Internet of Things

Technology, Communications and Computing

The series Internet of Things - Technologies, Communications and Computing publishes new developments and advances in the various areas of the different facets of the Internet of Things. The intent is to cover technology (smart devices, wireless sensors, systems), communications (networks and protocols) and computing (theory, middleware and applications) of the Internet of Things, as embedded in the fields of engineering, computer science, life sciences, as well as the methodologies behind them. The series contains monographs, lecture notes and edited volumes in the Internet of Things research and development area, spanning the areas of wireless sensor networks, autonomic networking, network protocol, agent-based computing, artificial intelligence, self organizing systems, multi-sensor data fusion, smart objects, and hybrid intelligent systems.

Indexing: *Internet of Things* is covered by Scopus and Ei-Compendex **

Debashis De · Samarjit Roy
Editors

Dew Computing

The Sustainable IoT Perspectives

Springer

Editors
Debashis De
Department of Computer Science
and Engineering
Maulana Abul Kalam Azad University
of Technology, West Bengal
Kolkata, India

Samarjit Roy
Department of Computer Science
and Engineering
D Y Patil International University
Pune, India

ISSN 2199-1073 ISSN 2199-1081 (electronic)
Internet of Things
ISBN 978-981-99-4589-4 ISBN 978-981-99-4590-0 (eBook)
https://doi.org/10.1007/978-981-99-4590-0

This Springer imprint is published by the registered company Springer Nature Singapore Pte Ltd.
The registered company address is: 152 Beach Road, #21-01/04 Gateway East, Singapore 189721, Singapore

Foreword

Dew computing fills the gap in ground-based services. At the beginning of the new millennium, in 2000, the speed of the processor was as fast as the speed of data transmission over the network, leading to a rapid evolution of distributed computing from the Grid to the Cloud. New service paradigms and a wide range of applications emerged. Service needs were met at lower application levels, and thus the *Dew-Fog-Cloud service hierarchy* emerged.

The term Dew Computing, which we introduced in 2015 (*Wang and Skala*), encompasses a broader concept and capabilities from IoT to Edge applications. It emerged as a result of the development of service applications as a widely deployed autonomous application, which is the first layer in the distributed service hierarchy. In the past period, the structured results of Dew Computing development have been presented at the DewCom conferences (2016–2023) by a number of authors gathered around *the IEEE Dew computing STC*.

As defined, the Dew computing is a paradigm for the software-hardware organization of on-premises computing in a Cloud computing framework in which the on-premises computing devices provide functions which are impartial of the cloud services and interoperate with cloud-induced services as well. Dew computing is an emerging paradigm that promotes the use of local computing resources to run various advanced user applications and reduce dependencies, but also to support remote computing resources provided by Edge, Fog, and Cloud computing environments to enhance user experience and capabilities.

The rapid development of IoT and mobile devices with or without Internet connectivity is the most common application area. The trend is that the most significant data processing around us is at the lowest probable level, directly linked with the physical ecosystem, and above all unambiguously controlling our immediate human environment. These devices, which are not in the Cloud or in the Fog or at the Edge or even at the mobile edge, are the foundation of the new Dew computing paradigm. Therefore, it is critical to reconnoitre the possibilities of Dew Computing and resolve the fundamental challenges of integrating the 'Dew' layer with the higher tier services in the Dew-Fog-Cloud hierarchy.

Dew computing differs from classical Cloud, Fog, and Edge computing in that it brings device-oriented services closer to the end user and adds autonomous processing functions that are independent of the Internet, but most importantly through simbiosis and communication with other devices to share data and information over the Internet or widely distributed hotspots. The difference is also expressed in terms of scalability, as Edge and Cloud providers can provision (nearly infinite) resources, while scalability in Dew computing must be realized at the device level rather than the server level.

This edited volume provides an overview of architectural and programming solutions in the conceptual and functional view of the Dew paradigm and their connections in overall solutions. The edition will help accelerate the adoption of Dew computing and promote the further development of ground-based service gaps, especially under the conditions of developing secure distributed service requests under the conditions of connectivity to a higher service hierarchy via the Internet or a wide-area hotspot network. At the beginning of the new millennium, in 2000, the speed of the processor was as fast as the speed of data transmission over the network, leading to a rapid evolution of distributed computing from the grid to the cloud.

Karolj Skala
Chair of IEEE DewCom Special
Technical Committee
Project Manager
Centre for Informatics and Computing
Ruđer Bošković Institute
Zagreb, Croatia

Preface

Dew computing depicts as an efficient computational schema which integrates the foundation of cloud computing with the proficiencies of end-user devices, such as the computing systems, smartphones, and so forth. It can heighten the technological practices for the users in contrast to solitary utilities of cloud computing. Dew computing tries to elucidate the key challenges relating to the cloud computing infrastructures, such as the much dependencies on Internet access. The key contributions of the dew computing are individuality and collaboration. Individuality signifies that the on-premises computational devices are capable to deliver such functionality mostly without the cloud services and the Internet connectivity. Collaboration defines as the applications must be proficient to link to the cloud services and orchestrate information transmission whenever appropriate. To illustrate the fundamental backbone of the Dew Computing and to resolve major technological challenges of the conventional cloud computing paradigms, this edited volume provides multiple fresh dimensions on Dew Computing in sustainable IoT perspectives.

This volume consists of three distinct sections. Section A primarily illustrates the architectural frameworks of the dew computing. This section contains four contributing chapters which highlight the demystification of the Dew Computing in Sustainable Internet of Things (IoT), Cache computing for Dew Devices at the Edge Networks, Dew Computing Monitoring System for Sustainable IoT, and the Security-Privacy aspects of authorized communications in Dew-assisted IoT Systems.

Section B emphasizes the Dew Computing platforms and Services. This section also consists of four distinctive chapters majorly focus on the implementation of Dew-inspired Matrix-Mesh communication protocol, Dew-assisted Blockchain schema for unreliable network, Dew-inspired Intrusion Detection system in the Edge of Things, and Machine Learning for Dew Computing.

The Dew Computing-inspired domains of applications are discussed in the Section C. This section comprises seven separate chapters, largely focusses on the Dew-assisted applications on the Sustainable Internet of Vehicular Things, Future Telerobotics or Teleoperations, Smart Healthcare, Dew-as-a-Service for intermittently connected Internet of Drone Things, Smart Aeroponics System in Agriculture

4.0, IoT-induced Smart Agriculture, and the Consumer Electronics for Sustainable Internet of Agricultural Things.

This edited volume embraces the comprehensive Dew Computing technologies and illustrations. It aspires to open up the perception of the Dew Computing that will fetch an efficient manifesto for the extent of industry-academia. The expansive Dew Computing subject matters differentiate this volume from the others. The substances include operative framework and system architectures of Dew Computing, Intelligent IoT analytics schema, Dew-induced sustainable IoT semantics, Knowledge extraction, Applications of CDEF (Cloud-Dew-Edge-Fog) computing paradigms in IoT, and Trustworthy Cyber-Physical systems. The aforementioned key-terms are tending to be implanted with the Dew Computing for the future generation automation.

This volume is also intended to create a centre of attention for the researchers who have been working specially in Information Technology, Computer Science and Engineering, and Electronics and Communication Engineering. This book is about fundamental and high-level conceptions regarding analytics-enabled Dew Computing paradigm in the context of sustainable IoT. We eagerly hope that this volume will serve as a beneficial guide for industry persons and assist the beginners to acquire emerging insights from rudimentary to advance in the contexts of sustainable IoT and CDEF Computing Technologies in broader spectrum of the scholarly insights and research expertise for working against the significant computational vulnerabilities.

Kolkata, India — Debashis De
Pune, India — Samarjit Roy

Contents

About the Editors

Prof. Debashis De earned his M.Tech. from the University of Calcutta in 2002 and his Ph.D. (Engineering) from Jadavpur University in 2005. He is a Professor and Director in the Department of Computer Science and Engineering at the West Bengal University of Technology (presently known as Maulana Abul Kalam Azad University of Technology, West Bengal), India, and an Adjunct Research Fellow at the University of Western Australia, Australia. He is a Senior Member of the IEEE, a Life Member of CSI, and a Member of the International Union of Radio Science. He worked as R&D Engineer for Telektronics and Programmer at Cognizant Technology Solutions. He was awarded the prestigious Boyscast Fellowship by the Department of Science and Technology, Government of India, to work at the Herriot-Watt University, Scotland, UK. He received the Endeavour Fellowship Award from 2008–2009 by DEST Australia to work at the University of Western Australia. He received the Young Scientist Award both in 2005 at New Delhi and in 2011 at Istanbul, Turkey, from the International Union of Radio Science, Head Quarter, Belgium. His research interests include wireless sensor networks, mobile cloud computing, green mobile networks, and nanodevice designing for mobile applications. He has published in more than 370 peer-reviewed international journals in IEEE, IET, Elsevier, Springer, World Scientific, Wiley, IETE, Taylor Francis, and ASP; more than 250 international conference papers; four research monographs in Springer, CRC, and NOVA; twenty books; and ten filed patents in his name. He has been listed in the Top 2% Scientist List of the world, Stanford University, USA. His research interest is Mobile Cloud Computing, IoT, and Quantum Computing. He is the Chairman of IETE and CSI Kolkata chapter.

Dr. Samarjit Roy is currently working as Assistant Professor at the School of Computer Science, Engineering, and Applications (SCSEA) at the D Y Patil International University, Pune, Maharashtra, India. He was a UGC Junior and Senior Research Fellow (UGC-SRF) at the Department of Computer Science and Engineering, Maulana Abul Kalam Azad University of Technology, West Bengal (formerly known as, West Bengal University of Technology), India. He has received a Ph.D. in Computer Science and Engineering on Intelligent Musical Pattern Recognition and Information Retrieval. He has served as Assistant Professor and Department Incharge at the Department of Computer Science and Engineering, Techno India Silli, Ranchi, Jharkhand. He completed his M.Tech. in Information Technology from Manipal Institute of Technology and B.Tech. in Information Technology from Government College of Engineering and Leather Technology, Kolkata, West Bengal. His research areas include Musical Pattern Recognition, Information Retrieval, Context-Aware Recommender Systems, Speech-Audio-Signal Processing, Cloud-Dew-Edge-Fog (CDEF) Computing, Soft Computing, Machine-Deep-Representation Learning, Internet of Music Things, Crowdsensing, and Data Analytics. He has published more than 50 research articles, books, book chapters, etc. in numerous reputed national and international journals, conferences, and with other international publication societies. He has achieved several international and national research recognitions, such as the Young Scientist Awards; Young Researchers Awards; Scientist of the Year 2020 from Illinois, USA; Socrates Award from Academic Union, Oxford, UK; Honorary Professor Project, Oxford; and doctoral fellowships from UGC India in Engineering and Technology section. He has organized and participated in several national and international conferences, workshops, symposiums, STTPs, and FDPs.

System Architecture

DewMetrics: Demystification of the Dew Computing in Sustainable Internet of Things

Samarjit Roy, Debadrita Panda, Byung-Gyu Kim, Palash Bairagi, Tamal Mondal, Sirshendu Arosh, Suprabhat Sinha, Debashis De, Yingwei Wang, Karolj Skala, and Davor Davidovic

1 Introduction

The personalization aspect is indispensable for the engrossment of the users in any computing genus. Dew computing is an emerging paradigm that inherits a flexible and super hybrid methodology to afford personal information to users' self-regulating Internetwork connectivity [1, 2]. Dew computing encompasses a set of innovative design attributes in terms of hardware and software prototypes that includes the dew computer, dew server, dew site, dew database, dew domain name

S. Roy (✉) · S. Sinha
School of Computer Science, Engineering, and Applications (SCSEA), Department of Computer Science and Engineering, D Y Patil International University, Pune 411044, Maharashtra, India
e-mail: samarjit.tech89@gmail.com

S. Roy · K. Skala · D. Davidovic
Center for Informatics and Computing, Ruđer Bošković Institute, Zagreb, Croatia

S. Roy
School of Computing and Information Technology, Eastern International University, Thu Dau Mot City, Vietnam

D. Panda
Department of Social Sciences, Technology, and Arts, Luleå University of Technology, Luleå, Sweden

B.-G. Kim
Department of Artificial Intelligence Engineering, Sookmyung Women's University, Seoul, South Korea

P. Bairagi
School of Commerce and Management (SCM), D Y Patil International University, Pune 411044, Maharashtra, India

D. De and S. Roy (eds.), *Dew Computing*, Internet of Things,
https://doi.org/10.1007/978-981-99-4590-0_1

system, dew domain name redirection, Software-as-a-Dew Product, Infrastructure-as-a-Dew service and super-hybrid-peer-to-peer network [3, 4]. The foremost indication behind dew computing is to minimize the dependency on obtainable Internetwork backhaul, thus reducing network traffic, indirect overall power consumption of the network system as well as data dependency over cloud-fog-edge services [4, 5]. Dew computing provides augmentations of real-time personalized services to users employing instantaneous web data that is surf-ready. Further, transparency and lose coupling with the Internetwork configuration are envisioned [4].

Most people refer to modern mobile and wireless ubiquitous solutions as the Internet of Things applications. The encroachments of the technology and establishment of cloud-based systems emerge the idea of a connected world over the Internet based on distributed processing frameworks [5, 6]. We illustrate in this contribution the dew computing architectural approach for sustainable IoT solutions and give an organizational outline of the dew server and its connections with IoT devices in the overall cloud-based solutions. Dew servers act as another computing layer in the cloud-based architecture for IoT solutions [7, 8], and we are going to demonstrate its specific goals and requirements. This is compared to fog computing and cloudlet solutions with an overview of the overall computing trends [8, 9]. The dew servers are analyzed from architectural and organizational aspects as devices that collect, process, and offload streaming data from the IoT sensors and devices, besides the communication with higher level servers in the cloud [10, 11].

The dew computing concept is on the edge of the Internet network means that the analyzed devices and systems will work only as a part of a general common integrated system, such as in the case of cyber-physical systems and various devices that act as an Internet of connected Things [5, 6]. The dew computing implementation in cyber-physical systems allows autonomous devices and smart systems, that can collaborate and exchange information with the environment, still, be independent of other external systems, or perform in a connected more complex cyber-physical system of systems [5].

The Internet of Things is a promising paradigm that integrates additionally a plethora of heterogeneous computational devices, incorporating the crowd, frameworks, additional system elements, and infrastructure [12]. Information sensing, modeling, retrieval, and distribution perform an emerging role in the Internet of Things network [13]. Dew computing is a challenging research issue, which needs to demonstrate its impact on the sensor data in the domain of parallel and distributed

T. Mondal · S. Arosh
Symbiosis Centre for Information Technology, Symbiosis International (Deemed University), Pune, Maharashtra, India

D. De
Department of Computer Science and Engineering, Maulana Abul Kalam Azad University of Technology, West Bengal, Kolkata, India

Y. Wang
School of Mathematical and Computational Sciences, University of Prince Edward Island, Charlottetown, Canada

computing [14]. The crowdsourcing paradigms are efficient to collect and analyze billions of information efficiently with a diminutive cost. In this promising paradigm, participated sensing devices sense information from the environment; transmit the sensor data to associated edge and fog computing devices through a dew repository, and eventually, the cloud data center stores the processed data for providing aggregated information and relevant services to the end-users [10].

Technological innovations have brought revolutionary modifications in terms of information storage and its accessibility both in subjective and business-related domains. The Internet facility helps everybody to fetch their information requirements without any location, or timeframe constraints. There are fundamentally three subdomains, i.e., cloud, edge, and fog computational paradigms that exist in brain-computer interfaces. Information can be stored and transmitted whenever required from the cloud data center. These Internet-enabled features untangle the optimum usage of physical IoT devices. Along with the immense benefits in the context of contemporary digital transformations, cloud data servers, and data centers may suffer from certain issues in terms of constricted bandwidth, latency, and cost [15].

1.1 Motivations

We depict that the cloud, edge, and fog computing paradigms flop principally to convey the Internet-free computing prototypes [2, 3], near-optimal scalable explanations, and reliable service concerns [14, 16]. To avoid these restraints in the conservative cloud, fog, and edge computing archetypes, the integrated or disseminated information storage is desirable to specify copious facilities to those who are availing the services, where the expedients will be within the Internet-dependent zone may be active or inactive genre [7, 10]. Hence, in the progressive mobile edge computing framework [11, 13, 15], the authors projected an emerging computational and session-based data repository platform, titled dew computing [4, 7]. The dew takes care of the existing peripherals in a collective network, although the surroundings are somehow Internet dependent [2, 3]. By adopting dew computing, diverged set of services, for example, distributed networks are strictly dependent on locations, facilities dependent on information dissemination, network latency, and cyber-physical combined interfaces, which are mostly decentralized and aid energy-efficient protocols [5, 8–10]. The projected dew-induced sustainable IoT paradigm also sets provisions in smart information sensing and confined data analytics [10, 11]. The incorporation of information sensing arrangements and widespread computing Intra-network-based apparatus occasioned the sustainable Internet of Things (IoT) advancement [10]. The sustainable IoT devices and paradigms are assiduous which specify voluminous applications and put together massive data volumes [12]. We can claim that dew computing is not a surrogate paradigm for depicting influences on the sustainable IoT, rather this newly emerged computing schema is an expanded augmentation over the conventional schemas of information processing and system evaluations.

1.2 Chapter Contributions

The significant contributions of DewMetrics are mentioned in the following:

(a) We depict a systematized study on the potential Dew computing framework aiming toward sustainable IoT applications.
(b) We attempt to focus on unfolding the concept of dew computing by illustrating the background infrastructures, system architectures, and a set of potential application areas where the minimized Internet-dependent schemas are significant enough.
(c) We discuss the agglomerative real-time case studies, such as the cache computing framework for the dew devices, the reduced Internet dependency-induced decision-making processes, the crises, and the humanitarian Internet of Music Things.
(d) This chapter provides a deep insight into upcoming tools and technologies which can be incorporated with the traditional dew computing archetype to acquire supplementary efficiency in the Internet-dependent distributed prototypes.

1.3 Chapter Organization

We organize the remaining DewMetrics chapter as follows: Sect. 2 presents the potential state-of-the-art on dew computing architectures, systems, and applications. We illustrate the summarized outline of dew computing in Sect. 3. We demonstrate a Dew Computing-induced prototype in Dealing with Crisis Situations in Sect. 4 along with the crisis management through the dew-cloud architecture. We explain a procedural schema in the context of reduced Internet-dependency-aware consumer decision-making procedures in Sect. 5. Dew computing-inspired Computational finance outline is presented in Sect. 6. We additionally propose a systematic summary to define Dew Computing paradigms for Quantum Machine Learning and Quantum Cryptography in Sect. 7. We append the influence of Dew computing in the congregating paradigms of Computational musicology and sustainable Internet of Things in Sect. 8. We discuss several topics of interest in Sect. 9 which have significant applicability in the dew-driven sustainable Internet of Things paradigms. Eventually, we conclude the chapter with the concluding remarks in Sect. 10.

2 Related Works

As we have highlighted Dew Computing is relatively a new post-cloud computing paradigm that was introduced in 2015. While the conventional cloud computing framework utilizes centralized servers for providing numerous services, dew computing utilizes on-premises computational devices to provide decentralized, contemporary cloud-friendly, and collaborative services to the end-users.

Researchers have illustrated a set of dew-induced system architectures for providing distributed and collaborative services and application domains.

In the paper [1], the authors have demonstrated the concept of newly emerged dew computing, the significant dew computing paradigms, and a set of future research challenges. Researchers projected dew computing as the ubiquitous and permeating computational framework to enable secretive networks to depict processed sensor data orchestration [2, 3]. Working principles, methodologies, large application paradigms, and a set of future research scopes had also been outlined in [2, 4]. Taxonomical representation has been depicted in the dew computing context that can be a significant component for the dew-cloud-induced future research direction [2, 3, 16, 17]. Authors have projected three diverged schemas where the future IoT applications may proceed for real-time data analytics [3], such as the multi-operations, cloud-centered IoT information processing, and contextual dew computing framework [6, 9, 16, 17]. Researchers have also depicted that service models, such as SaaS and SaaP can be associated with the dew-based framework [7]. Key advantages and limitations had additionally been outlined. Moreover, in [12] a schematic demonstration of web surfing has been presented when the Internet connectivity is lost or inactive which is a challenging real-time application scenario of dew computing. In [8], the preliminary demonstration of hierarchical connexions has been framed among the conventional computational artifacts, such as the cloud, fog, and dew, where they have proposed that the dew computing layer can be set up as the ground-level tier of edge, fog, and cloud computing hierarchy.

In [14] Information privacy-related concern has been represented in the accessible dew-cloud design. A cloud computing-inspired multi-agent scenario has been executed for postulating healthcare services using neural networks where the custom-made web resource has been acquired to shrink the access to large data volume to be processed through the cloud [8]. An illustration of the dew computing paradigm for cyber-physical systems has been introduced in [5] that encompasses autonomous systems for collaboration and exchanging system information. In [18], a dew-induced trace-driven validated framework has been projected to quicken resource management. The researchers have framed an efficient resource allocation paradigm to assign time-astute tasks by the appropriate allotment of Virtual Machines [16, 17]. Moreover, they have anticipated a comprehensive task distribution order among cloud-dew-edge-fog computation archives by evaluating performance time, transmission and service latency, and network usage. Authors proposed a paradigm in [19], titled, DC-Health, an emerging dew computing-enabled sustainable IoT healthcare solution for decisions with offline and ultra-low latency. The projected endeavor incorporates several healthcare peripherals and is provisioned in user-specific domains even when Internet connectivity is not active [20]. In the recent research contributions [21], the researchers have presented profound insight into machine learning-inspired diverge archetypes for detecting intrusion in edge-induced IoT networks. Paper [22] depicted the query optimization on graph database in a dew-cloud environment along with the performance metrics in terms of memory usage and execution time. Articles [23, 24] illustrate a set of future research challenges and projected solutions in the domains of next-generation IoT-driven fog computing and analytics. In [25], the researchers have proposed to allocate tasks in Dew computing paradigms using an artificial intelligent agent, named, Proximal Policy Optimization that acquires tasks in the simulated Dew ecosystem. The hierarchical artifact of cloud-fog-dew has been presented in [26] to overwhelm the restrictions of cloud computing in the real-time application environment, such as latency and resource management. They have verified the projected model using Non-dominated Sorting Genetic Algorithm II for analyzing the system scalability and a Mixed Integer Non-Linear Programming model for scheduling the real-time applications to lessen Internet data traffic and power dissipation. [27] demonstrates an archetype, titled as MedGini, which generates effective utilization of IoT and cloud-dew framework for monitoring a sustainable healthcare paradigm. They have used Wireless sensor nodes to monitor the dynamic bio-signals, the Gini index, and Shannon entropy for ensuring intelligent information synchronization in the cloud along with the evaluations of the cost and power consumption of the system. Mental health monitoring schema has been represented in [28] which was powered by a Lightweight Convolutional Neural Network. Demonstrations on the attack framework and required security characteristics for the dew computing paradigm have been presented in [29]. The authors have illustrated client–server security schemes and identified two specific criteria such as security point and key agreement protocols. The authors of the [30] have addressed the security issues, a reciprocated authentication framework for dew computing, which certifies authorized and secured session formation deprived of the necessity of a reliable server. Authors have implemented dew computing in cyber-physical systems that tolerate core appliances and smart peripherals which involved in information communication in the surroundings [31]. Their projected study aims at the demonstration of architecture to apply cyber-physical system-induced dew computing for setting up new characteristics and peripheral services and compared with additionally comparable system architectures. The chapter [32] illustrates an AI-induced self-monitoring potential healthcare application in edge and dew computing. He elaborated a use-case schema for serving as an AI-based cardiologist for the end-user expending a smart device and an electrocardiogram sensor along with the future challenges, advantages, and limitations of the proposed paradigm. A precise review of the blockchain implementations in cyber security perception and energy information protections of smart grids has been illustrated in [33] with the depiction of the key security issues of smart grid environments with big data and blockchain. In paper [34], a content-based Dew-Cloud-inspired framework has been modeled for enabling a hierarchical federated learning approach and to provide a higher level of data privacy with significant availability of Internet of Medical Things critical scenarios. Authors have depicted a blockchain and dew computing-inspired converged framework for the smart city application with the evaluations of performance metrics of the proposed model [35]. The researchers in [36] have highlighted the lacking of Saurabh et al.'s strategy in the context of forward security and user anonymity and introduced an authenticated key agreement (AKA) protocol, titled e-SMDAS. Researchers have stated that there are numerous solutions offered to enhance system performance in Cloud healthcare computing domains, such as Edge, Fog, Mist, and Dew computing [37]. The proposed article studied Fog computing schema, its benefits, and qualitative system architecture. Authors [38] presented an emerging distributed edge framework pointing to analyzing and storing the Internet of Things [39] and multimedia information in modern application scenarios. A systematized case study on healthcare cyber-physical systems has been illustrated along with the characteristics, the task of diverged technologies, and the implementation of cyber-physical medication Paradigms [40]. Researchers have discussed a dew-induced and blockchain-inspired Internet of Things, termed BCoT, that shows the convergence phenomena in the reformation of Industry 5.0 along with some typical future research challenges of industrial IoT applications [41]. In [42], the contribution illustrates a comprehensive study to realize scalable Blockchain storage systems for dealing with the ambiguity among the redundancy and decentralization characteristics, based on scalable Blockchain storage systems (SMBSS).

3 Dew Computing: The Infrastructure Outline

With the increase of end-users and readily available computing devices such as mobile, portable laptops, tablets, etc., dew computing has become a very useful technique to balance the load sharing between the end-user and the cloud. Most of the current services such as mobile apps, google documents, presentations, GitHub, and YouTube videos are on the cloud and users need steady Internet connections to access these services which might increase the load over the Internet which in turn leads to network congestion and the need for more bandwidth. However, there are many places where data and service availability, network latency, and throughput quality are limited and restricted [43]. Regions affected by natural calamities or disasters might face Internet loss and loss of other dedicated connectivity like GSM, PSN, etc. for a prolonged duration [44]. Dew computing architecture can be an alternative in certain situations where establishing direct communication to the cloud is improbable.

By the definition, Dew computing depends on two basic entities: Independence and collaboration. Independence relates to the mutually exclusive service to every user. Collaboration indicates synchronization to local and remote data. This computing includes six fundamental features Rule-based Data Collection, Synchronization, Scalability, Re-origination, Transparency, and Any Time Any How Accessibility. The model includes predefined policies to save end-users' data to its database which is addressed in Rule-based Data Collection. Synchronization deals with the organization of local and distributed data with integrity. The data stored needs to be minimum taking the mobile users into account and that is addressed by scalability. The data loss policy is managed by Re-origination. Dew transparency deals with data replication and distribution. Lastly and most importantly, the service should be available to everyone with or without the Internet which is tackled by Any Time Any How Accessibility. These methods will decrease the dependency on the cloud as well as reduce the load over the Internet connectivity. As of now, there are seven service models available in the literature [4, 5].

Infrastructure-as-dew (IAD) [7] demands that the end-user is supported by cloud services either by Dew virtual machines or by saving the data and setting the service from the local device to the cloud. The advantage of such a system is that data and devices will be completely independent of each other. If the device is damaged, another new device can take the place of the old device and the user just needs to log in to its account where all possible data can be accessed. Nowadays many mobile companies are providing such services. Web in Dew (WiD) [12] is a dew computing category where end-user hardware contains a fraction of duplicated or modified world wide web depending on their frequent usage or important service category. Since it is available locally, without an Internet connection also, the information can be browsed as per capacity. Since this is the exact copy of the original one, it satisfies the collaborative property too. Software in Dew (SiD) [45] is a dew computing category where the end-user owns a certain piece of software in his/her local devices. Additional information or portions of the application can be acquired as per requirement and further downloaded or real-time data transfer-based communications also can be possible. The best examples are the iPhone app store and google play store. For certain applications, a set of information is available offline (saved maps, train time table or standard non-dynamic fare charts of buses, trains, etc.), while the app goes online, additional activities can be done (e.g., downloading the further portion of the maps, know the current running status of trains, book tickets). Platform in Dew (PiD) [46] is a dew computing category where development and operational software are installed in the end user's hardware; the setting and application data are dynamically synchronized as per available connection. GitHub qualifies for such a service. Storage in Dew (STiD) [47] is a dew computing category where the stored data in end-user hardware is duplicated in the cloud service and get updated automatically. A common existing STiD service is Dropbox. The files/folders in Dropbox are available to users at any time with or without the Internet. Hence, it satisfies the independence feature. Moreover, these files/folders are automatically synchronized with cloud services when the Internet is available so that it satisfies the collaboration property. Database in Dew (DBiD) [48] is a dew computing category where a local database is kept which is a replication of the original cloud database. The local database can be synchronized on-roll, periodically, or dependent on an Internet connection. As discussed before, it follows both the properties of Dew computing. Data in Dew (DiD) [17] is a dew computing category where data such as email, messaging, calendaring, contact management, and scheduling are kept locally and for personal usage, and with the availability of the Internet, those are synchronized in the cloud with global sharing. The summary of the Dew service models has been depicted in Table 1.

Dew computing is distinguished by its minimal reliance on the Internet, extreme user control flexibility, and proximity to the end-user. Dew computing attempts to provide a variety of services, such as (1) densely distributed service, (2) awareness of one's location, (3) support for heterogeneity, (4) reduction of network-service latency, (5) facilitation of any time, anywhere access, and (6) quality of service that uses little energy. Infinite applications can benefit from its faultless security services, including

Table 1 Summarized representation of the essential dew computing services

Services	Illustrations
Infrastructure-as-dew (IAD)	User is supported by cloud services either by dew virtual machines or by saving the data and setting the service from the local device to the cloud
Web in dew (WiD)	End-user hardware contains a fraction of duplicated or modified local world wide web depending on their frequent usage or important service category. It won't require Internet connectivity
Software in dew (SiD)	End-user owns a certain piece of software in his/her local devices. Additional information or portions of the application can be acquired on the requirement and downloaded or real-time data transfer-based communications also can be possible
Platform in dew (PiD)	Development and operational software are installed in the end user's hardware; the setting and application data are dynamically synchronized as per the available connection
Storage in dew (STiD)	Dew computing category where the stored data in end-user hardware is duplicated in the cloud service and get updated automatically
Database in dew (DBiD)	Dew computing category where a local database is kept which is a replication of the original cloud database
Data in dew (DiD)	Dew computing category where data such as email, messaging, calendaring, contact management, and scheduling are kept locally and for personal usage, and with the availability of the Internet, those are synchronized in the cloud with global sharing

smart retail, e-healthcare, real-time data analytics, localized industrial automation, and smart sensing and actuation (Fig. 1).

With this knowledge, the importance of dew computing in distressed areas can be apprehended easily. Dew computing can be helpful for those end-users dealing with the absence or limitation of Internet connectivity. Now, the question lies in how the dew computing services, models, and architectures can be found resourceful in dealing with crisis response situations. That is, how the offline and online services of Dew computing can be utilized simultaneously in IoT devices and applications for effective data flow to retrieve real-time crisis information. In a crisis, it might be possible that (a) the Internet connectivity gets disrupted in the affected regions due to loss of communication infrastructures or (b) the connectivity might present in some small pockets. In such circumstances, the Dew computing infrastructure should be helpful in storing, processing, and analyzing the locally generated data in offline mode and sending the information periodically whenever Internet connectivity is available to the users' devices.

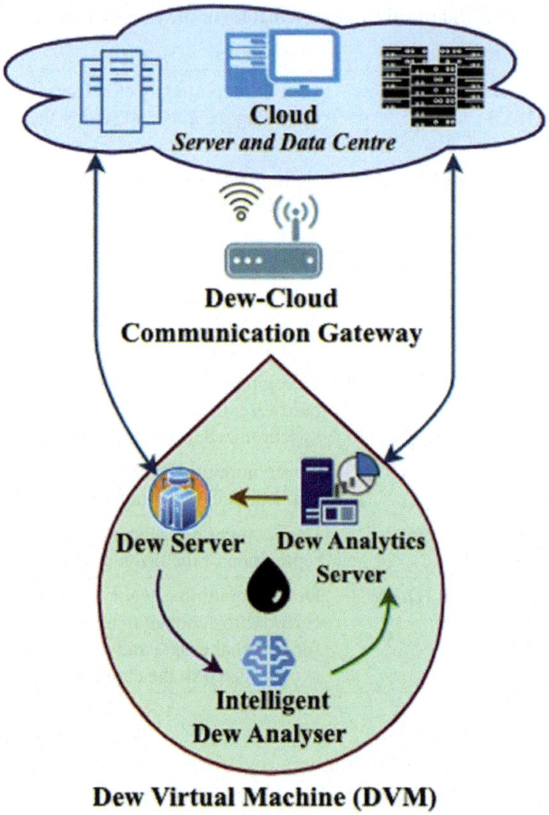

Fig. 1 Representation of the dew-cloud integration infrastructure

4 Crisis Environments and Significance of Data Flow for Response

Acquisition, analysis, and processing of the large volume of raw data germinated from various data generation platforms, i.e., Web applications, Online Social Networks [49, 50] like Twitter, WhatsApp, Facebook, etc., SMS generated in GSM services and real-time testbeds and applications (portable devices and applications used for establishing communication in opportunistic environments) for extraction of comprehensive information remain a considerable challenge. With the advent of the enormous amount of raw data generated from these data sources, different sectors (E-services, Business, Agriculture, Disaster Management, Health, Education, etc.) are currently focusing on data mining steps for extracting comprehensive knowledge from the data. From the "Crisis Management" viewpoint, one of the prime goals is to perform effective crisis response and recovery which solely relies upon real-time analysis and processing of situational data and proper dissemination of information acquired from data. The crisis management operation also involves the participation of diverse stakeholders like Government Personnel, NGOs, Rapid Action Forces, etc. consuming data from heterogeneous sources for real-time situation analysis and resource management. Therefore, the overall information and communication technology (ICT) plays a critical role in coordination among the stakeholders through a constructive flow of information [51]. Any ICT-based crisis response can be characterized in two ways (also depicted in Fig. 2): (a) Handling the data flow through deploying the hybrid Ad Hoc communication infrastructure with the help of low-cost portable devices (Smartphones, Single Board Computers, UAVs, etc.) and protocol stacks [52, 53] when the existing communication services like GSM, Internet, Satellite, etc. are limited in use and (b) Utilization of data received through news reports, press releases, social media posts (Facebook, Twitter, WhatsApp, etc.) through the Internet, emergency calls, etc. when the existing communication persists into crisis zones or regions [54, 55]. These diverse data sources contribute to real-time damage and require assessment for obtaining effectual crisis retaliation and recuperation.

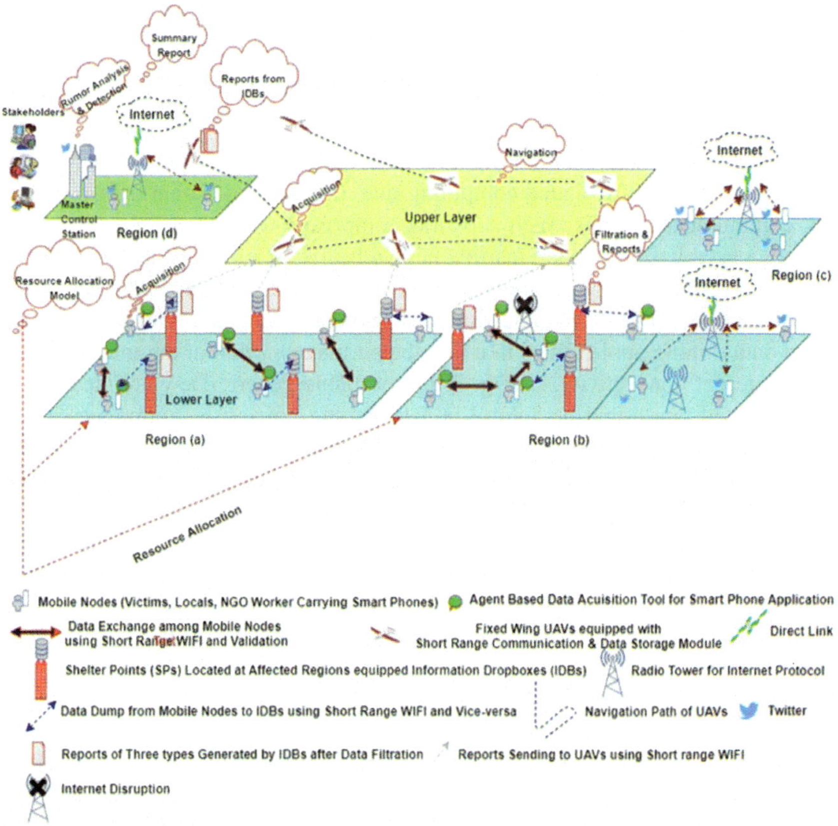

Fig. 2 Framework of data-driven ICT-based crisis management

From the above study, it can be comprehended that data dissemination sources can be multimodal and subject to the status of conventional network infrastructure in any crisis response situation. For developing overall ICT-based services, equivalent utilization of such data sources both for opportunistic and traditional network environments (i.e., GSM, Internet, etc.) is required. Depending upon the status of the crisis regions in terms of the Internet accessibility, both offline and online data sources should be made useful intermittently for real-time information extraction. Besides, through exploiting the data acquisition and mining steps, proper collection, the realization of association rules, and processing of such data are also essential for representation in a standard periodic report format that satisfies the interests regarding the "damage and needs assessment" of crisis management stakeholders. Moreover, such regular reports should also be made utilizable for designing real-time decision support tools for resource management. In Sect. 4.1, the utilization of the Dew computing component in crisis management has been discussed in greater detail.

4.1 Utilization of Dew Computing in Crisis Situations

Dew computing models can be considered ubiquitous, pervasive, and convenient ready-to-go, plug-in facilitated computing that empowers personal hybrid peer-to-peer communication links. Such networks comprise low-cost scalable devices like PC, laptops, tablets, high-end Smartphones, and software protocol stacks. The prime objective of such an arrangement is to efficiently collect, process, store, and utilize raw data in the offline communication environment. Such communication can help to run data mining applications in a distributed manner without the intervention of stable Internet connectivity. In Fig. 3, the probable generic Dew architecture has been depicted that might be helpful in establishing post-crisis communication for information flow. The Dew cloud architecture has been used in several application domains like healthcare, cyber-physical systems, air quality monitoring, disaster management, and many more. Since Dew is a newborn baby, therefore, it's yet to be applied to such real-life application domains mentioned above. Here, in this section, we specifically discuss the usage of Dew architecture for crisis mitigation.

4.2 Dew-Cloud Architecture for Crisis Management

The intermittent integration of Dew clients like wireless sensor networks (WSN), Smartphones with applications, Single board computers, etc. with dedicated cloud architecture can be considered as an ICT-based data communication and mining tool which can further augment the crisis management services. In [56], a 3D-based cloud platform has been built utilizing the wireless censored data for an effective disaster response, i.e., optimal allocation of resources like (robots, drones, firefighters, etc.).

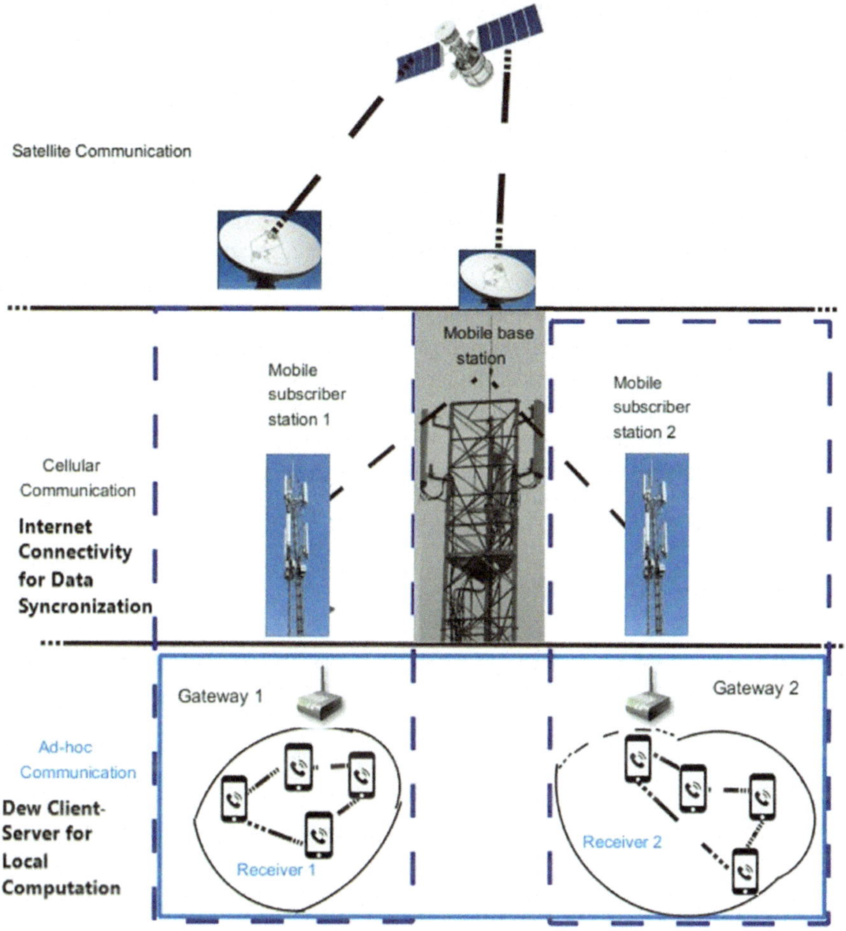

Fig. 3 Proposed dew computing-based communication architecture

The system's purpose is to be used as a training environment for a rescue team to develop various rescue plans before they are applied in real emergency situations. The proposed cloud architecture combines 3D data streaming and sensor data collection to build an efficient network infrastructure that meets the strict network latency requirements for 3D mobile disaster applications. As compared to other existing systems, the proposed system is truly complete. In [57], proposed a scheme called MGRA for the allocation of computing nodes which takes into account different issues like the users' mobility, inefficacy in resource allocation, and handling of failure situations. Experimentation was carried out using a Dew computing testbed comprising low-cost android devices connected with Wi-Fi Direct protocol. MGRA exhibited significant improvement in terms of time for application completion, amount of battery usage, and time required for recovering from failure as compared to present-day approaches.

4.3 Internet of Drone Things

UAVs can play a major role in crisis management in terms of data collection, communication establishment, surveillance, etc. These few design frameworks have been proposed for crisis management through utilizing IOT devices and drones in the past literature. In [58], a distributed architecture of clusters of UAVs and IoT devices located on the ground is proposed which will monitor various disaster management applications and do real-time surveillance. Here, each UAV has been considered as a Dew client which involves the computation of real-time route prediction, object detection, obstacle avoidance, etc. The authors in [59] proposed a Dew cloud computing framework adjoined with the UAV networks known as "DewDrone" for providing opportunistic network connectivity for smart cities, rural sectors, etc. Also, a robust disaster management framework has been considered for the design of an IoT infrastructure with UAVs to generate an Internet of drone things for establishing communication between electricity service and smart city.

4.4 Dew Robotics Applications

In [60], a cloud robotics application has been developed named Dew Robotics which allows the development of solutions that do not rely completely on the Cloud but whose computational capabilities are distributed among different devices. Dew Robotics exploits the intelligence of edge devices to change the operating conditions of our applications and adapt them to the system's status.

5 Reducing the Dependency on Internetwork During Consumer Decision-Making

From the inception to the latest, most frameworks related to consumer decision-making are fully interconnected and hugely interdependent. If any problem or frustration arises at any level, it propagates to all the consequent steps and finally hampers consumer decision-making. Especially when purchases are made on online platforms, consumers are the utmost sufferers. Internet connection loss or Internetwork-free orientation completely exhausts the information and decision-making flow of these frameworks. Although cloud computing comes with a huge opportunity for users, like universal access and scalability, some challenges are present, like all resources located remotely, far from the user's machine and control. For this, if the Internet connection goes off, all access to those resources is completely lost. This is the thrust area where dew architecture intervenes [1]. In the cloud-dew architecture, installed websites are always available to the users irrespective of

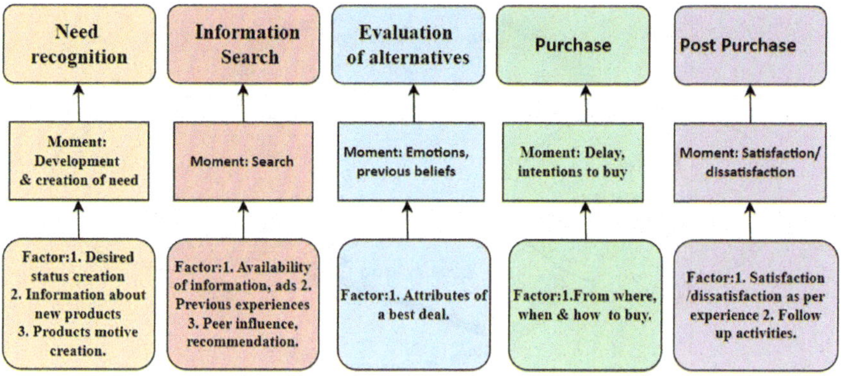

Fig. 4 Traditional framework of consumer decision-making

the Internet connection. When the Internet connection resumes, it starts synchronizing with the cloud server. In the beginning, dew computing was started as a web application [4]. Dew computing is characterized as a software organization paradigm for personal computers through this definition. Through this architecture, the local computers offer rich capability independent of the cloud services, and local computers collaborate with cloud services when the network connection resumes [5, 6]. Numerous frameworks related to consumer decision-making have been present in the literature since 1960. Most of these frameworks are also applicable to online consumer decision-making. One of the traditional online consumer decision-making frameworks has been presented in Fig. 4.

This traditional framework has been developed in [61]. There are a total of five steps present in the framework. Along with the steps, the moment and all the factors are also present. The moment is representing the thrust area of the step and the factors are broadly capturing the responsible criteria that are driving the concerned step. All the steps are interdependent and hugely dependent on the Internet while making online purchase decisions. For example, All the responsible factors for need recognition, i.e., information about new products, are mostly driven by online platforms, the social media can influence desired status creation. Next, if the information search online is satisfactory, the evaluation of alternatives will be complete. If the evaluation of the alternatives is not complete, the purchase experience will not reach a satisfactory level. Each step has a huge dependency on the network connection. Suppose the Internet connection fluctuates or becomes off frustrations of consumer increases. In such a situation, dew computing through reducing the interdependency on the Internet widens a huge scope for the users. Not only traditional or root decision-making models but also the latest or modified decision-making models have shown strong dependency on the Internet connection if the consumer purchases online. One of the latest another framework of consumer decision-making is McKinsey's dynamic model of consumer decision [62]. Figure 5 shows this framework. This framework also consists of highly interconnected steps and depends on the network connection.

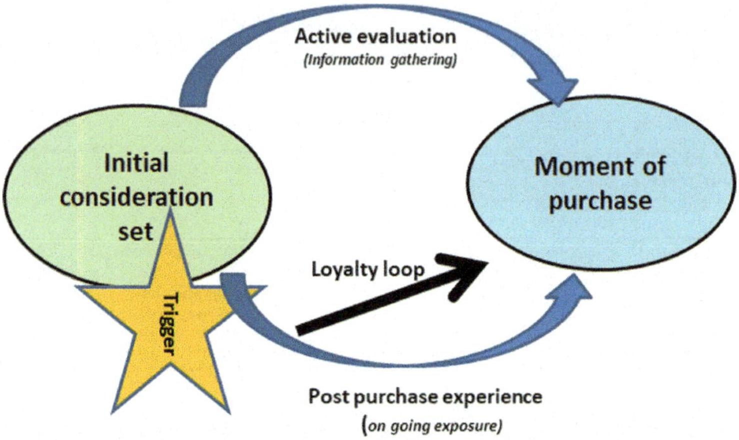

Fig. 5 McKinsey's dynamic model of the consumer decision journey

Just like the previous framework, the loss of an Internet connection creates frustrations. Here also, dew computing can be a solution through which customer loyalty can increase significantly.

5.1 Intervention of Dew Computing

There are some other areas where dew computing can enhance performance such as:

(a) The operational efficiency of retail businesses.
(b) Improvised and customized consumer experiences.
(c) Helps to determine consumer trends.
(d) Provide hassle-free consumer services to those places/retail stores where Internetwork connection is not present/disrupting.

In this regard, the advantage of dew computing can be utilized by integrating the decision-making models in the cloud-dew architecture. The interlinkage between cloud-dew architecture and consumer decision-making traditional funnel is presented in Fig. 6.

The Dew server is an essential part of the proposed framework. The server must be equipped with many adaptive technologies. The four main co-servers are Database, Mobile Information, Application, and POP/IMAP message protocols. There are some pre-installed websites present on the dew server. When a user searches some website contents, the dew server provides all the responses. Internet connection is not at all an issue. After searching, the dew server creates a dual copy of the visited website on the user's local machine. This dual copy is called Dew Site. The dew scripts generally take care of the modifications of dew sites. The supervisor of the dew scripts is called the dew analyzer. The user can access the website or modify the contents through

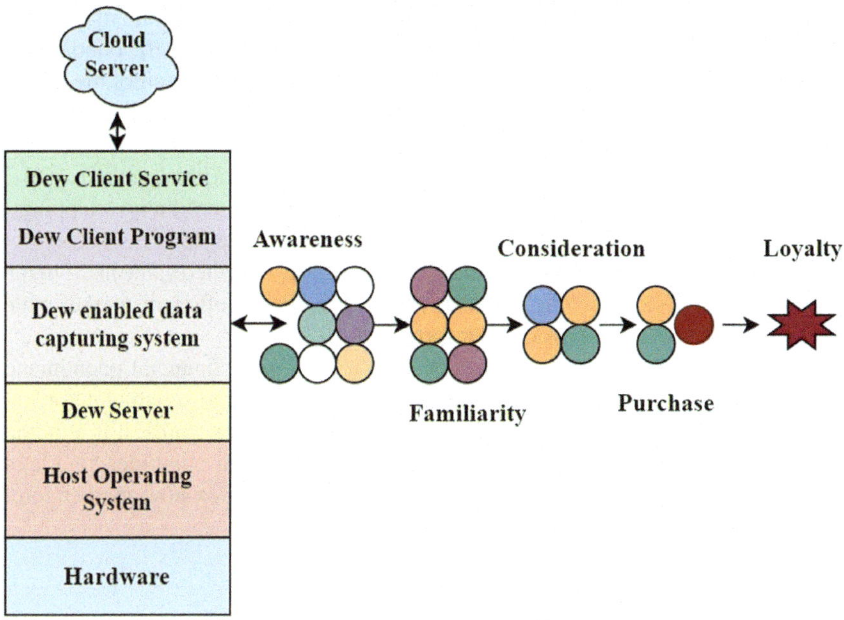

Fig. 6 Demonstration of the projected dew-assisted consumer decision-making paradigm

the dew server without any Internet presence. Once the Internet connection resumes, the dew server synchronizes with the cloud server.

6 Dew in Computational Finance Theories

The process of computational finance with the behavioral components causes complex problems. Agent-based computational finance can be defined as bottom-up approaches that use computer technology to simulate financial environments [63]. The decision-makers called agents, interact with individual rules of behavior and the trading mechanism [64]. In this model financial market is the interacting group of learning, heterogeneous and bounded-rational agents, including the financial market's efficiency and rationality, Time series data remain a curious puzzle to understand and Financial data provides a wealth of price and volume of data that can be analyzed [65]. So, several models of the Artificial Finance Markets (AFM) have increased the discussion on the research literature and problems of the models. According to [14, 15] agent-based computational finance (ACF) can be classified into three most important components such as few-type model, a dynamic model under learning, and a many-type model and its emergence. In this model, agents usually follow two base strategies called technical analysis and fundamental analysis.

In the few-type model, technical analysis follows a historical pattern and trend in data to predict future trends, in which the future replicates itself. Fundamental analysis is more complicated in determining the trends, as it determines the internal value of the securities. If the agent finds bias in its intrinsic value and the face value, then they find an opportunity to develop a path strategy [66].

A dynamic model under learning includes the generic algorithm methods [63], in which optimal risk and risk-free rate are considered to measure the overlapping of price in the exchange rate and its conversion to a single value [67].

The emergence and many-type model determines the expected value in a dynamic environment and confirms that the markets evolve into an efficient market which means price reflects on all available information.

Based on the above context we can call an agent of the financial phenomenon a fundamental/rational agent, noise agent, and technical agent to simulate or integrate the financial data in the computational process. In other words, it is a method that guides the science of complexity based on multi-agent computer simulation technology with financial theories [68]. This model indicates how market forces influence and impact stock prices.

6.1 Relational Agent's Decision Model

In finance, we believe that "investors behave rationally and they are rational agents" most of the time [69, 70]. These rational agents' are the new computational financial agent (CFA) that are evolving as autonomous interacting system agents.

The trade network game model and the computational decision in finance are broad categories in three strategic decision functions [71].

(a) First, the agent's decisions are heterogeneous in a character who interacts with other agents by initializing their initial data and the behavioral pattern, i.e., making a protocol to communicate with other traders, for trade partner search and matching or to trade interaction.
(b) Second, the agent predicts and makes the association or relationship between the price and the number of different asset classes based on the information gathered, attributes, beliefs, and preferences of the agents.
(c) Third, creating a new model of agent behavior and their interaction by gathering new information, determining trade behavior, update in investors' preferences, and modifying the fitness level.

6.2 Technical Agents

The technical agent is usually called momentum traders, they use historical market data to forecast the trend of price. They buy (sell) when the price goes up (down). They follow a simplified way of trading in technical analysis and traders follow a hard behavior [63, 72]. In this model T_i is assigned each momentum trade at the beginning of the simulated random walk model, following a normal distribution of the time-series data, which is subjected to the Markowitz portfolio model. This model also used in genetic networking programming (GNP) or the Genetic algorithm (GA) Model and the reinforcement algorithm model defined by [63].

6.3 Noise Trader/Agent

Noise agents are irrational investors who do not follow the common trading mechanism, technical analysis methods, and portfolio optimization methods. Such traders do not have access to inside information about the stock. Irrational behavior acts as noise traders with an assumption of expecting biases in stock trading [63, 73, 74]. The bias follows a normal distribution with a constant variance in their expected price.

6.4 Reinforcement Learning

Reinforcement learning is the framework or deterministic approach in financial computation that tells the agent is the software portfolio manager performing trading actions in the financial market or in the financial environment [67]. These financial comprises all financial assets and the expectations of the participants toward them. It describes how an agent can learn an optimal action policy in a sequential decision process [66]. It creates a feedback loop between the learning system and the experience gained from the environment. It treats the time t period as:

(a) The agent can take an action ($a_t \in A$).
(b) The agent obtained a reward as ($r_t \in R$) for short-run.
(c) The State becomes ($s_{(t-1)} \in S$).

Starting from using online platforms for investing to capturing financial information through system programming or mathematical formula, everything is dependent on an Internet connection. Loss or disturbance of Internet connection hampers real-time data capturing, data fetching, and transactions and creates huge dissatisfaction among investors. It increases the risk factor among the investors, which may restrict them from investing. Under such circumstances, dew computing can be an excellent solution. The dew server resides in the investor's local machine and consists of four components, such as (1) Dew Server, (2) Dew data capturing system, (3) Dew Client program, and (4) Dew Client Service Application. This server helps serve the requested service to the investors by correlating the local with the remote data. The dew server keeps track, based on the configuration of the investors or self-motivated, of all the visited websites (investment platforms) and creates a dual copy. One or multiple dew sites can be mapped with one dew server. At first, an

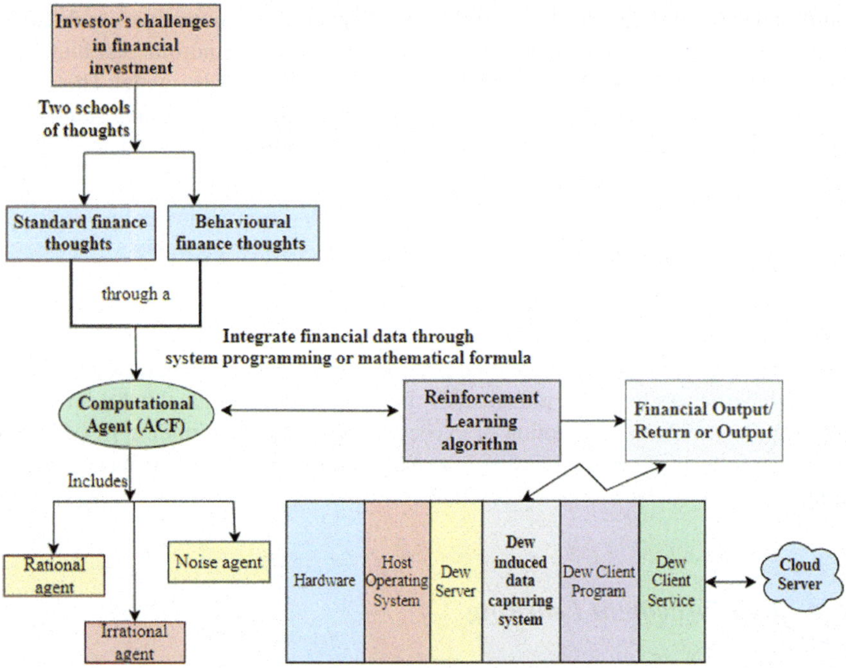

Fig. 7 Interconnection between ACF and cloud-dew architecture

investor login into a website as well as the dew site. The dew site does systematic rule-based data collection. Now if the Internet connection disrupts, the dew script analyzer starts and generates the required files. Next, an acknowledgment was sent to the dew client program and the updating of the master mapping table took place. With the resume of the Internet connection, the synchronization took place and any ordinary user can access the dew investor's data in read mode. The following flowchart presents all the interconnections between the dew-cloud architecture and agent-based computational finance (ACF). Figure 7 depicts the interconnection paradigm in the dew-cloud-assisted ACF.

7 Dew Computing Paradigms for Quantum Machine Learning and Quantum Cryptography

Machine learning and dew computing both are the most powerful emerging technologies in the present era. These technologies contribute a crucial role in the development of modern science and have advantages in their domains. But when they merge it produces a significant advantage. Machine learning algorithms are primarily used to learn and improve machines from experience. It provides the capability to the machines so that they can learn automatically without any human intervention or assistance and adjust actions according to necessity. There are different types of machine learning algorithms and many algorithms have already been developed. Most of these algorithms required a huge amount of storage to execute, which is a primary point of concern to most people. This is the place where dew computing walks in. It is the outsourcing technology, which enables us sufficient space to access applications and data remotely through Internet connectivity. Besides this, the technology has a tremendous advantage in terms of flexibility, availability, accessibility, cost-effectiveness, and so on [17, 16]. These things make this field an obvious choice for every sector. With the advancement of time machine learning algorithms are becoming more powerful and complex and demanding better computational power than every previous day. Besides this, the security of the system is also facing challenges every day. Quantum technology offers a way that the introduction of this technology can address the aforementioned issues.

7.1 Quantum Machine Learning in Dew Computing

Quantum machine learning [75] is such a domain that connects two fields machine learning and a new kind of computing device known as a Quantum Computer. This domain mainly focused to find the complex models in machine learning, which can't be calculated using classical computing methods. It also concentrates to find a way to invent and apply quantum software that enables the machine learning faster than classical computers. Machine learning is mainly used to train machines using data. To minimize the training time CPUs and GPUs are used in the classical computing method. But if the data is too large it will take a huge time to train a machine. Here quantum machine learning comes forward. It can compute multiple states at the same time. Besides this quantum machine learning uses quantum data for operation and during the processing, it follows the laws of quantum physics. For this reason, it is found that quantum machine learning algorithms can exponentially reduce the training time of a machine. It also improves the learning capacity and efficiency of a machine [76]. It is found that quantum computing devices (Quantum Computers) can produce patterns that are very difficult for classical computing devices (Classical Computers). This is another reason which shows that machine learning based on quantum computing can outperform the domain based on classical computing.

Quantum machine learning algorithms execute on quantum computing devices (Quantum Computers). This is the main challenge in this field. Because quantum computers are not fully developed yet. It is also not possible to set up these currently available quantum computers everywhere or at very frequent intervals. Because there are some serious challenges with these quantum computers. One of the major challenges is quantum computers need very low temperatures and isolation. Normal temperature and interaction with other particles cause the increment of the decoherence rate of computer qubits very fast. It will cause the loss of the quantum properties of computer qubits which will further be pointed toward the loss of data stored in the computer qubits. On the other side, rotations can affect the computer qubits, which can cause a crucial error in the circuit. Another challenge comes in this field with the development of algorithms. The development of quantum algorithms is a very critical task because, during the development, the developer has to be very concerned about the background physics. These challenges can impact quantum machine learning very deeply. Dew computing offers a partial solution to these challenges. Few quantum computers can be developed with high capacity and proper precaution. These few computers can be used from anywhere using Internet connectivity through dew computing technology. This emerging technology also offers proper resource management of accessible quantum computers.

7.2 Quantum Cryptography in Dew Computing

The security of the domain of dew computing has become a point of concern with the advancement of technology. In the classical cryptographic method, the primary drawback is that there is no way to exchange the keys with full security. It is impossible to address whether the exchanged key has been revealed or not in the journey. An asymmetric key encryption method has been introduced to solve this problem but this method also has its drawback. It is a very slow method, and it can't be applied to a large size of data. Besides this, the development of quantum technology directly challenges the modern cryptographic technique. Computes based on quantum technology can decrypt classical encryption in a limited time. For any classical modern computers, this time amount is impractical. The introduction of quantum technology in cryptography can address these problems [77, 78]. The no-cloning theorem is the fundamental concept of the quantum cryptographic technique. According to this theorem, quantum states can't be copied. Since quantum states are used in the quantum cryptographic technique so, as per the no-cloning theorem they can't be intercepted during transmission. If anyone tries to intercept information, it will be detected, and the signal will be destroyed. Quantum cryptography is not a new technique as a whole, it doesn't develop from scratch. The working principle of this cryptographic technique is similar to the classical asymmetric cryptographic system. The only difference is that a quantum Cryptographic system uses the properties of quantum physics for the transmission of key. This method is termed Quantum Key Distribution (QKD).

7.3 Quantum Key Distribution (QKD)

Quantum Key Distribution (QKD) is the central technique of quantum cryptography. This technique is used to generate and share random keys between the sender and receiver. There are different types of schemes for QKD, i.e., BB84 [79], Silberhorn, Decoy state, KMB09, E91, etc. Among these BB84 is the most commonly discussed

Table 2 Demonstration of quantum key-distribution indexes

Basis	0	1
+	↑	→
x	/	\

Table 3 Representation of quantum key-distributions

1	0	1	1	0	0	1	0	1	1	Random code
+	x	+	+	x	+	x	x	+	x	Alice's basis
→	/	→	→	/	↑	\	/	→	\	Polarized photons sent by Alice
x	x	+	x	+	+	x	+	+	+	Bob's measurement basis
\	/	→	\	→	↑	\	→	→	↑	Polarized photons measured by Bob
w	r	r	w	w	r	r	w	r	w	Classical channels discussion
	0	1			0	1		1		Final shared secret key

QKD protocol scheme. In this protocol, binary information is encoded on photons using their different properties (i.e., polarization, spin, etc.) and sent these encoded photons to the receiver.

Let Alice and Bob agree to share their secret key using QKD (BB84). At first, Alice choose any random bit as her secret key and two bases (rectilinear basis [+] and diagonal basis [x] as mentioned in Table 2) to encode the secret key. Then she encodes the secret key on photons by the preparation of the polarization state of it using the considered basis randomly. After that, she sent the encoded photons to Bob using quantum channels. As Bob does not know the sequence of encoding basis, after receiving the encoded photons he measures these photons using any random basis. After completing the measurement, Bob contacts Alice by any classical channel and discusses the sequence of his basis of measurement. Now Alice compares her basis sequence with Bob's basis sequence and discards the mismatch basis. After removing the mismatched basis, the matched basis's information formed the communication secret key between Alice and Bob (Table 3).

According to the no-cloning theorem, quantum states can't be copied. So if anyone tries to intercept the encoded photons between transmissions, then he has to measure these. As others (except Alice) don't know the basis sequence of encoding photons, when he tries to measure these photons the information in the state will be destroyed.

7.4 Application of QKD in Dew Computing

Kerberos [80, 81] is a widely used secure authentication technique in cloud computing including dew. In this technique, the Key Distribution Center (KDC) plays a significant role to provide the secure key to the user. The user uses this secure key to access the cloud server safely. Applying QKD in KDC introduces a tremendous advancement of security in this technique over the classical approach. QKD can also be used in asymmetric cryptographic systems for different applications.

8 Dew Computing in Sustainable Internet of Music Things

The usability of Information and Communications Technology in music education displays the signs of existence to provoke conventional music teaching and learning strategies. This projected contribution will assert the advancement of new conceptions of music teaching–learning, music composition, and reconstruction paradigms through the application's congenital applications to the Information and Communication framework. The device-oriented music composition and reconstruction in the Internet-dominant era is usually participant-specific musical information recognition and analytical strategies which incorporates the devices, i.e., the mobile phones, sound sensor devices, and the Cloud-Dew-Edge-Fog (CDEF) peripherals. These smart devices intensify the succession of the Internet of Things (IoT), largely, the Internet of Everything (IoE), where they sense the information from the environment in a collective concern as end-users require. Data sensing through sensors, framing, information retrieval, and dispensation accomplish a significant benefaction on the IoT network. The projected contribution will analyze a set of fruitful music reconstruction and composition applications in the IoT context, which we titled the "Internet of Music Things (IoMT)". The proposed framework will illustrate that the involved individuals who will have the information sensing devices, will be qualified for musical data sensing and composition tasks, which may be shareable for experience within the group and recoup the required information for advanced music analytics and reconstruction operations of common attentiveness. The Internet of Things is a propitious framework that amalgamates the plethora of diversified computational gadgets, assimilating the crowd, substructures, supplementary system ingredients, and organizational structures.

8.1 System Outline: Dew-Induced Internet of Music Things

We demonstrate the system outline of the Dew-induced IoMT into two distinct subsections. The outline of the anticipated architecture is presented in subsection 8.1.1. We depict the modeling assumptions in the Sect. 8.1.2 of estimated dew-induced schema for efficient IoMT.

8.1.1 The System Architecture

This subsection deals with a five-layered ordered dew-induced IoMT schema. Figure 8 consists of five assorted layers, (a) Data Sensing layer, (b) Dew computing layer, (c) Edge computing layer, (d) cloud computing layer, and (e) Application layer. Distinct layers of the projected and ordered framework are illustrated in the following:

(a) Data Sensing Layer: The data sensing layer is the lowermost layer of the provided dew-induced IoMT archetype. This layer comprises numerous virtual segments that embrace the music sources and physical sound sensors. Sound sensing peripherals are capable to sense and accumulate the unprocessed and pre-processed musical data from the musical environments [15]. The data sensing layer also comprises the connected and active microcontrollers for communicating with the associated computing systems and additional microcontrollers to sense the information transmission.
(b) Dew computing Layer: The dew computing layer accumulates the ordinary sound sources, physical peripherals, sensors, and connected microcontrollers. It involves the agglomerative and virtual dew clusters. Dew clusters comprise a minimum of three crucial components, such as the (i) dew server, which enables interactions with upper layered computing devices and sporadic synchronization of information; (ii) dew analyzer amasses the sensor information and explores local computations; (iii) dew storage stores pre-processed sensor data for Internet-independent settings [10].
(c) Edge computing Layer: Proposed system transmits the pre-processed sensor information to the immediate next layer from the dew layer when the Internet connectivity is ON. This layer is titled the edge computing layer. Modules of this layer are the decision-making edge instances, that consist of computing tools, storage, and edge-cloud communication gateways [10]. The available devices are proficient for information evaluations, processed information accumulation, and information transmission to the higher layer. It additionally contains the communication networks for interacting within themselves and/or among edge and cloud computing layers.
(d) Cloud computing layer: The adjacent next to the edge layer in the projected architecture hierarchy in Fig. 8, the cloud computing layer encompasses data centers and essential services. This layer necessitates music resource managing and handling musical compositional activities for the level of data aggregation assignments. It comprises cloud resources dealing with the competencies of predetermination and authorized service excellence of edge computing implementations.
(e) Application Layer: The uppermost layer in the proposed dew-music architecture approaches the processed musical presentations from the cloud data center. End-users in the application layer are efficient in assembling musical compositions, combinations, and reconstruction assignments as the audiences prefer. Eventually, the uppermost layer of the hierarchy grasps the user-requester performances

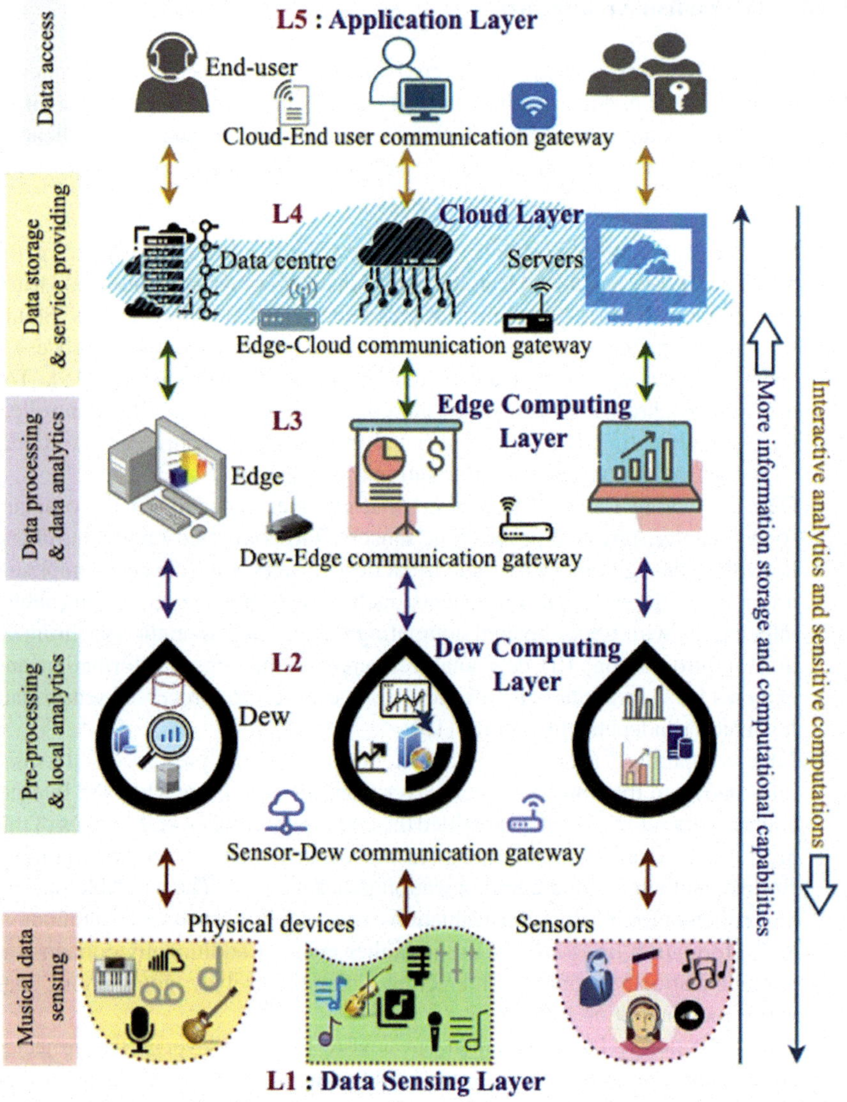

Fig. 8 Representation of the anticipated dew-induced sustainable internet of music things schema

that effects the dew-cloud-based musical information approachability to convey inventive and smart music arrangements of the music auditors and the composers as well.

8.1.2 System Assumptions for Dew-Driven IoMT Schema

We are to depict numerous methodical assumptions centered on the real-time dew-cloud computing consequence. We denote the useful hypotheses over the projected scheme state-of-the-art in the following:

(a) We categorize all peripherals and nodes in the data sensing layer into two all-inclusive segments: the sources of music pieces and partaken in sound sensing equipment.
(b) We assume the musical performances as the sound repository in our projected endeavor and a set of vocal and instrumental music enactments. We also undertake that all information demands pre-processing.
(c) Sensing peripherals are smart gadgets and real-time sound sensors that appreciate optimal pre-processed and unprocessed information. We additionally assume that sensors are positioned in outright Geospatial locations.
(d) In Dew peripherals assemble the sensor information as the restricted sources to evaluate the locale and when the Internet connectivity status is inactive.
(e) Edge computing gadgets are disseminated in a tier within a distributed network to accomplish computational assignments based on the crowdsourcing policy.
(f) Cloud server represents the accumulated music analyzer in which the information summarization and storage of managed information-allied undertakings are achieved.
(g) The end-users in the application layer are authenticated in accessing information from the cloud data center. The users, music composers, and audience are capable to approach public data centers for music information retrieval, qualitative music composition, and reconstruction of the music pieces.

8.2 Connection Strategies for Scheming an Efficient Dew Computing Framework for IoMT

As we discussed earlier, musical information is diffused to associated computational devices with the assistance of a cloud server. The networks can be overcrowded and integrated peripherals may be overburdened owing to viscous bottlenecks. To reduce, the transmission-related latency and to enhance the performance, information is scattered to the cloud server as well as the edge devices. Dew computing conglomerates the underlying perception of the cloud and edge computing, which feeds access to the cloud data, and sustains duplicates on the local repositories. This suggestive schema in the anticipated framework is essentially contingent upon diminishing the duration of data transmission, latency, system power dissipation, and energy depletion. We assume the projected system can be Quality of Service (QoS) and Quality of Experience (QoE), evaluating dual-fold aspects: (a) Reduce the response time of the system, lessen the information transportation time within intermediate computing nodes, and shrink system the service latency that can make a system energy-efficient; and (b) Minimize the intermediate gaps between the sensor devices to the active dew nodes, dew nodes to the adjacent edge, and integrated edge to the cloud data center, that may correspondingly expose toward developing an effective dew-induced IoMT framework.

9 Topics of Interest: Application Domains and Future Research Challenges of Dew Computing-Assisted Sustainable Internet of Things

We illustrate numerous application domains of the Dew Computing in the Sustainable IoT perspectives in the following, although the projected dew computing paradigm can conceptually be depicted in the contexts of large-scale human-centric Internet of Things application domains.

9.1 The Dew-Assisted IoT in Computational Musicology

The research in the context of computational musicology is typically an interdisciplinary research area where systematic analytics, information representation, and humanized creation interact. The technological advancements draw on and significantly contribute to our understanding, perceptions, and visions of the physics and psychophysics of sound. Computational Musicology is always an interdisciplinary or to some extent, a multidisciplinary domain for methodical music analysis, musical patterns identification, and humanized musical information retrieval which is a comprehensive amalgamation of Person Intelligence (PI), Artificial Intelligence (AI), typical programming languages, algorithms, representation learning; additionally, the psychology, music theory, acoustics, signal processing, multimedia information systems, engineering, physics, performance practice, library science, applied mathematics, statistics, and so forth. Systematized technology is perpetually confounding and altering the landscape of human beings' musical experiences as music creators, participants, music listeners, and consumers. Music theorists, musicologists, vocal and instrumental performers, and music composers are habitually meeting with real-world challenges as musical compositions are largely dependent on human perceptions. Apart from the conventional methodologies to retrieve elementary musical components, we apply heuristic and representation learning approaches to recognize or extract the higher level musical features from symbolic musical information, intending to automate or expedite real-world musical tasks, in terms of music generation, composition, and substantial music reconstruction. In the remote and scattered musical performances concern, the dew-assisted IoMT framework can be significant enough to compose specific and smart musical compositions without much-reliant on Internetwork connectivity. Projected phenomena will additionally illustrate the major implications for the music teaching–learning and composition in the subordinate smart classroom with IoT, inclusive of the significance of rigorous proposition, crowd inclusion, paradigms of learning objectives, methodical and pervasive music learning, evaluation, and system integration.

9.2 Dew Computing in Sustainable Internet of Music Things

In the dew-music computational paradigm, future endeavors can be involved in: (a) the illustration of participatory crowdsensing in CDEF framework (b) bandwidth allotment; (c) service and resource virtualization in Internet of Sounds; (d) Performance metrics in low-resource settings.

9.3 Sustainable and Humanized Internet of Music Things

In the contexts of the humanistic-care inspired IoMT schema, we may highlight the following scopes for the further researches, as (a) Dew-driven music crowdsourcing architecture, and (b) modeling of dew-assisted humanized music information fusion.

9.4 Dew-Osmosis: Convergence of Dew and Osmotic Computing in Human-Centric IoT

It is a promising Internet-independent technology, which requires demonstrating parallel and distributed computing over the heterogeneous network. The integration framework of dew and osmotic computing in the context of the Internet of Music Things (IoMT) can be illustrated for framing heterogeneous musical data migration within a dynamic and distributed infrastructure. This work aims to intellectualize, how audiences can be benefited from the osmotic computing-based dew-induced IoMT system analytics. The scope of the future research of the projected Dew-Osmosis computational paradigm may be (a) illustration of participatory and opportunistic crowdsensing and crowdsourcing perceptions, (b) energy-efficient dew-inspired osmotic computing paradigm, (c) performance visualization and virtualized service monitoring for IoT applications, (d) cluster computing in the context of heterogeneous and distributed systems, and (e) dew-osmotic convergence architecture for Industrial IoT (IIoT) applications.

9.5 Communication and Computational Intelligence in Dew-Assisted IoT

Computational intelligence assessments, correctness, and meticulousness at distinct layers in IoT paradigms are verbalized by computational intelligence strategies, which are capable of data collection, the interconnection among the devices and Internet, data processing, intelligent analytics, and decision-making, exclusive of direct human interactions with the systems.

9.6 Cloud-Dew-Edge-Fog (CDEF) and Sustainable Internet of Musical Things: The Convergence

The traditional cloud computing paradigm along with the extended performance-driven computational strategies such as the dew, edge, and fog computing frameworks provide the efficient and smart mechanisms of musical compositions in the domain of the (a) Remote musical performance control and organization, (b) Remote Recording environment, (c) Auto-tuning of the smart musical instruments and Remote Live Mixing, (d) Generative Music and Algorithmic Composition within the shared network.

9.7 Dew-Assisted Big Data Analytics and Industrial Automation

The Network infrastructure in the age of Big data, Industry 4.0, sustainable IoT, and Artificial Intelligence enhance the restrictions of information availability, system scalability, strategic reliability, optimum bandwidth, and less latency. This stemmed from an emerging schema of cloud computing offering centralized information storage, efficient resource balancing, and software-based consequences promoted in the as-a-Service policy. Conversely, there are still a lot of QoS-aware use cases, such as sufficient information and service availability, network delay, jitter, and throughput requirements to be additionally addressed. Limited outcomes to these challenges are solved by the Fog and the Edge computing, offering computing and information closer to the end-user network peripherals. Although these conceptions do not optimally address the concerns of offline information availability and Inter-network latency. A hypothetical response to those challenges could be the conception of dew computing—a supplementary tier in the obtainable client–server framework, functioning on end-users devices to succeed at the optimum level of information synchronization among the data in the cloud and the local devices, that makes a system reliant in the typical Cloud-Fog-Edge schema in the context of network connectivity. Dew-assisted paradigm can address the efficient system modeling in the context of the industrial automation and big data analytics evaluated in the cloud server, with local accomplishment by lessening the network delay and affording offline information availability, specifying the significant data synchronization among the local and cloud databases.

9.8 Influence of Dew Computing on Human–Computer Interaction

Human–Computer Interaction is the key finding procedure of interactions of human beings with computing systems, specifically, as it communicates the technological innovations. The User-centered design, UI, and UX are coalesced with the Human–Computer Interaction to afford the intuitive technology and outcome. Human–Computer Interaction analysts reflect on how to establish and adopt computing systems that can provide satisfactory products to human beings. Nowadays, when we are majorly dependent on the Internet, human-centric applications and outcomes should be offered to the end-users without much interruptions of the Internetwork, and hence, the expected outcome can be achieved shortly. In this scenario, the projected dew computing schema can play a significant role to provide services and to maintain the Quality of Experience.

9.9 Integration of Dew and Affective Computing

Recently, affective computing has acquired admiration and has been harnessed extensively in numerous domains, including marketing, e-learning management, financial and economic behaviors, smart healthcare, assistive tools, and human–machine interface design. Artificial Intelligence plays a pivotal role in the context of designing affective computing schemas and is mainly utilized for decision-making systems. Advancements in artificial intelligence, CDEF, and IoT research have empowered researchers to articulate cost-efficient and robust tools for a diversity of application scenarios. Specifically, the arrival of the machine, deep, and representation learning strategies have crafted it feasible to implement proficient effective computing artifacts for potential healthcare, humanitarian, agricultural, and additional sustainable applications.

9.9.1 Dew-Assisted Sonification in Internet of Humanitarian Things

Sonification refers to the utility of non-speech audio to put across information or perceptualized data. To date, the researchers have carried out a few findings on the applicability of the Internet of Things (IoT) paradigm to the context of interactive sonification. IoT has the promise of facilitating the emergence of emerging outlines of interactive sonification which are the outcome of the shared expertise of interactive sonification artifact by both the presenters performing gestures locally within the infrastructure and with the set of remote users. The dew-assisted interactive sonification in the IoT context may have a large impact on smart healthcare and the Internet of Medical Things. Researchers can additionally utilize the interactive sonification in the Internet of Behaviors, Internet of Sounds, and Internet of Music Things which can be remotely controlled and can be capable to sense the analyzing data based on the implementation requirements.

9.9.2 Dew-Inspired Tactile Internet of Things

One of the foremost Tactile Internet applications is the sustainable Tactile Internet of Things (IoT) which enables human beings to control the desired systems remotely and construct a new-fangled generation of cyber-physical systems, with the significant assistance of haptic technologies. The future research challenges of the dew-inspired Tactile Internet of Things can be the (a) Remote Presence: The interactions of Human-Agent to Human-Agent in the contexts of virtual and augmented reality, (b) Telerobotics: the interactions from Human-Agent to the Working Robot, (c) Remote Robotics: The task flows from a Human-Agent to the Robot-Agent, (d) Extended Reality: A potential task transformation from the Human-Agent to the Virtual Agent.

9.9.3 Cloud-Dew-Edge-Fog-Roof (CDEFR) Computing Strategies for Internet of Risky Things

Along with the traditional computing paradigms, such as cloud, edge, and fog computing, new-age computing technologies like the dew and roof have been introduced to enhance the service providing schemas in IoT applications. If dew computing highlights (a) lessening the network latency, (b) affording offline data availability, and (c) specifying qualitative data synchronization between the local and cloud databases, simultaneously the roof computing emphasizes as an emerging federated computing and networking schema to operate the constrained tools in IoT which offers (i) connectivity, (ii) context assembling, (iii) data and service management, and (iv) majorly the issues of security and privacy.

Conventional IoT conveys computerization and connectivity to billions of performing devices worldwide. We are concerning vehicles, drones, healthcare systems, smart homes, and smart cities. Although the major threats in these disciplines are attacks, the security on the Internet significantly relies on the security of Internet-assisted devices. By specifying visibility in real-time systems and at the granular level, the research challenges should be (a) to securely connect Inter-networks the authorized "risky" IoT devices and (b) to recognize and retort to all vulnerability topics, which can cause when the unauthorized and "risky" devices for communicating through the cache-driven network.

10 Conclusions

In this chapter, we have illustrated a personalized dew-cloud-based sustainable Internet of Things framework. The dew computing schema enables contextual collaboration between the local computing resources and the contemporary cloud infrastructures. The local analyzers stipulate the essential data transmission and analytics functionalities being independent of the integrated Internetwork when in offline mode, synchronize the processes automatically, and update the modifications performed during offline mode with the cloud when the Internet connectivity is restored. This chapter depicts a set of dew-supported diverged IoT and consumer behaviors-oriented paradigms with concepts, system architectures, and systematized service-oriented outlines. Additionally, we have discussed numerous application-oriented case studies and future research challenges in dew-assisted IoT perspectives. We have incorporated the dew computing schema into perception-driven information sensing, data synchronization, system scalability, and distributed information accessibility individualities. Likewise, the dew computing-enabled devices embrace the answerability of human–computer interactions that exist at the derived level in the projected dew-edge-cloud-based computing archetype. We can measure the dew-assisted system performance evaluations, in terms of the network latency, information transference, service, and transportation time delay, system and integrated information entropy, power dissipation, energy consumption, bandwidth allocation and service virtualization, jitter, analysis of QoS and QoE-aware system depiction, resource provisioning, secured communication outlining, scalability, and so forth.

References

1. Olabisi, D., Abubakar, S.K., Abdullahi, A.T.: Demystifying dew computing: concept, architecture and research opportunities. Int. J. Comput. Trends Technol. **70**, 39–43 (2022). https://doi.org/10.14445/22312803/IJCTT-V70I5P105
2. Ray, P.P.: An introduction to dew computing: definition, concept and implications. IEEE Access **6**, 723–737 (2018). https://doi.org/10.1109/ACCESS.2017.2775042
3. Ray, P.P.: Minimizing dependency on internetwork: is dew computing a solution? Trans. Emerg. Telecommun. Technol. (2018). https://doi.org/10.1002/ett.3496
4. Wang, Y.: Definition and categorization of dew computing. Open J. Cloud Comput. **3**, 1–7 (2016). urn:nbn:de:101:1-201705194546
5. Gushev, M.: Dew computing architecture for cyber-physical systems and IoT. Internet Things (Elsevier) (2020). https://doi.org/10.1016/j.iot.2020.100186
6. Gusev, M.: A dew computing solution for IoT streaming devices. In: Information and Communication Technology, Electronics and Microelectronics (MIPRO) , pp. 387–392. IEEE (2017). https://doi.org/10.23919/mipro.2017.7973454
7. Wang, Y., LeBlanc, D.: Integrating SaaS and SaaP with dew computing. In: IEEE International Conferences on Big Data and Cloud Computing (BDCloud), pp. 590–594 (2016). https://doi.org/10.1109/bdcloud-socialcom-sustaincom.2016.92
8. Axak, N., Rosinskiy, D., Barkovska, O., Novoseltsev, I.: Cloud-fog-dew architecture for personalized service-oriented systems. In: 9th International Conference on Dependable Systems,

Services and Technologies (DESSERT), pp. 78–82. IEEE (2018). https://doi.org/10.1109/dessert.2018.8409103
9. Frincu, M.: Architecting a hybrid cross layer dew-fog-cloud stack for future data-driven cyber-physical systems. In: Information and Communication Technology, Electronics and Microelectronics (MIPRO) , pp. 399–403. IEEE (2017). https://doi.org/10.23919/mipro.2017.7973456
10. Roy, S., Sarkar, D., De, D.: DewMusic: crowdsourcing-based internet of music things in dew computing paradigm. J. Ambient. Intell. Humaniz. Comput. **12**, 2103–2119 (2021). https://doi.org/10.1007/s12652-020-02309-z
11. Roy, S., Mukherjee, A., De, D.: IoHMT: a probabilistic event-sensitive music analytics framework for low resource internet of humanitarian musical things. Innov. Syst. Softw. Eng. **1–24**,(2022). https://doi.org/10.1007/s11334-022-00499-7
12. Wang, Y., Skala, K., Rindos, A., Gusev, M., Shuhui, Y., Yi, P.: Dew computing and transition of internet computing paradigms. ZTE Commun. **15**(4), 30–37 (2019). https://doi.org/10.3969/j.issn.1673-5188.2017.04.004
13. Roy, S., Sarkar, D., De, D.: Entropy-aware ambient IoT analytics on humanized music information fusion. J. Ambient Intell. Hum. Comput. **11**(1), 151–171 (2020). https://doi.org/10.1007/s12652-019-01261-x
14. Patel, H., Suthar, K.: A novel approach for securely processing information on dew sites (Dew computing) in collaboration with cloud computing: an approach toward latest research trends on Dew computing. In: Engineering (NUiCONE), pp. 1–6 (2017). https://doi.org/10.1109/nuicone.2017.8325622
15. Roy, S., Sarkar, D., Hati, S., De, D.: Internet of music things: an edge computing paradigm for opportunistic crowdsensing. J. Supercomput. **74**, 6069–6101 (2018). https://doi.org/10.1007/s11227-018-2511-6
16. Khan, F.A., Shaheen, S., Asif, M., Rahman, A.U., Imran, M., Rehman, S.U.: Towards reliable and trustful personal health record systems: a case of cloud-dew architecture based provenance framework. J. Ambient Intell. Hum. Comput. **10**(10), 3795–3808 (2019). https://doi.org/10.1007/s12652-019-01292-4
17. Khan, M.S.H., Roy, P., Khanam, F., Hera, F.H., Das, A.K.: An efficient resource allocation mechanism for time-sensitive data in dew computing. In: 2019 International Conference of Artificial Intelligence and Information Technology (ICAIIT) , pp. 506–510. IEEE (2019). https://doi.org/10.1109/icaiit.2019.8834633
18. Hirsch, M., Mateos, C., Rodriguez, J.M., Zunino, A.: DewSim: a trace-driven toolkit for simulating mobile device clusters in Dew computing environments. Softw. Pract. Exp. (Wiley) (2019). https://doi.org/10.1002/spe.2696
19. Medhi, K., Ahmed, N., Hussain, M.I.: Dew-based offline computing architecture for healthcare IoT. ICT Express **8**(3), 371–378 (2022). https://doi.org/10.1016/j.icte.2021.09.005
20. Ray, P.P., Dash, D., De, D.: Internet of things-based real-time model study on e-healthcare: device, message service and dew computing. Comput. Netw. (2018). https://doi.org/10.1016/j.comnet.2018.12.006
21. Kaura, S., Bhardwaj, D.: A comprehensive review on intrusion detection in edge-based IoT using machine learning. Intell. Commun. Technol. Virtual Mob. Netw. **615–624**,(2023). https://doi.org/10.1007/978-981-19-1844-5_48
22. Alyas, T., Alzahrani, A., Alsaawy, Y., Alissa, K., Abbas, Q., Tabassum, N.: Query optimization framework for graph database in cloud dew environment. CMC-Comput. Mater. Contin. **74**(1), 2317–2330 (2023). https://doi.org/10.32604/cmc.2023.032454
23. Bhatia, J., Italiya, K., Jadeja, K., Kumhar, M., Chauhan, U., Tanwar, S., Bhavsar, M., Sharma, R., Manea, D.L., Verdes, M., Raboaca, M.S.: An overview of fog data analytics for IoT applications. Sensors **23**(1), 199 (2023). https://doi.org/10.3390/s23010199
24. Roman, R., Lopez, J., Mambo, M.: Mobile edge computing, fog: a survey and analysis of security threats and challenges. Fut. Gen. Comput. Syst. **78**, 680–698 (2018). https://doi.org/10.1016/j.future.2016.11.009

25. Sanabria, P., Tapia, T.F., Toro Icarte, R., Neyem, A.: Solving task scheduling problems in Dew computing via deep reinforcement learning. Appl. Sci. **12**(14), 7137 (2022). https://doi.org/10.3390/app12147137
26. Javadzadeh, G., Rahmani, A.M., Kamarposhti, M.S.: Mathematical model for the scheduling of real-time applications in IoT using Dew computing. J. Supercomput. **78**(5), 7464–7488 (2022). https://doi.org/10.1007/s11227-021-04170-z
27. Karmakar, A., Banerjee, P.S., De, D., Bandyopadhyay, S., Ghosh, P.: MedGini: Gini index based sustainable health monitoring system using dew computing. Med. Novel Technol. Dev. **100145**,(2022). https://doi.org/10.1016/j.medntd.2022.100145
28. Podder, T., Bhattacharya, D., Majumdar, A.: Dew computing-inspired mental health monitoring system framework powered by a lightweight CNN. In: Disruptive Technologies for Big Data and Cloud Applications, pp. 309–319. Springer, Singapore (2022). https://doi.org/10.1007/978-981-19-2177-3_31
29. Braeken, A.: Authenticated key agreement protocols for dew-assisted IoT systems. J. Supercomput. **1–21**,(2022). https://doi.org/10.1007/s11227-022-04364-z
30. Rana, S., Obaidat, M.S., Mishra, D., Mishra, A., Rao, Y.S.: Efficient design of an authenticated key agreement protocol for dew-assisted IoT systems. J. Supercomput. **78**(3), 3696–3714 (2022). https://doi.org/10.1007/s11227-021-04003-z
31. Raza, H., Amjad, M., Muneer, S.: IoT based cyber-physical system in automobile devices with dew computing architecture. J. NCBAE **1**(1) (2022)
32. Gusev, M.: AI cardiologist at the edge: a use case of a dew computing heart monitoring solution. In: Artificial Intelligence and Machine Learning for EDGE Computing, pp. 469–477. Academic Press (2022). https://doi.org/10.1016/B978-0-12-824054-0.00020-4
33. Alorf, A.: Blockchain and Dew computing for secure energy trading in smart grids: a profit-aware approach. SSRN 4088660
34. Singh, P., Gaba, G.S., Kaur, A., Hedabou, M., Gurtov, A.: Dew-cloud-based hierarchical federated learning for intrusion detection in IoMT. IEEE J. Biomed. Health Inform. (2022). https://doi.org/10.1109/JBHI.2022.3186250
35. Hati, S., De, D., Mukherjee, A.: DewBCity: blockchain network-based dew-cloud modeling for distributed and decentralized smart cities. J. Supercomput. **78**(6), 8977–8997 (2022). https://doi.org/10.1007/s11227-021-04203-7
36. Ma, Y., Ma, Y., Cheng, Q.: Cryptanalysis and enhancement of an authenticated key agreement protocol for dew-assisted IoT systems. Secur. Commun. Netw. **2022**,(2022). https://doi.org/10.1155/2022/7125491
37. Kaur, N., Mittal, A., Kumar, A., Kumar, R.: Healthcare monitoring through fog computing: a survey. ECS Trans. **107**(1), 7689 (2022). https://doi.org/10.1149/10701.7689ecst
38. Debauche, O., Mahmoudi, S., Guttadauria, A.: A new edge computing architecture for IoT and multimedia data management. Information **13**(2), 89 (2022). https://doi.org/10.3390/info13020089
39. Lee, K., & Man, K.L.: Edge computing for internet of things. Electronics **11**(1), 1239 (2022). https://doi.org/10.3390/electronics11081239
40. Verma, R.: Smart city healthcare cyber physical system: characteristics, technologies and challenges. Wirel. Pers. Commun. **122**(2), 1413–1433 (2022). https://doi.org/10.1007/s11277-021-08955-6
41. De, D., Karmakar, A., Banerjee, P.S., Bhattacharyya, S., Rodrigues, J.J.: BCoT: introduction to blockchain-based internet of things for industry 5.0. In: Blockchain based internet of things, pp. 1–22. Springer, Singapore (2022). https://doi.org/10.1007/978-981-16-9260-4_1
42. Fan, X., Niu, B., Liu, Z.: Scalable blockchain storage systems: research progress and models. Computing **104**(6), 1497–1524 (2022). https://doi.org/10.1007/s00607-022-01063-8
43. Maheshwari, S., Raychaudhuri, D., Seskar, I., Bronzino, F.: Scalability and performance evaluation of edge cloud systems for latency constrained applications. In: 2018 IEEE/ACM Symposium on Edge Computing (SEC), pp. 286–299. IEEE (2018). https://doi.org/10.1109/SEC.2018.00028

44. Jarwan, A., Sabbah, A., Ibnkahla, M., Issa, O.: LTE-based public safety networks: a survey. IEEE Commun. Surv. Tutor. **21**(2), 1165–1187 (2019). https://doi.org/10.1109/COMST.2019.2895658
45. Ahammad, I., Khan, M.A.R., Salehin, Z.U.: Software-defined dew, roof, fog and cloud (SD-DRFC) framework for IoT ecosystem: the journey, novel framework architecture, simulation, and use cases. SN Comput. Sci. **2**, 1–51 (2021)
46. Brezany, P., Ludescher, T., Feilhauer, T.: Cloud-dew computing support for automatic data analysis in life sciences. In: 2017 40th International Convention on Information and Communication Technology, Electronics and Microelectronics (MIPRO), pp. 365–370. IEEE. (2017). https://doi.org/10.23919/MIPRO.2017.7973450
47. Suwansrikham, P., Kun, S., Hayat, S., Jackson, J.: Dew computing and asymmetric security framework for big data file sharing. Information **11**(6), 303 (2020). https://doi.org/10.3390/info11060303
48. Ageed, Z.S., Zeebaree, S.R., Sadeeq, M.A., Ibrahim, R.K., Shukur, H.M., Alkhayyat, A.: Comprehensive study of moving from grid and cloud computing through fog and edge computing towards dew computing. In 2021 4th International Iraqi Conference on Engineering Technology and Their Applications (IICETA), pp. 68–74. IEEE (2021). https://doi.org/10.1109/IICETA51758.2021.9717894
49. Bramer, M.: Principles of Data Mining, vol. 180, p. 2. Springer, London (2007). https://doi.org/10.1007/978-1-4471-7493-6
50. Chen, M.S., Han, J., Yu, P.S.: Data mining: an overview from a database perspective. IEEE Trans. Knowl. Data Eng. **8**(6), 866–883 (1996). https://doi.org/10.1109/69.553155
51. Li, T., Xie, N., Zeng, C., Zhou, W., Zheng, L., Jiang, Y., et al.: Data-driven techniques in disaster information management. ACM Comput. Surv. (CSUR) **50**(1), 1–45 (2017). https://doi.org/10.1145/3017678
52. Saha, S., Nandi, S., Paul, P.S., Shah, V.K., Roy, A., Das, S.K.: Designing delay constrained hybrid ad hoc network infrastructure for post-disaster communication. Ad Hoc Netw. **25**, 406–429 (2015). https://doi.org/10.1016/j.adhoc.2014.08.009
53. Hazra, K., Shah, V.K., Bilal, M., Silvestri, S., Das, S.K., Nandi, S., Saha, S.: A novel network architecture for resource-constrained post-disaster environments. In: 2019 11th International Conference on Communication Systems & Networks (COMSNETS), pp. 328–335. IEEE. https://doi.org/10.1109/COMSNETS.2019.8711166
54. Zhang, C., Fan, C., Yao, W., Hu, X., Mostafavi, A.: Social media for intelligent public information and warning in disasters: an interdisciplinary review. Int. J. Inf. Manag. **49**, 190–207 (2019). https://doi.org/10.1016/j.ijinfomgt.2019.04.004
55. Mondal, T., Pramanik, S., Pramanik, P., Datta, K. N., Paul, P. S., Saha, S., & Nandi, S.: Emergency Communication and Use of ICT in Disaster Management. Emerging Technologies for Disaster Resilience: practical Cases and Theories, pp. 161–197 (2021). https://doi.org/10.1007/978-981-16-0360-0_10
56. Sharma, K., Anand, D., Sabharwal, M., Tiwari, P.K., Cheikhrouhou, O., Frikha, T.: A disaster management framework using internet of things-based interconnected devices. Math. Probl. Eng. **2021**, 1–21 (2021). https://doi.org/10.1155/2021/9916440
57. Savyanavar, A.S., Ghorpade, D.V.R.: Resource allocation scheme for Dew computing paradigm using mobile grids. Int. J. Innov. Technol. Explor. Eng. **199–203**,(2019). https://doi.org/10.35940/ijitee.H6545.078919
58. Mukherjee, A., De, D., Dey, N., Crespo, R.G., Herrera-Viedma, E.: DisastDrone: a disaster aware consumer internet of drone things system in ultra-low latent 6G network. IEEE Trans. Consum. Electron. (2022). https://doi.org/10.1109/TCE.2022.3214568
59. Mukherjee, A., De, D., Dey, N.: Dewdrone: dew computing for internet of drone things. IEEE Consum. Electron. Mag. **12**(1), 52–57 (2021). https://doi.org/10.1109/MCE.2021.3139306
60. Stanco, G., Botta, A., Ventre, G.: Dewros: a platform for informed dew robotics in ros. In: 2020 8th IEEE International Conference on Mobile Cloud Computing, Services, and Engineering (MobileCloud), pp 9–16. IEEE (2020). https://doi.org/10.1109/MobileCloud48802

61. Belch, G.E., Belch, M.A.: Advertising and Promotion: an Integrated Marketing Communications Perspective, 6th edn. McGraw-Hill, New York
62. Court, D., Elzinga, D., Mulder, S., Vetvik, O.J.: The consumer decision journey. McKinsey Q. 3(3), 96–107 (2009)
63. Chen, Y., Xu, Z., Yu, W.: Agent-based artificial financial market with evolutionary algorithm. Econ. Res.-Ekonomska Istraživanja **35**(1), 5037–5057 (2022). https://doi.org/10.1080/1331677X.2021.2021098
64. Sachs, J., Meng, Y., Giarola, S., Hawkes, A.: An agent-based model for energy investment decisions in the residential sector. Energy **172**, 752–768 (2019)
65. Anufriev, M., Radi, D., Tramontana, F.: Some reflections on past and future of nonlinear dynamics in economics and finance. Decis. Econ. Financ. **41**, 91–118 (2018). https://doi.org/10.1007/s10203-018-0229-9
66. Charpentier, A., Elie, R., Remlinger, C.: Reinforcement learning in economics and finance. Comput. Econ. 1–38 (2021)
67. Jiang, Z., Xu, D., Liang, J.: A deep reinforcement learning framework for the financial portfolio management problem (2017). arXiv:1706.10059. https://doi.org/10.48550/arXiv.1706.10059
68. Wang, Y., Xu, T., Niu, X., Tan, C., Chen, E., Xiong, H.: STMARL: a spatio-temporal multi-agent reinforcement learning approach for cooperative traffic light control. IEEE Trans. Mob. Comput. **21**(6), 2228–2242 (2020). https://doi.org/10.1109/TMC.2020.3033782
69. Panda, B., Leepsa, N.M.: Agency theory: review of theory and evidence on problems and perspectives. Indian J. Corpor. Gov. **10**(1), 74–95 (2017)
70. Kumala, R., Siregar, S.V.: Corporate social responsibility, family ownership and earnings management: the case of Indonesia. Soc. Responsibil. J. **17**(1), 69–86 (2021). https://doi.org/10.1108/SRJ-09-2016-0156
71. Adhikari, R.: Foundations of computational finance. Math. J. **22**, 1–59 (2020). https://content.wolfram.com/uploads/sites/19/2020/08/Adhikari-1.pdf
72. Tran, M., Pham-Hi, D., Bui, M.: Parameter optimization for trading algorithms of technical agents. In: 2022 RIVF International Conference on Computing and Communication Technologies (RIVF), pp. 689–694. IEEE (2022)
73. Zubillaga, B.J., Vilela, A.L., Wang, C., Nelson, K.P., Stanley, H.E.: A three-state opinion formation model for financial markets. Phys. A **588**, 126527 (2022). https://doi.org/10.1016/j.physa.2021.126527
74. Granha, M.F., Vilela, A.L., Wang, C., Nelson, K.P., Stanley, H.E.: Opinion dynamics in financial markets via random networks. Proc. Natl. Acad. Sci. **119**(49), e2201573119 (2022)
75. Biamonte, J., Wittek, P., Pancotti, N., Rebentrost, P., Wiebe, N., Lloyd, S.: Quantum machine learning. Nature **549**(7671), 195–202 (2017)
76. Phillipson, F.: Quantum machine learning: benefits and practical examples. In: QANSWER, pp 51–56 (2020)
77. Pandya, M.: Securing Cloud-The Quantum Way (2015). arXiv:1512.02196
78. Sharma, G., Kalra, S.: A novel scheme for data security in cloud computing using quantum cryptography. In: Proceedings of the International Conference on Advances in Information Communication Technology & Computing, pp. 1–6 (2016)
79. Bennett, C.H., Brassard, G.: Quantum cryptography: public key distribution and coin tossing (2020). arXiv:2003.06557
80. Baliello, C., Basso, A., Giusto, C.D.: Kerberos protocol: an overview. Distributed Systems, Italy, Fall (2002)
81. Fan, K., Li, H., Wang, Y.: Security analysis of the kerberos protocol using BAN logic. In: 2009 Fifth International Conference on Information Assurance and Security, vol. 2, pp. 467–470. IEEE (2009)

Cache Computing for Dew Devices at the Edge Networks

Falguni Adhikary, Swarup Kumar Paul, M. S. Obaidat, Debashis De, and Abhijit Das

1 Introduction

Computers are used as controlling and monitoring tools, especially in the industrial IoT sector. Data is born at the edge by millions of IoT devices. The processor speed, size of Random-Access Memory (RAM) and wrong orientation of cache memory in the processor are the main factors behind the speed and performance of a computer.

A cache is a small amount of computer memory that allows data to be temporarily stored in a computing environment. It is an expensive, high-speed semiconductor memory, used to improve the accessibility of recently used or frequently used data. It has two parts: a directory that contains the addresses and remaining data line that stores the data. Cache memory performance has already been improved in the current decade based on speed, latency, and energy consumption. In addition, incorporating cache memory within the processor instead of using it externally enhances computer performance. The web browser, operating system, and CPU generally use cache memory [1, 2].

Grid computing, cloud computing, fog computing, and edge computing depend closely on Internet connectivity. In contrast, dew computing can work in the absence of the Internet. Here, cache memory plays an important role. Cache memory is enough instead of main memory when the same information is required. Thus, cache

---

F. Adhikary · S. K. Paul · A. Das (✉)
RCC Institute of Information Technology, Kolkata, India
e-mail: ayideep@yahoo.co.in

M. S. Obaidat
King Abdullah II School of Information Technology, University of Jordan, Amman 11942, Jordan

School of Computer and Communication Engineering, University of Science and Technology, Beijing, China

D. De
Maulana Abul Kalam Azad University of Technology, Kolkata, West Bengal, India

D. De and S. Roy (eds.), *Dew Computing*, Internet of Things,
https://doi.org/10.1007/978-981-99-4590-0_2

computing is highly demanded on devices like tablets, smartwatches, smart lights, among others.

1.1 Motivation

Dew computing is independent as the service is available from local devices without the Internet and collaborative as devices connect to cloud servers and synchronize the data as needed. Dew computing has two factors—independence, which provides independent services to the user and collaboration is responsible for synchronization between local and remote data. Dew computing enhances the capabilities of end-devices over the concept of cloud computing [3].

Grid, cloud, fog, and edge computing [4] rely heavily on Internet connectivity, but dew computing can work without it. If anyone needs repeated information, Cache Computing (CC) is sufficient instead of main memory. Therefore, cache computing is widely used in Dew Devices (DD) such as smart mobiles, tablets, computers, smartwatches, and various Internet of Things (IoT) applications in medical science, smart agriculture, etc.

1.2 Chapter Contributions

The key contributions in this chapter are summarized below:

(a) A cache computing paradigm is presented to enhance the system performance of computational devices when the cache memory is incorporated inside the processor.
(b) Dew device integration with the computing devices' processor through cache has shown significant enhancements in performance, speed, data transmission and service latency, energy dissipation, and power consumption.
(c) We have demonstrated the directory that contains the addresses and remaining data line that stores the information enriches the system performance metrics of the dew-enabled devices over the conventional concept of cloud computing.

1.3 Chapter Organization

The remaining chapter is organized as follows: Sect. 2 illustrates the application paradigms integrated with the dew computing peripherals. Section 3 depicts the convergence schema of the dew with cache computing. The cache system outline is presented in Sect. 4. The taxonomical demonstration is represented in Sect. 5. Section 6 illustrates the cache computing case studies in memory mapping, data caching, output caching, and distributed caching. The recent advancements in real-time cache computing measures are given in Sect. 7. The simulator descriptions are discussed in Sect. 8. Finally, the future enrichments of dew computing in the machine learning contexts are proposed in Sect. 8. Eventually, this chapter is concluded with concluding remarks and scope for future research.

2 Dew Devices

According to sources, Web in Dew (WiD), Storage in Dew (STiD), Platform in Dew (PiD), Software in Dew (SiD), and Database in Dew (DBiD) can be classified in dew computing. Cache computing-enabled devices can reduce user latency and improve the Quality of Experience (QoE). Dew devices can also perform their activities without communicating with cloud servers. Next, we list some of the real-time applications that demonstrate the systems dew computing.

2.1 Real-Time Image Analysis at Edge

With the advancement in dew computing, image processing frames are now processed in real time with no delay or frame loss from the image sensor. We can use our mobile phone or other devices to translate the preloaded images from one language to another. This eliminates processing delays and ensures that new records are synchronized quickly [5] using Dew Computing (Fig. 1).

2.2 Wearable E-healthcare

Wearable e-healthcare can compare our current heart condition and pulse rate to preloaded vital signs and transfer information to the devices such as smart-watches for emergencies [6]. This gadget acts as dew device and helps patients to track their medical expertise at home or outside (Fig. 2).

2.3 Social Networking Using Rich Cache Enabled Network Devices

Social networking plays an essential role in the world of information. User can employ their dew devices, such as mobile phones, tablets, or laptops, to upload

Fig. 1 Real-time image analysis using dew computing even at resource constrained Internet connection

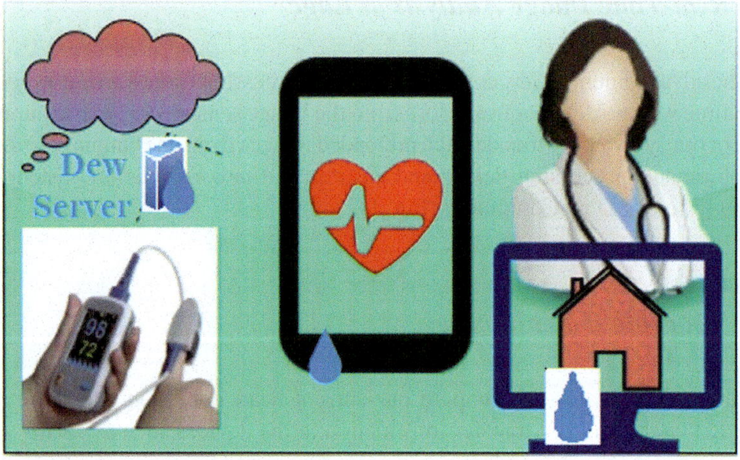

Fig. 2 Cache-based wearable e-healthcare with dew server during unreliable internet connection

or download data from their local copy of an individual's profile, even without an Internet connection [7] (Fig. 3).

Fig. 3 Dew computing-based real-time social networking

2.4 *Edge Content Caching for IoV in Smart City*

Smart transportation is an approach that integrates intelligent technology with sensors, automation, and high-speed networks into transportation systems to reduce congestion in cities and make transportation safer, more efficient, and less expensive than the Internet of Vehicle (IoV) [8], which is an emerging area in this filed. A dew-equipped ambulance can communicate with the nearest dew server to pre-clear traffic with pre-approval [9] (Fig. 4)

2.5 *Vehicular Ad-Hoc Network*

Today, huge traffic congestions have become a problem for large cities. Decentralized intelligent transportation can solve the problem using two key technologies: Vehicle to Infrastructure (V2I) and Vehicle to Vehicle (V2V) [8, 10]. Information can be exchanged with roadside units via Wi-Fi hotspots considering V2I. The latter is a decentralized approach to collecting real-time data from vehicle movements. Dew computing plays an important role in these cases for real-time processing at Edge networks (Fig. 5).

Fig. 4 Edge content caching for IoV in smart city for smart transportation

Fig. 5 Vehicular ad-hoc network

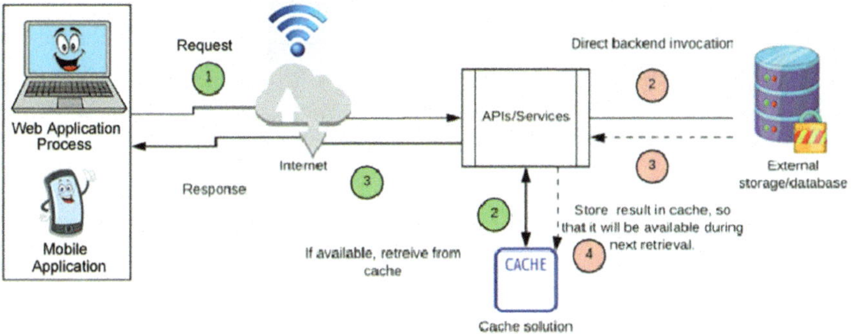

Fig. 6 Integration of dew devices

2.6 *High Definition (HD) Games*

These standalone and online games can be played seamlessly using dew-enabled devices like mobile, tablets, and so on.

3 Integration of Dew Devices

As the name suggests, dew is at the lowest level of the ground [11]. Similarly, the lowest level of the hierarchy is closely related to life in the real world. Dew-enabled devices allow services from local devices without an Internet connection. Once Internet connectivity is restored, the dew architecture will synchronize data with the cloud server [12]. Here, dew computing acts as an additional layer between local and cloud computing [13]. The primary purpose of cache computing is to improve the performance of data retrieval. Therefore, CC is necessary for DD to improve data access performance, scalability, reliability, efficiency, and synchronization latency (Fig. 6).

4 Cache Architecture

Cache memory is a buffer between the CPU and the main memory [14] that is used to improve CPU access speed (Fig. 7).

The CPU is used to look up data in cache memory when needed, and if not found, it goes to the main memory to access the data. A cache hit is performed when the requested data is found, and the search query is also satisfied. However, if the requested data is not found in main memory, it can be treated as a cache miss[15]. So,

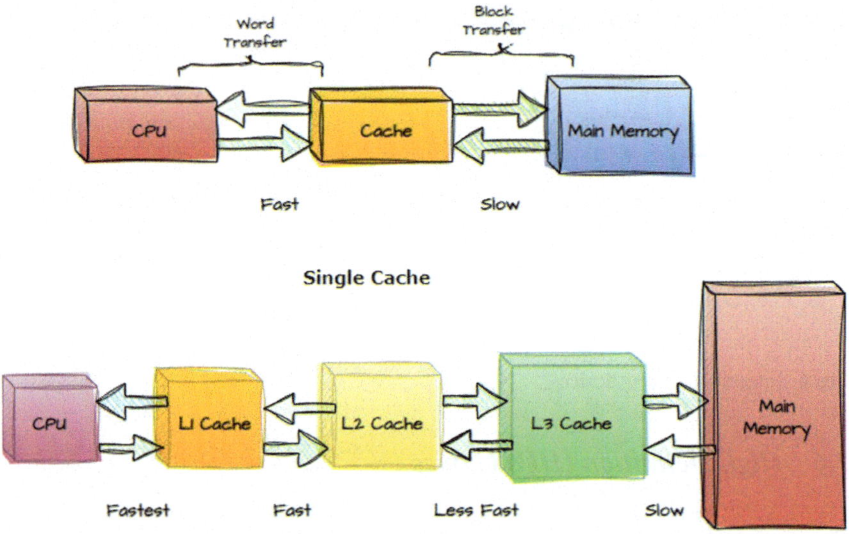

Fig. 7 Block diagram of cache memory

Hit Rate (Ratio) = Number of Hits/(Number of Hits + Number of Misses). There are 41 cache hits and 2 cache misses with a hit ratio of 41/(41 + 2), or 0.9535.

A cache hit ratio above 90% means that most requests are satisfied by the cache, but below 80% is treated as underperforming (Fig. 8).

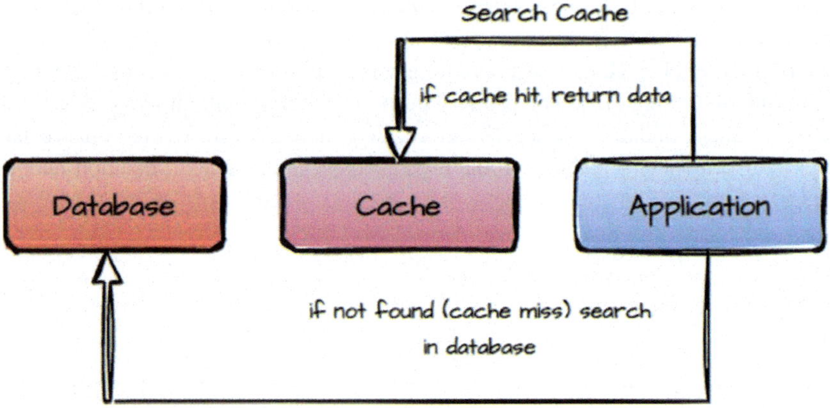

Fig. 8 Cache hit & cache miss

4.1 Taxonomy of Cache Memory

Cache memory can be classified into three tiers based on speed and capacity: L1, L2, and L3. The L1 is integrated into the processor and can store information. L1 is fast, small, and built into the CPU. The CPU consists of two sections used as the input cache and the data cache as the output cache.

The L2 can be embedded in or separated from the microprocessor chip. It can hold data of L1 or higher. L3 is built on the motherboard of the computer. It is a specialized memory and it improves the performances of L1 and L2. L3 is twice faster compared to the Dynamic Random-Access Memory (DRAM) (Fig. 9).

The instructions and data are stored in level 3. Next, the user must choose which cores to send instructions and data for execution and decide to assign cores. This level has separate caches that can handle instructions and data equally. For example, L1 consists of two parts, commands and data. Therefore, the instruction portion of L2 can be stored in the instruction portion of L1, and the data portion of L2 can also be stored in the data portion of L1. Now the CPU can fetch data and instructions one by one [16].

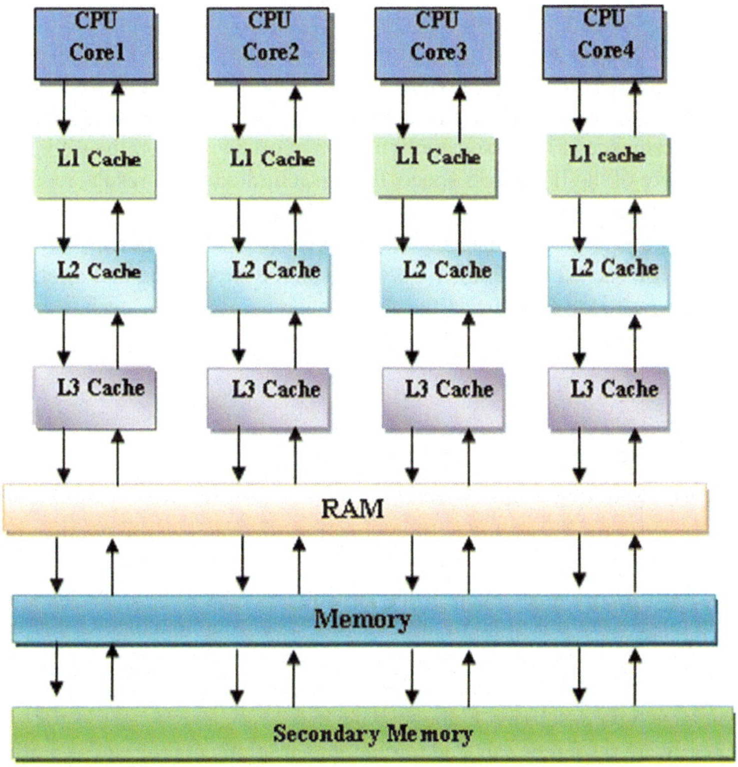

Fig. 9 Taxonomy of cache memory

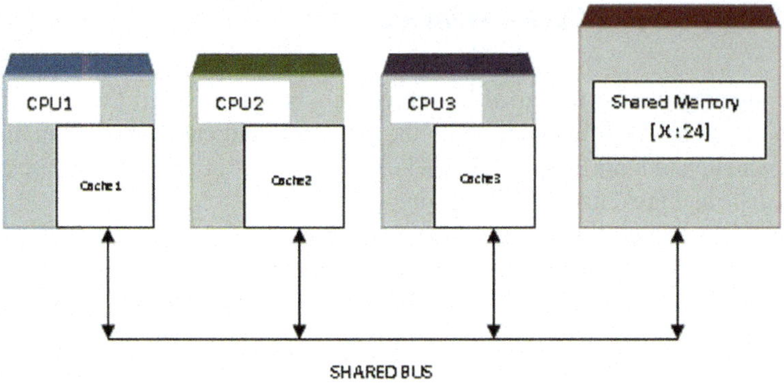

Fig. 10 Cache coherence

5 Cache Memory Design for Network Processors

Due to the enormous dynamics behind IoT and its applications, data network bandwidth requirements are increasing at an alarming rate for network devices, considering the development of microchips. However, during the execution of a program, more than caching is needed as having a low locality of packet address stream compared to the data reference stream.

To increase the range of valid Internet Protocol (IP) addresses and improve the cache's performance [17], a fixed configuration of cache is generally used. However, despite the size of the IP address space, there is a limited set of results regarding the number of outgoing interfaces for network devices.

Dew devices can access data on the network in a very efficient way of cache computing. The cache coherency log is based on the snoopy log and directory [18]. The Snoopy protocol establishes the data coherence between caches and shared memory across the bus [19] (Fig. 10).

6 Real-Time Systems' Cache Memory

In real-time systems, the accuracy of applications depends on the produced results. With fixed deadlines, you can effectively plan your computer resources to meet the deadlines [20]. However, designers and programmers need to know the execution time to complete the real-time requirements of their applications. Cache memory access times are unpredictable because execution times can vary. Therefore, more is needed for hard real-time systems [21]. To improve the processor's performance, cache memory was developed in the average cases.

The hardware resources can lead to random variations in program execution time though this is usually not allowed. Real-time systems are often forced to take a design-to-worst-case approach, resulting in poor processor utilization. Thus, hard real-time

Fig. 11 Real-time systems application of cache computing

systems are limited to circumstances where even the worst-case representation is reliable. Real-time systems must deliver results in time for applications such as airbags, armed tank motion, supertanker rotation, and toasters. CC improves the accessibility of data in DD by reducing latency (Fig. 11).

7 Cache Computing

Today, data-centric applications have strong relevance for specialization. Modern processors provide 40–60% area with a cache for storing and retrieving data. Active compute units transform elements in cache sub-arrays without transferring data. Such transformations unlock massive data-parallel computing power. Caching can work even without an Internet connection. The application cache can retrieve recently accessed or used data which helps to save resources. Several algorithms are used for cache maintenance—Least Used, Not Recently Used, and Recently used. Programs in the computer reside permanently in secondary storage.

The process is divided into pages of equal size. The main memory is also divided into frames of similar size. Frame size corresponds to page size. It is also a similar organization of main memory and cache memory. Here, the parts of the main memory are named as frame and the parts of the cache memory are line, i.e., same as the block size of main memory.

The smallest addressable memory unit is a word. 1 word = 1 byte. Let's assume that main memory size is 64 words and cache size =16 words, and line size = 4 words.

The number of blocks = 64/4 = 16 blocks.

0	0 1 2 3	⇐ Blocks
1	4 5 6 7	
2	8 9 10 11	
3	12 13 14 15	
4	16 17 18 19	
5	20 21 22 23	
6	24 25 26 27	
7	28 29 30 31	
8	32 33 34 35	
9	36 37 38 39	
10	40 41 42 43	
11	44 45 46 47	
12	48 49 50 51	
13	52 53 54 55	
14	56 57 58 59	
15	60 61 62 63	

Main memory
64 words

So, number of line in cache = 16/4 = 4, i.e., 0, 1, 2, 3.

Block Numbers

Lines →

0	4	8	12
1	5	9	13
2	6	10	14
3	7	11	15

The 0th block maps onto the 0th line, the 1st block maps onto the 1st line, the 2nd block onto the 2nd line, and 3rd block goes on to the map on the 3rd line. However, in the 4th block, there is no cache line, so, here, Round Robin is applicable to resolve the issues, i.e., the 4th block map onto the 0th line, and so on.

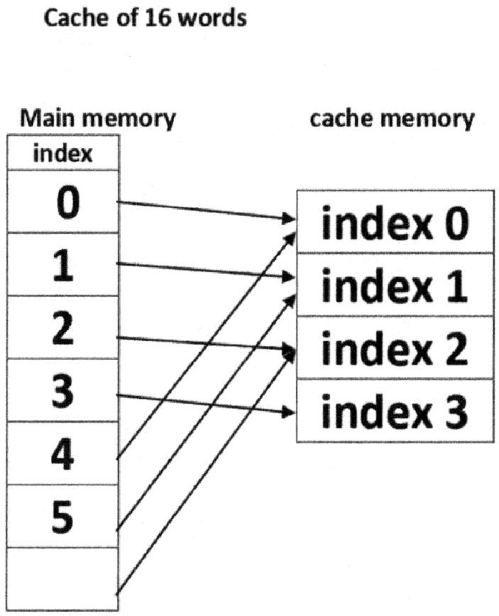

Now people are dealing with a large amount of data that needs to be recommended to store cloud services, but due to lack of Internet connectivity, data cannot access uninterruptedly. To overcome the issues, DD is used as DD contributing to deploying the fragmented working file to the cloud storage services.

7.1 *Data Caching*

When data is fetched from the database, it may take some time due to constraints. If the data set is used frequently, we recommend storing it on the application server, as the data can be accessed faster from the data cache. Caching and de-caching are handled at the application and server levels. The report query is sent to the database upon access, and the result set is visualized. Automatic caching reduces the report loading time. Holistic approach can retrieve the result set from the cache server when the user opens the report instead of a new query from the database (Fig. 12).

7.2 *Application/Output Caching*

Output caching stores the outcome of the pages, hypertext protocol response (HTPR), among others, in the memory. It can download the web page very fast and also reduce the overhead of the server. It caches raw HTML rather than a raw data set. Caching can happen at the module level, but generally at the HTML level. We can reduce 50%

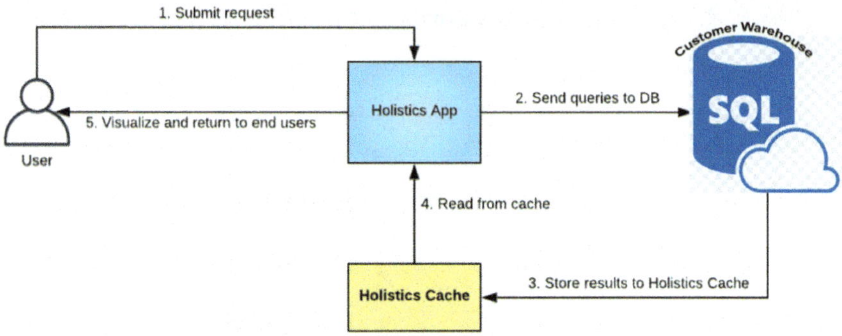

Fig. 12 Data caching

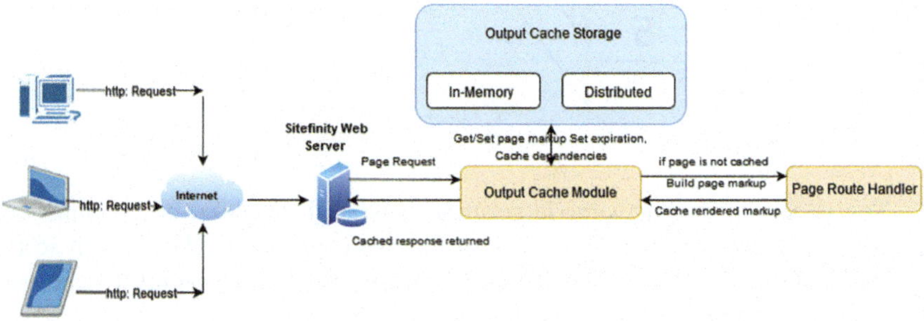

Fig. 13 Output caching

the time of page loading by using this method. The output cache is the important caching layer as incoming requests reach it first, and content is served instantly. When applying the output cache on the page level, Sitefinity CMS does each personalization segment of the page. The widget level has inherent dynamic behavior and performs as cache substitution.

There is only one default cache profile for all pages, though the cache profile can be changed on the page level. If the page is not found on the output cache Sitefinity CMS reads the output cache.

Each consecutive request for the same page is served from the cache unless someone has modified it; page has expired or output cache is disabled (Fig. 13).

7.3 Distributed Caching

A distributed cache is a technique that combines random-access memory from different computer networks into a single memory data store and uses it as a data

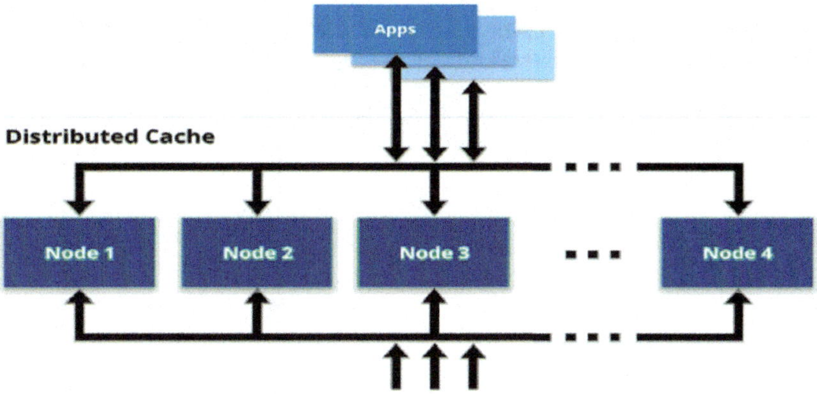

Fig. 14 Distributed caching

cache to improve scalability, performance, resilience, and cost-effectiveness [22] (Fig. 14).

For extensive system data, data can be cached on web servers and thus stored on multiple distributed servers. New servers can be easily added to the server pool so that cache never runs out of space. YouTube and Amazon also use this mechanism to deliver content faster. The Dew server provides services to a single client rather than multiple clients (Fig. 15).

7.4 A Recent Advancement in Cache Memory

In modern computer systems, the memory hierarchy consists of registers inside the processor [23], cache memory inside and outside the processor, and virtual memory inside the hard disk drive. This makes computers efficient which improves speed, latency, and hit ratio performance. The tech world has created its walls for storing data. Several advancements have been made to strengthen the wall performance, such as the Hybrid Memory Cube (HMC) [24]. It is a revolutionary high-speed 3D Dynamic Random-Access Memory (DRAM) architecture [25] that combines high-speed through-silicon-via stacked logic layers with bonded memory chips to increase memory performance significantly. Technology companies like SAMSUNG and MICRON have joined forces to make this technology better and more efficient and integrate various applications.

The alliance is called the Hybrid Memory Cube Consortium. This technology offers 15x better performance than DDR3 modules and consumes 70% less energy per bit than DDR3. Secondary storage on solid-state devices (SSDs) is much faster, and random-access storage performs better. Smart Response Technology (SRT) puts in place the Intel Z68 chipset that permits SATA solid-state devices (SSDs) to act much faster as a hard drive cache (Fig. 16).

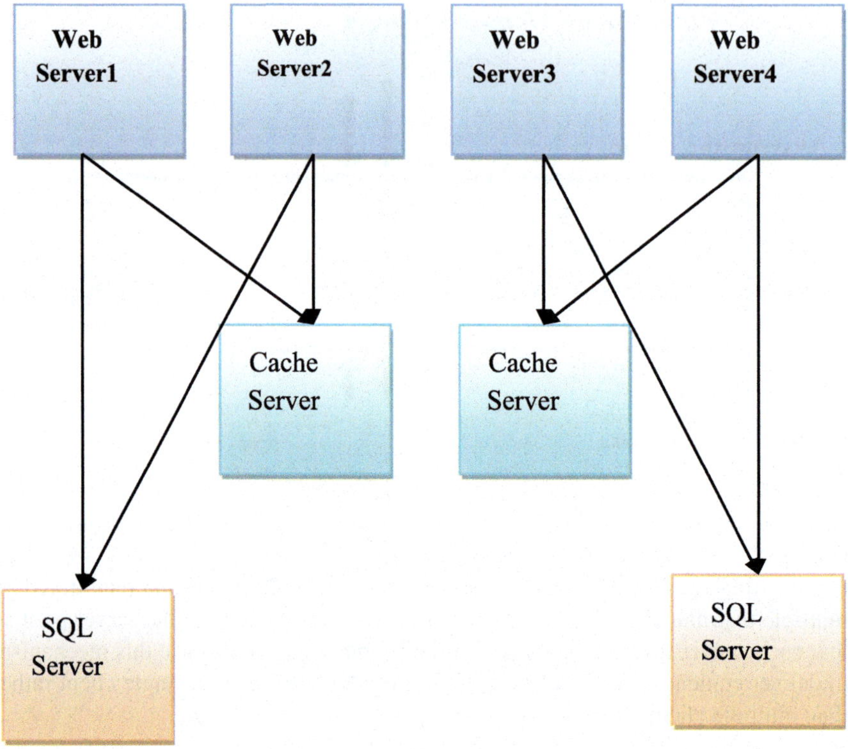

Fig. 15 Distributed cache server

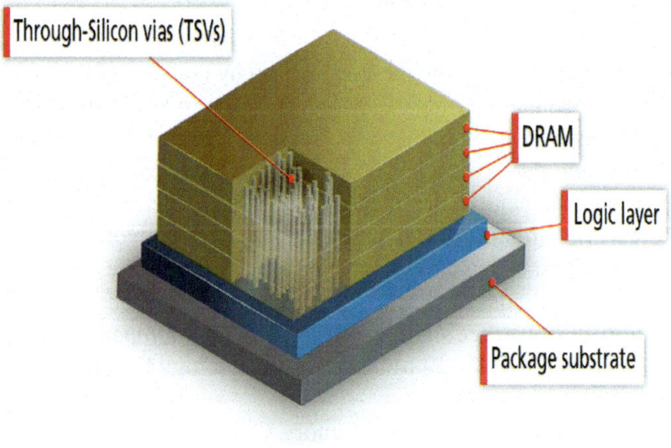

Fig. 16 Hybrid memory cube

7.5 *Execution Cache Memory Model*

Execution Cache Memory (ECM) is an analytical resource-based model that runs on multi-core processors to predict the runtime serial code [26]. The model decomposes the total execution time and summarizes it by machine model. ECM also works with parallel or sequential code. Instruction execution and data transfer are two essential resources. The sequential loop in building the ECM involves four steps.

- Derive Vi, i.e., data volume per loop iteration across the path *i*.
- Ti = Vi/bi overall data paths.
- Then calculate an "in-core" runtime prediction.
- Build a prediction T_{ECM} using a machine model.

In the parallel loop, the model assumes that the contributions come from the resources like pipelines, and shredding out scalable caches that are divided into several cores.

7.6 *Cache Simulator*

The Cache simulator is a tool used to implement and evaluate new ideas in the domain for understanding and finding micro-architectural bottlenecks. Cache hierarchy is an integral part of the processor. Some simulators support the whole processor, though others may only support cache hierarchy. It can be used on the existing processor where hardware performance is insufficient or unavailable. Cache simulators can be categorized based on time and function. The functional simulator is used to analyze statistical evaluation and functional correctness to trace execution and the timing simulator is used to look at the timing frame to the completion of each execution. In the general cache, a simulator is responsible for simulating the user-level execution [27]. The full system simulator can run on one or more operating systems to detain all instructions for execution. The simulator can work in three modes execution, emulation, and trace-driven (Table 1).

Table 1 Comparison of some cache simulators

Name of simulator	Type of simulator	Level of execution
Sniper	Timing	User level
Tejas	Timing	User level & full system
MacSim	Timing	User level
Sniper	Timing	User level
Gem5	Timing	User level & full system
Manifold	Timing	User level & full system
SMPCache	Functional	Not applicable

Cache simulators can be validated based on hardware performance, performance measured on another simulator or analytical models.

Simulation trees are simulated from top to bottom in the forest as requests for memory addresses. The simulation starts from the smallest cache configuration to the largest cache configuration. Using the FIFO replacement policy speeds up the level 1 cache memory.

Dew maintains a table that tracks cache misses for various caches. The number of simulation cache configurations depends on the size of the table.

7.7 *Machine Learning in Cache Computing*

With the tremendous growth of mobile technology in recent years, from multimedia technology to online gaming, machine learning technology, and upcoming 5G networks will play a key role [28] in optimizing cache performance across Dew devices. These required efficient data delivery means provide minimal latency. Cache computing helps retrieve frequently used data from backhaul traffic. Machine learning (ML) is used to solve computer science problems in areas such as bioinformatics, traffic lights, image processing, traffic prediction, wireless sensor networks (WSNs), and resource management. Dew devices can be used to train data-driven ML techniques to ensure optimal resource allocation and reduce cache compute time complexity. Recurrent Neural networks (RNN) and Short-term minimal memory (LSTM) are commonly used to improve cache replacement machine learning techniques in reinforcement learning. RNN solve the cache replacement problem and generate new policies from previously accessed data. An LSTM learns the sequential data to compute the outputs. For cache replacement, the history of cache accesses is sequential data. Hawkeye and Glider are two of the most effective PC-based predictors. Still, having a cheap alternative strategy with little overhead and hardware upgrades is essential. The target features of the reinforcement learning repository (RLR) are first split by training a reinforcement learning (RL) agent and hill-climbing analysis. The final access type for feature lines, line hits, is chosen based on neural network weights. The feature selection process is fully automated, allowing RL agents to adapt to dynamic changes in access patterns. These constrained characteristics are then used to design replacement policies by assigning priority levels to cache lines.

8 Conclusions

Cache computing on dew devices is developed to improve the endpoint capabilities over contemporary cloud computing concepts. This chapter discusses how cache computing could be integrated with dew devices to enhance data scalability, reliability, and accessibility without Internet access or applications. Synchronization with all the data in dew devices with different cache computing enabled on heterogeneous platforms is possible using IoT interoperability.

References

1. Kumar, S., Singh, P.K.: An overview of modern cache memory and performance analysis of replacement policies. In: 2016 IEEE International Conference on Engineering and Technology (ICETECH), pp. 210–214, Coimbatore, India (2016). https://doi.org/10.1109/ICETECH.2016.7569243.
2. Agarwal, A., Simoni, R., Hennessy, J., Horowitz, M.:An evaluation of directory schemes for cache coherence. In: The 15th Annual International Symposium on Computer Architecture. Conference Proceedings, pp. 280–289, Honolulu, HI, USA (1988). https://doi.org/10.1109/ISCA.1988.5238
3. Ray, P.P.: An Introduction to Dew Computing: definition, Concept and Implications. IEEE Access **6**, 723–737 (2018). https://doi.org/10.1109/ACCESS.2017.2775042
4. Skala, K., Davidovic, D., Afgan, E., Sovic, I., Sojat, Z.: Scalable distributed computing hierarchy: cloud, fog and dew computing. Open J. Cloud Comput. **2**(1), 16–24 (2015). https://doi.org/10.19210/1002.2.1.16
5. Hauet, A., Kruger, A., Krajewski, W.F., Bradley, A., Muste, M., Creutin, J.D. Wilson, M.: Experimental system for real-time discharge estimation using an image-based method. J. Hydrol. Eng. **13**(2) (2008). https://doi.org/10.1061/(ASCE)1084-0699(2008)13:2(105)
6. Partha, P.R., Dinesh, D., De, D.: Internet of things-based real-time model study on e-healthcare: device, message service and dew computing. Comput. Netw. **149**(11), 226–239 (2019). https://doi.org/10.1016/j.comnet.2018.12.006
7. Chen, C., Wang, C., Qiu, T., Atiquzzaman, M., Wu, D.O.: Caching in vehicular named data networking: architecture, schemes and future directions. IEEE Commun. Surv. Tutor. **22**(4), 2378–2407 (2020). https://doi.org/10.1109/COMST.2020.3005361
8. Ray, P.P.: Minimizing dependency on internetwork: is Dew computing a solution? Trans. Emerg. Telecommun. Technol. (Wiley) **30**(1). https://doi.org/10.1002/ett.3496
9. Ghosh, S., De, D.: DewCityGame: Dew computing-based 5G IoT for smart city using coalition formation game. IETE J. Res (2022). https://doi.org/10.1080/03772063.2022.2120916
10. Yang, F., Wang, S., Li, J., Liu, Z., Sun, Q.: An overview of internet of vehicles. China Commun. **11**(10), 1–15 (2014). https://doi.org/10.1109/CC.2014.6969789
11. Wang, Y., Pan, Y.: Cloud-dew architecture: realizing the potential of distributed database systems in unreliable networks. In: Proceedings of the 21st International Conference on Parallel and Distributed Processing Techniques and Applications (PDPTA 2015) , pp. 85–89. Las Vegas, USA (2015)
12. Gushev, M.: Dew computing architecture for cyber-physical systems and IoT. Internet Things **11** (2020). https://doi.org/10.1016/j.iot.2020.100186
13. Agarwal, A., Li, H., Roy, K.:DRG-cache: a data retention gated-ground cache for low power. In: Proceedings 2002 Design Automation Conference (IEEE Cat. No.02CH37324), pp. 473–478, New Orleans, LA, USA (2002). https://doi.org/10.1109/DAC.2002.1012671
14. Puaut, I.: Cache analysis versus static cache locking for schedulability analysis in multitasking real-time systems. In: Proceedings of the 2nd International Workshop on Worst-Case Execution Time Analysis, in Conjunction with the 14th Euromicro Conference on Real-Time Systems, Vienna, Austria (2002)
15. Thiebaut, D.: On the fractal dimension of computer programs and its application to the prediction of the cache miss ratio. IEEE Trans. Comput. **38**(7), 1012–1026 (1989). https://doi.org/10.1109/12.30852
16. Byna, S., Chen, Y., Sun, X.-H.:A taxonomy of data prefetching mechanisms. In: 2008 International Symposium on Parallel Architectures, Algorithms, and Networks (i-span 2008), pp. 19–24, Sydney, NSW, Australia (2008). https://doi.org/10.1109/I-SPAN.2008.24

17. Rodriguez, P., Spanner, C., Biersack, E.W.: Analysis of web caching architectures: hierarchical and distributed caching. IEEE/ACM Trans. Netw. **9**(4), 404–418 (2001). https://doi.org/10.1109/90.944339
18. Hou, T., Feng, G., Qin, S., Jiang, W.: Proactive content caching by exploiting transfer learning for mobile edge computing (2018). https://doi.org/10.1002/dac.3706
19. Chaiken, D., Fields, C., Kurihara, K., Agarwal, A.: Directory-based cache coherence in large-scale multiprocessors. Computer **23**(6), 49–58 (1990). https://doi.org/10.1109/2.55500
20. Chiueh, T.-C., Pradhan, P.: Cache memory design for network processors. In: Proceedings Sixth International Symposium on High-Performance Computer Architecture. HPCA-6 (Cat. No.PR00550), pp. 409–418 (2000). https://doi.org/10.1109/HPCA.2000.824369
21. Milligan, M.K., Cragon, H.G.: The use of cache memory in real-time systems. Control Eng. Pract. **4**(10), 1435–1442 (1996). ISSN 0967-0661, https://doi.org/10.1016/0967-0661(96)00154-2
22. Chang, Z., Lei, L., Zhou, Z., Mao, S., Ristaniemi, T.: Learn to cache: machine learning for network edge caching in the big data era. IEEE Wirel. Commun. **25**(3), 28–35 (2018). https://doi.org/10.1109/MWC.2018.1700317
23. Banday, M.T., Khan, M.: A study of recent advances in cache memories. In: 2014 International Conference on Contemporary Computing and Informatics (IC3I), pp. 398–403, Mysore, India (2014). https://doi.org/10.1109/IC3I.2014.7019786
24. Santos, P.C., Alves, M.A.Z., Diener, M., Carro, L., Navaux, P.O.A.: Exploring cache size and core count Tradeoffs in systems with reduced memory access latency. In: 2016 24th Euromicro International Conference on Parallel, Distributed, and Network-Based Processing (PDP), pp. 388–392, Heraklion, Greece (2016). https://doi.org/10.1109/PDP.2016.55
25. Jeddeloh, J., Keeth, B.:Hybrid memory cube new DRAM architecture increases density and performance. In: 2012 Symposium on VLSI Technology (VLSIT), pp. 87–88 (2012). https://doi.org/10.1109/VLSIT.2012.6242474
26. Hager, G., Eitzinger, J., Hornich, J., Cremonesi, F., Alappat, C.L., Röhl, T., Wellein, G.: Applying the executioncache-memory model: current state of practice. In: Proceedings of ACM (SC18), Dallas, Texus USA (2018)
27. Brais, H., Kalayappan, R., Panda, P.R.: A survey of cache simulators. ACM Comput. Surv. **53**(1), 1–32 (2021), Article No. 19. https://doi.org/10.1145/3372393
28. Sethumurugan, S., Yin, J., Sartori, J.:Designing a cost-effective cache replacement policy using machine learning. 2021 IEEE International Symposium on High-Performance Computer Architecture (HPCA), pp. 291–303, Seoul, Korea (South) (2021). https://doi.org/10.1109/HPCA51647.2021.00033

DewMonitor: Dew Computing Monitoring System for Sustainable IoT

Amiya Karmakar, Pritam Ghosh, Matías Hirsch, Partha Sarathi Banerjee, Debashis De, Mateos Cristian, and Zunino Alejandro

1 Introduction

The development of conventional industries like healthcare, agriculture, electricity, education, and transportation have benefited from the Internet of Things (IoT) devices' reconfigurability, networking, task automation, and control abilities [1–5]. However, the volume of data generated by IoT devices presents significant difficulties for the system's energy sustainability and storage, communication, processing, and security. Therefore, the idea of "green sensing and communication" has grown significantly to overcome these difficulties [5–12]. This article explores the available green sensing and communication approaches for dew-based IoT to create sustainable systems for various IoT-based applications. Furthermore, sustainable monitoring and analysis of I-Dew help users and owners use I-Dew's functionalities without compromising the desired service quality.

Major contributions:

- By introducing a new monitoring and analysis layer between the IoT device and higher-level servers in the cloud, this article provides an operational and analytical view of Dew Computing's IoT device solution.
- DewMonitor makes the environment sustainable for Dew computing

A. Karmakar · D. De
Maulana Abul Kalam Azad University of Technology, Haringhata, West Bengal, India

P. Ghosh
Iswar Chandra Vidyasagar Polytechnic, Jhargram, India

M. Hirsch · M. Cristian · Z. Alejandro
ISISTAN-UNCPBA-CONICET, Tandil, Argentina

P. S. Banerjee (✉)
Kalyani Government Engineering College, Nadia 741235, India
e-mail: psbanerjee.kgec@gmail.com

D. De and S. Roy (eds.), *Dew Computing*, Internet of Things,
https://doi.org/10.1007/978-981-99-4590-0_3

- This research examines how the Dew Computing system and IoT devices function in various scenarios.

2 Dew Computing—From the Perspective of IoT

This paper offers an operational and analytical view of Dew Computing's IoT device solution by adding a new monitoring and analysis layer between the IoT device and higher-level servers located in the cloud [13]. A mobile phone or processing center for intelligent homes can serve as the dew monitoring layer's autonomous computer and communication device. Based on the DewMonitor's analytical output, stakeholders can fine-tune their Dew Computing system.

DewMonitor makes the items in the I-Dew sustainable by bringing analytics closer or integrating it. This idea returns computation to the data sources rather than dumping it into the clouds [14].

The IoT sensor is the component that transforms environmental signals into data so that it may be processed to produce pertinent information about a particular physical, chemical, or biological parameter [15–18].

We are using the Internet of Things sensor to sense the occurrence of a specific event, a status change, or other digital data in addition to measuring natural signals. The three fundamental tasks of an IoT sensor are sensing signal conversion to digital data and digital data transmission [18, 19].

The dew server is a piece of hardware that sits between the IoT sensor and high-level servers in the architecture hierarchy.

It conducts fundamental processing and regulating operations while interacting with IoT sensors and high-level servers. IoT device control, Internet communication management, data collecting, storage, transmission, data processing, data visualization, IoT device administration, and user interface are the core categories for the functions.

High-level servers are typical cloud servers that carry out final data processing by distributing web services and applications to the target audience [20–22].

Combining the introduction of intermediate fog and cloud servers with the intermediate dew computing level is not advisable. In our approach, we have kept the dew servers close to the IoT sensors and cloudlets and fog servers representing high-level servers in front of the clouds. Many fog computing scenarios consider a smartphone to be the user. In our concept, this device serves as a dew server to other IoT sensors close to the smartphone (as in the typical ubiquitous computing scenario) [23–25].

Advantages of using Dew monitoring layer

- Create user interface (UI) mockups: Analysts might utilize DewMonitor to better comprehend how a dew device interacts with a system.
- Define system usage patterns: It is crucial to understand dew device behaviors and trends within the system during runtime. This assists in concentrating on crucial system components that are useful for understanding the efficient use of Dew environment.

```
[root@localhost system]# cat node-exporter.service
[Unit]
Description=Prometheus node exporter
After=network.target

[Service]
Type=simple
ExecStart=/usr/local/bin/node_exporter
[Install]
WantedBy=multi-user.target
```

Fig. 1 Service file of the node exporter

```
[root@localhost system]# systemctl status node-exporter.service
● node-exporter.service - Prometheus node exporter
   Loaded: loaded (/etc/systemd/system/node-exporter.service; disabled; vendor >
   Active: active (running) since Sun 2022-09-25 13:08:02 EDT; 2s ago
 Main PID: 3241 (node_exporter)
    Tasks: 3 (limit: 19438)
   Memory: 4.8M
   CGroup: /system.slice/node-exporter.service
           └─3241 /usr/local/bin/node_exporter
```

Fig. 2 Status of the node exporter service

- Perform a feasibility analysis: Feasibility studies aid in evaluating dew system viability and potential effects on the overall dew environment.

Dew server is the centralized analytics hub that relates to distributed dew device network. To analyze the dew device performance throughout the operational time, we installed a node exporter service in the dew system [50]. This service helps the dew system to send the infrastructure-related data to the target destination. Figure 1 shows the service file of the node exporter, and it starts the process by executing ExecStart, and Fig. 2 shows the status of the node exporter service. Node exporter also helps to acquire data for unreliable networks. Stakeholders can monitor the system's performance and outage by using the node exporter's functionality. In this work, we further created some alert channels and alert systems based on predefined thresholds to provide an alert to stakeholders about the system. Monitoring the system's chaos can also be achieved with the power of DewMonitor.

3 I-Dew Integration

Without the dew computing layer incorporated in the overall architecture of the solution, the application of direct cloud connection, fog computing, and cloudlets is not conceivable.

A reliable power source and Internet connection are required to connect IoT devices to the cloud directly. Only immovable IoT sensors and devices can handle the more critical energy requirements of direct connection, making this viable. I-Dew is a term that describes this interaction between IoT and Dew Computing devices [25–29].

This research examines how the Dew Computing system and IoT devices function in various scenarios. For example, they will need more energy resources to transport data to the local network to communicate with the remote server. Therefore, the suggested fix includes a dew monitoring layer to assess the ecosystem's performance in vital data gathering, storing, processing, and communicating to enable sustainable I-Dew [30–34].

The dew computing monitoring layer works by simulating a corresponding dew server, which could be a standalone intelligent home processing and communication box or a mobile smartphone [35–38].

We have displayed a dashboard that shows the dew server in analytical form. Recommending the appropriate dew server system solution over the existing one is explained.

The newly defined dew server system realizes an additional I-Dew category of the already existing dew computing categories.

The idea allows the devices monitoring the system to analyze and organize data more sustainably [39–44].

4 DewMonitor Architecture

Independence and cooperation are the central tenets of dew computing. Independence requires that the local device be capable of operating without a constant Internet connection. In order to collaborate, an application must be able to connect to a cloud service and synchronize data as needed [45–47].

A Dew Cloud Monitoring layer is required to set up a DewMonitor architecture. The data analytics server (DAS), the dew nodes, and the DewMonitor visualizer are the minimum number of components that make up the DewMonitor layer, which is an isolated environment for processing collected data generated from the sensing layer. Figure 3 illustrates the overall architecture of DewMonitor.

Dew server (DS): The DS functions on the Dew node like a cloud service. It communicates with the cloud service and regularly syncs content with it.

Dew analytics server (DAS): The dew analytics server gathers data on how the distributed dew devices are being utilized and visualizes all the real-time data. This layer aids in automatic alert generation based on preset thresholds.

DewMonitor Analyzer (DMA): After receiving information on usage patterns from the DAS, the DMA uses the information to modify and adapt the dew server to the user's needs to improve that user's experience.

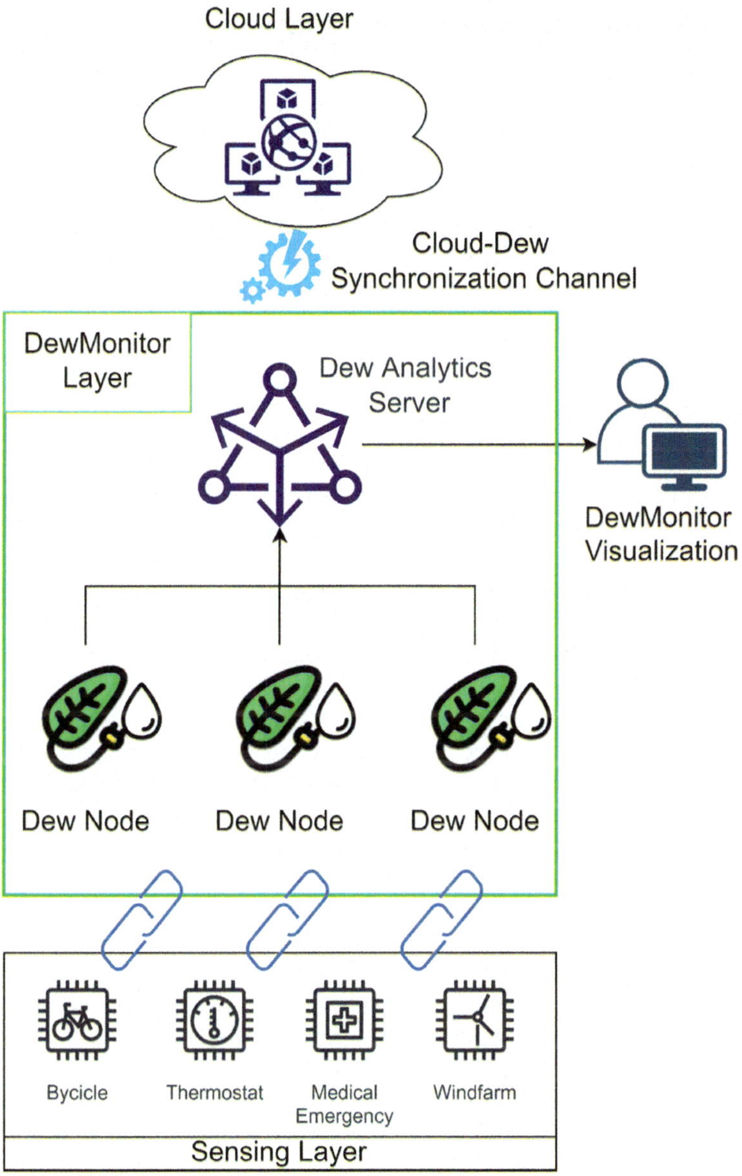

Fig. 3 DewMonitor architecture

5 Dew Computing Device Monitoring and Analysis

This analysis has been performed based on Data synchronization task on dew layer. Monitoring was started when there was no internet connectivity. When Internet connectivity was restored, Dew monitor analyzed all the system performance before and after successful internet tunnel creation. After closely monitoring the DewMonitor, it is observed that upon gaining Internet connection there is a jump in system resource and network utilization. The resource utilization trend is analyzed and illustrated in this section. DewMonitor helps the system admin to scale up the resource vertically which means adding greater computational power. It also helps admin in horizontal scaling, also known as "scaling out," to increase the system's capacity by adding more instances to the Dew environment and distributing the processing and memory burden over several devices.

Advantages of DewMonitor over the legacy system

- If a user is unable to access their personal data because they do not have an active Internet connection, using a DewMonitor provides them increased control as well as flexibility. The data are initially kept as a local copy at the DewMonitor, which serves as the basis for instantiation of the Internet being synced with the master copy that is kept on the cloud side.
- The primary benefit of using DewMonitor is that it allows users to bridge resource analysis that takes place at a distance from their own premises. This is especially helpful in situations in which the user's ability to control and access data is entirely dependent on connectivity. Another benefit is that it is applicable to end devices that have a poor processing speed in addition to limited data storage and computational capability. This describes the majority of IoT devices.
- In the event that a user is unable to access their personal data because they do not have an active Internet connection, using a DewMonitor provides them increased control as well as flexibility (Fig. 4).

CPU analysis:

Stakeholders may evaluate how intensively running programs are being handled by checking the Dew layer's CPU utilization. The corresponding value shows how much of a processor core's total working time is used for data processing. The CPU may be used up to 100% of the time. In other words, CPU consumption may be a good predictor of the processor's present level of stress and, if required, the amount of remaining capacity. It aids in dew system optimization and developing environmentally friendly dew computing, as presented in Figs. 5, 6, 7, and 8.

The available operating time and the actual performance duration are compared to arrive at this calculation. Users will encounter lengthy load and save times, and, in the worst case, programs may start to freeze because an excessive number of processing commands overworks the processor if CPU usage is too high.

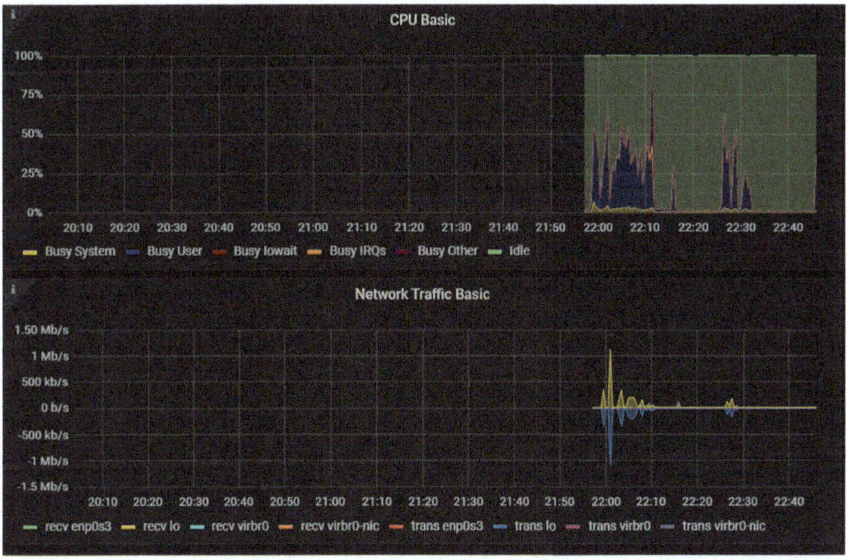

Fig. 4 CPU and network traffic analysis

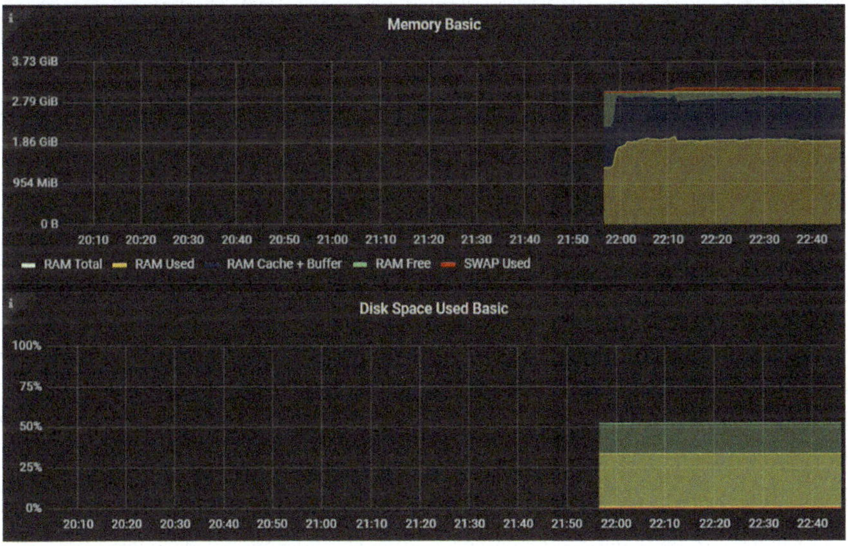

Fig. 5 Memory and disk space used analysis

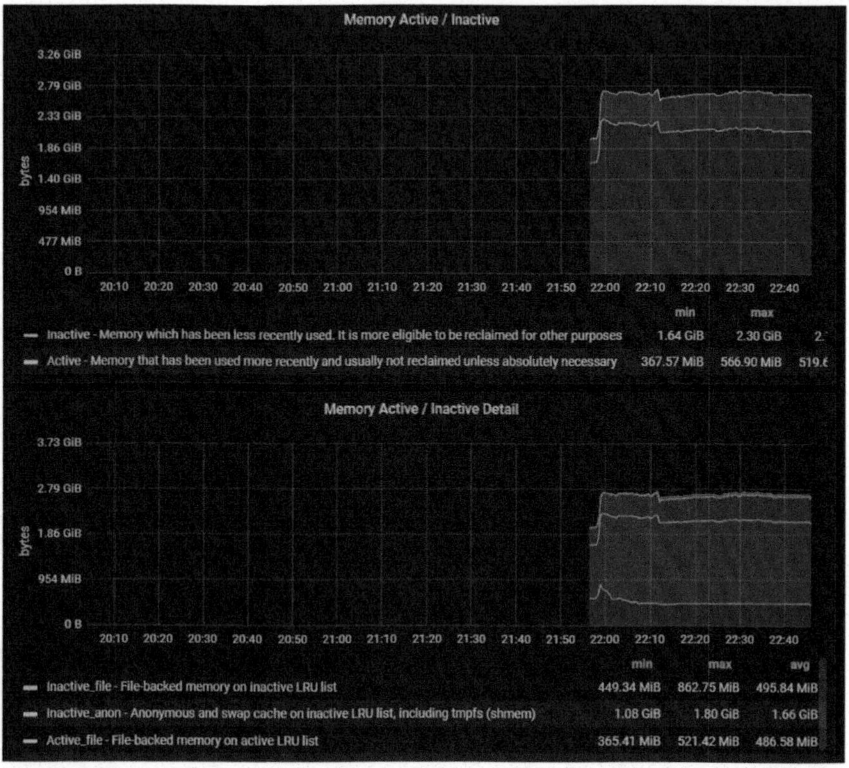

Fig. 6 Memory active/inactive analysis

5.1 Network Analysis

DewMonitor utilizes the network bandwidth efficiently by performing network analysis. This helps to achieve vertical and horizontal auto-scaling and prevent network congestion. It all helps the admin to create an efficient network load balanced architecture. Dew monitor also suggests which layer load balancing is required for dew server.

Dew network analysis offers unique perceptions of how an organization function. For example, employees frequently solve problems, make decisions, and share information through informal social networks. According to our research, an employee's position in the organization's informal network is five times more likely to predict their success than their intelligence, education, or experience.

However, these informal networks may diverge significantly from what executives might anticipate depending on formal organizational structures or established commercial procedures. As a result, there is a risk that executives make decisions without considering how people interact. In effect, they are wearing blinders.

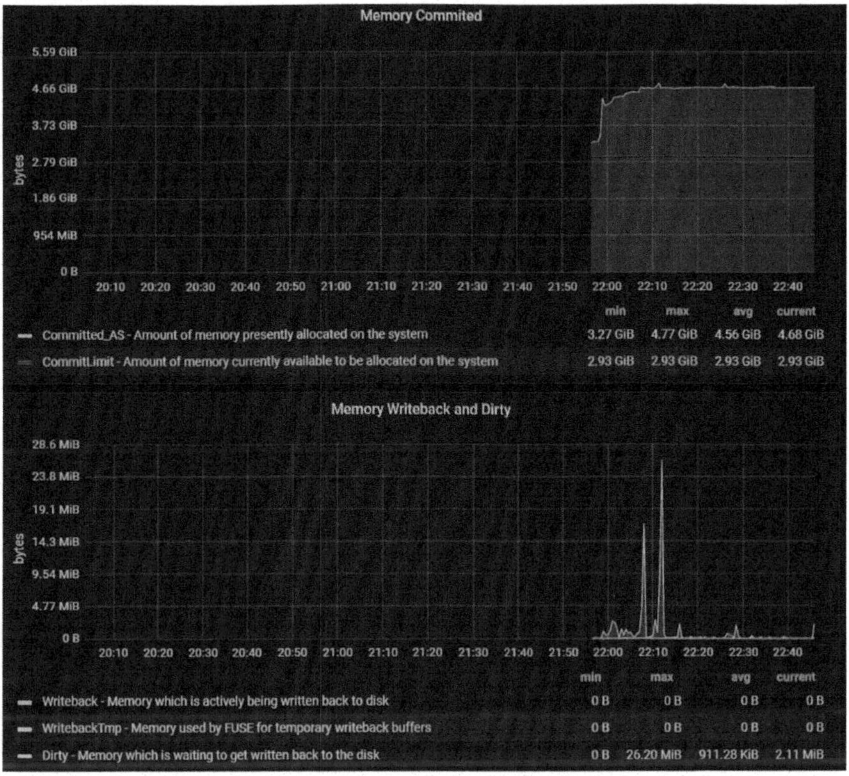

Fig. 7 Memory committed and memory writeback and dirty analysis

The organizations are scanned using network analysis. To demonstrate what is happening, it analyzes and depicts informal networks. The resulting insights can be applied to boost innovation, enhance collaboration, recognize employees who may be at risk, optimize organization design, change culture, and train staff, among other things. DewMonitor assists stakeholders in network analysis, as presented in Figs. 9, 10, and 11.

5.2 *Memory Analysis*

Because we can determine which issues can be solved using a certain quantity of each computing resource, computational resources are helpful. Doing so may assess an algorithm's efficiency and evaluate whether it is the best option for addressing the problem.

Whenever memory use approaches 100%, someone reports a machine with strange behavior. In other words, the host displays extremely high levels of memory use

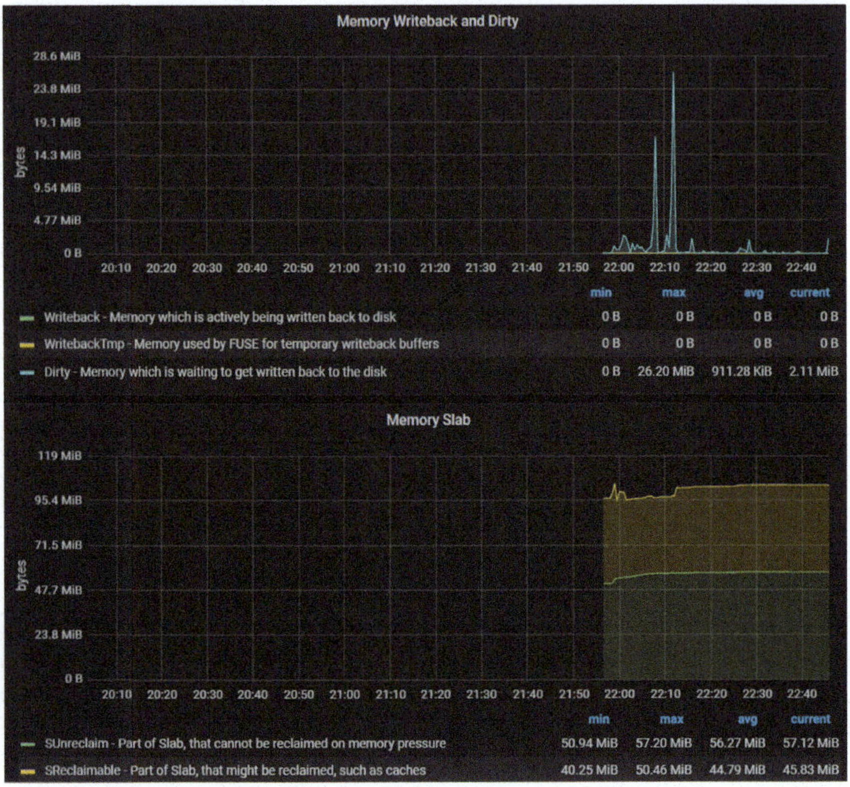

Fig. 8 Memory writeback and dirty and memory slab analysis

anytime running processes consume the physical memory capacity. High memory usage impacts the functionality of actual running tasks and interaction responsiveness. It is difficult to troubleshoot this issue. This issue is ultimately connected to how the kernel handles big transparent pages. The defragmentation of memory pages increases the system burden when the option is enabled for all processes. DewMonitor helps to prevent this type of situation, as presented in Figs. 12, 13, and 14.

6 Requirement of Dew Layer Monitoring

When the primary line of business apps, their associated databases, and email system are not operating correctly, stakeholders will be notified via dew monitoring. In addition, DewMonitor can trend usage, performance, and growth. These are all crucial for capacity planning, ensuring that the SLA is met, and identifying issues before they result in outages.

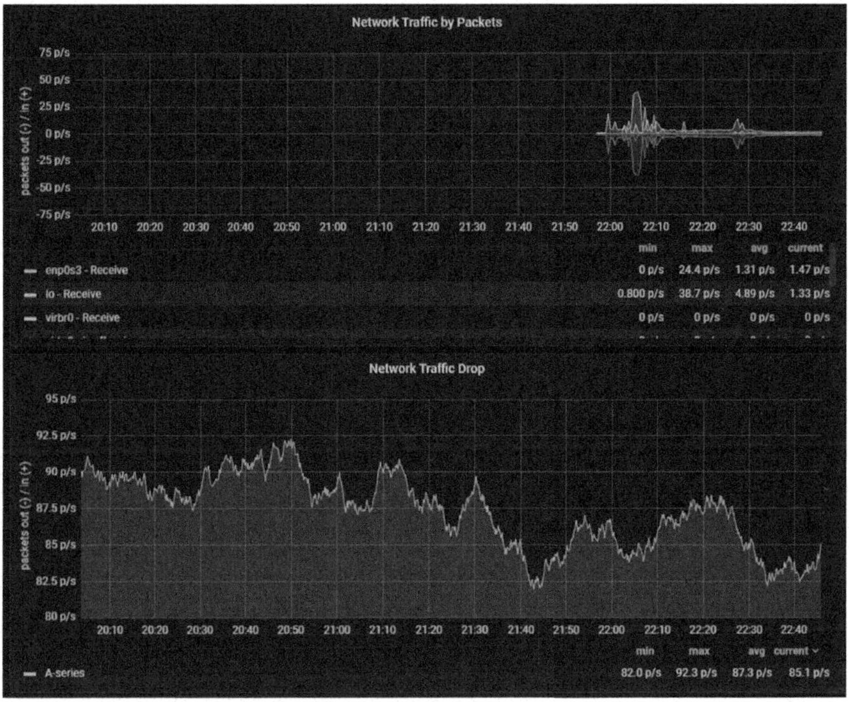

Fig. 9 Network traffic by packets and network traffic drop analysis

DewMonitor is used to manage and monitor software application performance and availability. Monitoring identifies and treats complicated application performance issues to maintain the intended level of service. IT measurements are converted into business meaning or value via DewMonitor.

7 Dew Monitoring to Ensure Sustainable Computing

7.1 Economical Sustainability Analysis of DewMonitor

DewMonitor optimizes the utilization of resources. The system is, therefore, by nature, inexpensive. Consequently, compared to other modern precision systems, DewMonitor requires a lower initial expenditure for installation. It is constructed with high-quality components, but everything has a shelf life. DewMonitor can function perfectly without additional resources, lowering maintenance costs. DewMonitor will lower the expense of the infrastructure. Higher optimization translates into more sales. As a result, the stakeholder will also gain financially, and the Return on Investment will be quicker (RoI).

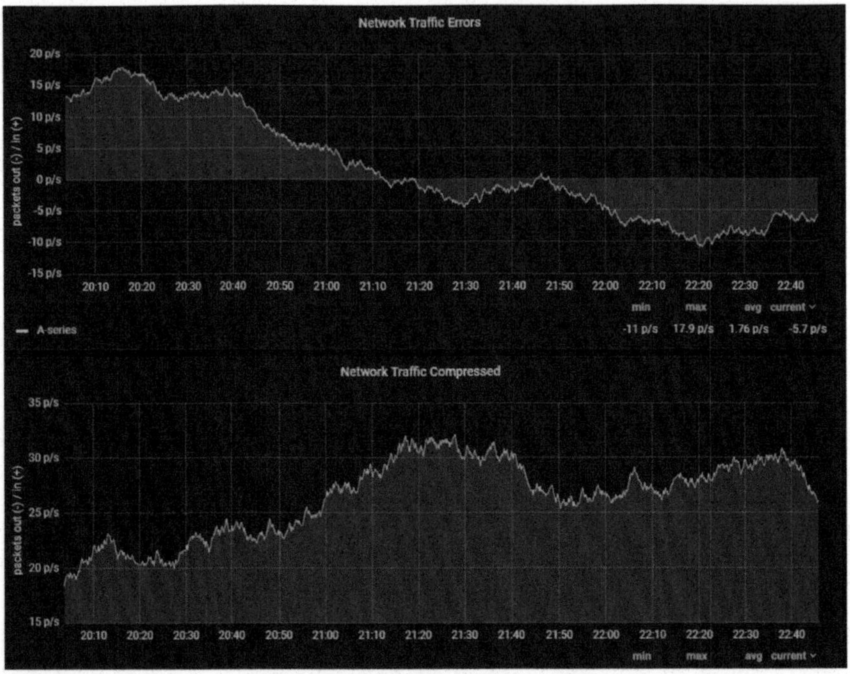

Fig. 10 Network traffic errors and network traffic compressed analysis

7.2 Social Sustainability of I-Dew

All human hierarchies can effectively apply and use DewMonitor. DewMonitor fulfills its goal for many socioeconomic strata, regardless of their social status.

7.3 Environmental Sustainability of I-Dew

DewMonitor is an intelligent system that analyzes the system resource panel using the dew computing idea to reduce the excessive usage of computer resources. Using fewer computing resources is undoubtedly a greener strategy. DewMonitor's carbon impact during use will be more negligible as a result than with other products. DewMonitor thoroughly measures the system, after all. I-Dew has less environmental impact than other gadgets since I-Dew uses fewer resources compared to similar solutions presently.

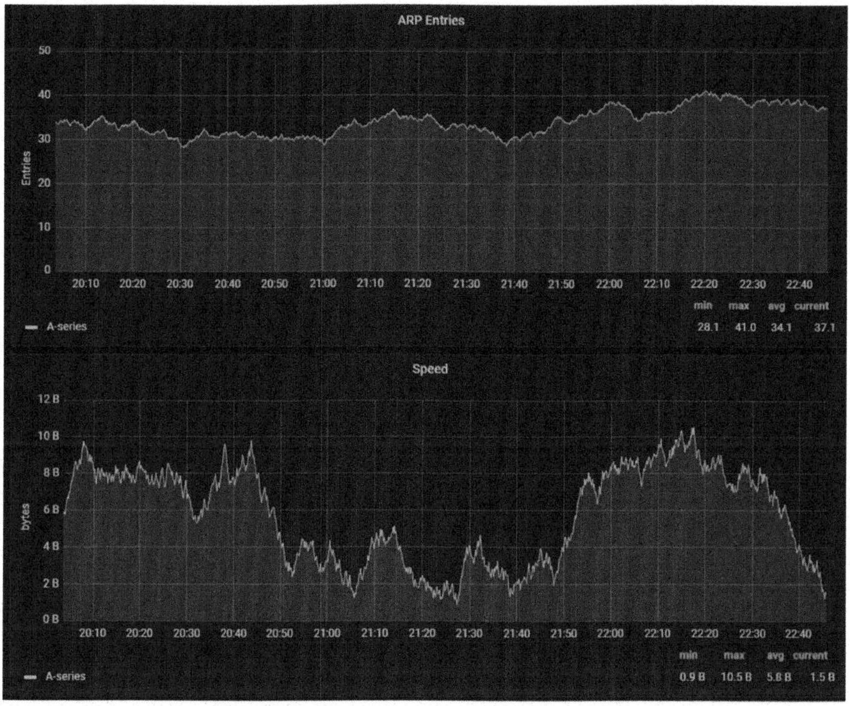

Fig. 11 ARP entries and speed analysis

8 Limitation of DewMonitor

Theoretically sound as they are, current cloud-dew architecture ideas have some flaws. For starters [48–52], the issue of data consistency across various client nodes and distant servers has never been entirely resolved [53, 54]. We anticipate a solution to this problem to be challenging but completely doable, like the difficulties encountered in cache consistency. Although fully localizing cloud behavior seems excellent in principle, there are overwhelming obstacles to overcome in practice. For instance, a client system is unlikely to have the storage and computing power needed to replicate cloud behavior, even if it serves only one user.

9 Conclusions

The performance and availability of software applications are managed and tracked using DewMonitor. DewMonitor is set up in a local machine referred as the Dew analytics server. This server is connected with the distributed dew device via local wireless network tunnel. This analytics server is responsible for gathering data on

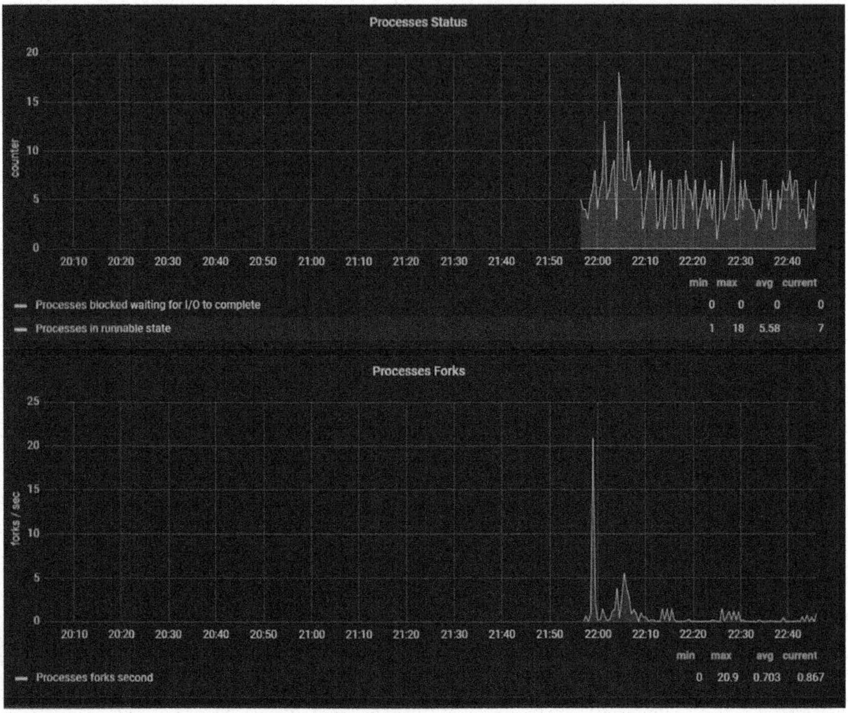

Fig. 12 Processes status and processes forks analysis

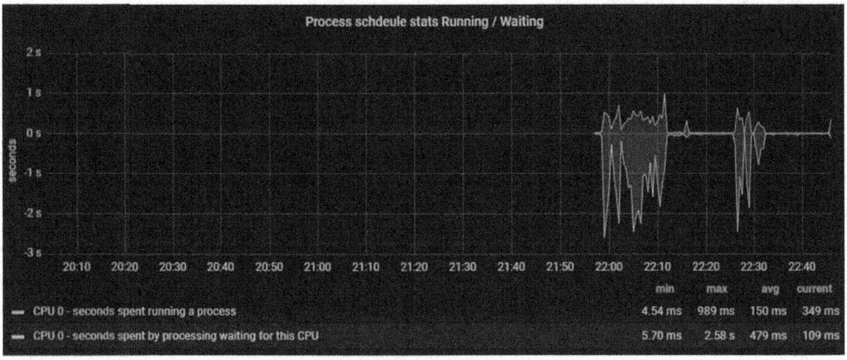

Fig. 13 Processes schedule stats running/waiting analysis

how the distributed dew devices are being utilized and visualize all the real-time data. This layer aids in automatic alert generation based on preset thresholds. In order to maintain the expected level of service, DewMonitor is utilized to locate and resolve complex application performance issues. DewMonitor transforms IT measurements into commercial meaning or value. DewMonitor may also help organizations to

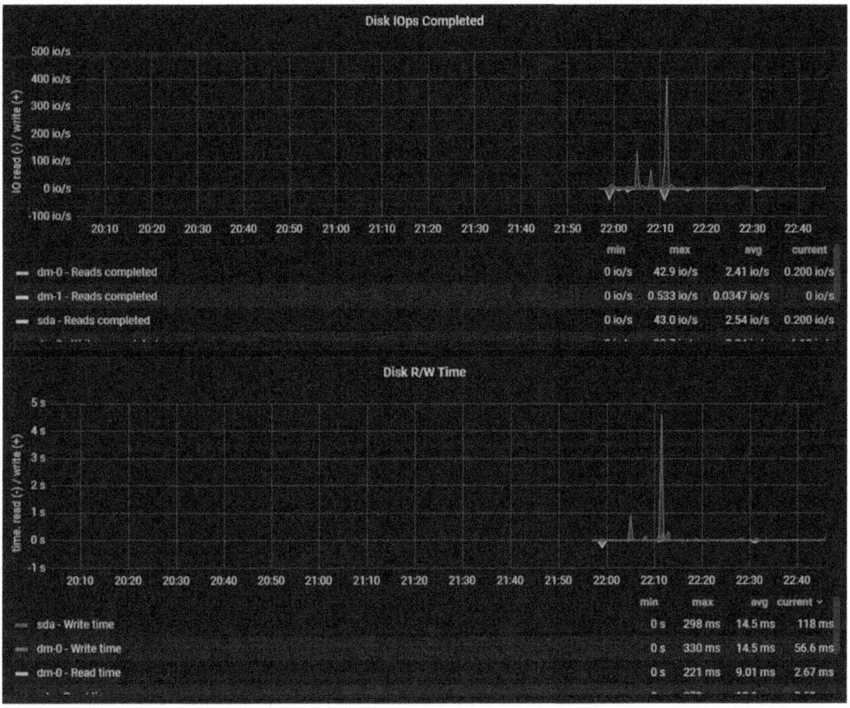

Fig. 14 Disk IOps and disk R/W time analysis

enhance growth and performance. These are essential for capacity planning, ensuring the SLA is satisfied, and spotting problems before they become outages. This paper examines the green sensing and communication options currently available for dew-based IoT to develop sustainable systems for various IoT-based applications. I-Dew customers and owners can use its functions without sacrificing the appropriate level of service quality for sustainable monitoring and analysis.

References

1. Wang, Y.: Definition and categorization of dew computing. Open J. Cloud Comput. (OJCC) **3**(1), 1–7 (2016)
2. Chandra, A., Weissman, J., Heintz, B.: Decentralized edge clouds. IEEE Internet Comput. **17**(5), 70–73 (2013)
3. Ray, P.P.: An introduction to dew computing: definition, concept and implications. IEEE Access **6**, 723–737 (2017)
4. Rindos, A., Wang, Y.: Dew computing: the complementary piece of cloud computing. In: 2016 IEEE International Conferences on Big Data and Cloud Computing (BDCloud), Social Computing and Networking (SocialCom), Sustainable Computing and Communications (SustainCom) (BDCloud-SocialCom-SustainCom), pp. 15–20. IEEE (2016)

5. Albreem, M.A., Sheikh, A.M., Alsharif, M.H., Jusoh, M., Mohd Yasin, M.N.: Green internet of things (GIoT): applications, practices, awareness, and challenges. IEEE Access 9, 38833–38858 (2021). https://doi.org/10.1109/ACCESS.2021.3061697
6. Sarkar, S., Debnath, A.: Green IoT: design goals, challenges and energy solutions. In: 2021 6th International Conference on Communication and Electronics Systems (ICCES), pp. 637–642 (2021). https://doi.org/10.1109/ICCES51350.2021.9489167
7. Jörke, P., Wietfeld, C.: How green networking may harm your IoT network: impact of transmit power reduction at night on NB-IoT performance. In: 2021 IEEE 7th World Forum on Internet of Things (WF-IoT), pp. 753–758 (2021). https://doi.org/10.1109/WF-IoT51360.2021.9596046
8. Shah, K., Narmavala, Z.: A survey on green internet of things. In: 2018 Fourteenth International Conference on Information Processing (ICINPRO), pp. 1–4 (2018). https://doi.org/10.1109/ICINPRO43533.2018.9096789
9. Sharma, N., Panwar, D.: Green IoT: advancements and sustainability with environment by 2050. In: 2020 8th International Conference on Reliability, Infocom Technologies and Optimization (Trends and Future Directions) (ICRITO), pp. 1127–1132 (2020). https://doi.org/10.1109/ICRITO48877.2020.9197796
10. Wang, Y.: Cloud-dew architecture. Int. J. Cloud Comput. **4**(3), 199–210 (2015)
11. Constantinou, S.: Green planning of IoT smart environments. In: 2021 22nd IEEE International Conference on Mobile Data Management (MDM), pp. 267–268 (2021). https://doi.org/10.1109/MDM52706.2021.00055
12. Verma, G., Prakash, S.: A comparative study based on different energy saving mechanisms based on green internet of things (GIoT). 2020 8th International Conference on Reliability, Infocom Technologies and Optimization (Trends and Future Directions) (ICRITO), pp. 659–666 (2020). https://doi.org/10.1109/ICRITO48877.2020.9197848
13. Rindos, A., Wang, Y.: Dew computing: the complementary piece of cloud computing. In: 2016 IEEE International Conferences on Big Data and Cloud Computing (BDCloud), Social Computing and Networking (SocialCom), Sustainable Computing and Communications (SustainCom), pp. 15–20, Atlanta, Georgia, USA (2016). https://doi.org/10.1109/BDCloud-SocialCom-SustainCom.2016.14
14. Wang, Y., LeBlanc, D.: Integrating SaaS and SaaP with Dew computing. In: 2016 IEEE International Conferences on Big Data and Cloud Computing (BDCloud), Social Computing and Networking (SocialCom), Sustainable Computing and Communications (SustainCom), pp. 590–594, Atlanta, Georgia, USA (2016). https://doi.org/10.1109/BDCloud-SocialCom-SustainCom.2016.92
15. Zlatanov, N.: The data center evolution from Mainframe to Cloud, Researchgate (2016)
16. Baun, C., Cocos, H.-N., Spanou, R.-M.: Performance aspects of object-based storage services on single board computers, Christian Baun, Henry-Norbert Cocos, Rosa-Maria Spanou. Open J. Cloud Comput. (OJCC) **4**(1), 1–16 (2017)
17. Zalazar, A.S., Ballejos, L., Rodriguez, S.: Security and compliance ontology for cloud service agreements. Open J. Cloud Comput. (OJCC) 4(1), 17–25 (2017)
18. Sojaat, Z., Skalaa, K.: The dawn of Dew: Dew computing for advanced living environment. In: 2017 40th International Convention on Information and Communication Technology, Electronics and Microelectronics (MIPRO), pp. 347–352 (2017). https://doi.org/10.23919/MIPRO.2017.7973447
19. Oparin, G.A., Bogdanova, V.G., Gorsky, S.A., Pashinin, A.A.: Service-oriented application for parallel solving the parametric synthesis feedback problem of controlled dynamic systems. In: 2017 40th International Convention on Information and Communication Technology, Electronics and Microelectronics (MIPRO), pp. 353–358 (2017). https://doi.org/10.23919/MIPRO.2017.7973448
20. Gordienko, Y., Stirenko, S., Alienin, O., Skala, K., Sojat, Z., Rojbi, A., Lopez Benito, J.R., Artetxe Gonzalez, E., Lushchyk, U., Sajn, L., Llorente Coto, A., Jervan, G.: Augmented coaching ecosystem for non-obtrusive adaptive personalized elderly care on the basis of cloud-fog-dew computing paradigm. In: 2017 40th International Convention on Information and Communication Technology, Electronics and Microelectronics (MIPRO), pp. 359–364 (2017). https://doi.org/10.23919/MIPRO.2017.7973449

21. Brezany, P., Ludescher, T., Feilhauer, T.: Cloud-Dew computing support for automatic data analysis in life sciences. In: 2017 40th International Convention on Information and Communication Technology, Electronics and Microelectronics (MIPRO), pp. 365–370 (2017). https://doi.org/10.23919/MIPRO.2017.7973450
22. Crnko, N.: Distributed database system as a base for multilanguage support for legacy software. In: 2017 40th International Convention on Information and Communication Technology, Electronics and Microelectronics (MIPRO), pp. 371–374 (2017). https://doi.org/10.23919/MIPRO.2017.7973451
23. Gusev, M.: A dew computing solution for IoT streaming devices. In: 2017 40th International Convention on Information and Communication Technology, Electronics and Microelectronics (MIPRO), pp. 387–392 (2017). https://doi.org/10.23919/MIPRO.2017.7973454
24. Podbojec, D., Herynek, B., Jazbec, D., Cvetko, M., Debevc, M., Kožuh, I.: 3D-based location positioning using the Dew computing approach for indoor navigation. In: 2017 40th International Convention on Information and Communication Technology, Electronics and Microelectronics (MIPRO), pp. 393–398 (2017). https://doi.org/10.23919/MIPRO.2017.7973455
25. Frincu, M.: Architecting a hybrid cross layer dew-fog-cloud stack for future data-driven cyber-physical systems. In: 2017 40th International Convention on Information and Communication Technology, Electronics and Microelectronics (MIPRO), pp. 399–403 (2017). https://doi.org/10.23919/MIPRO.2017.7973456
26. Brezany, P., Khan, F.: CloudonScanning-Dew data provenance framework. In: Oral Presentation, DEWCOM 2017, Opatija, Croatia (2017)
27. Lipic, T., Skala, K.: The key drivers of emerging socio-technical systems: a perspective of dew computing in cyber-physical systems. In: Oral Presentation, DEWCOM 2017, Opatija, Croatia (2017)
28. Wang, Y.: An attempt to model Dew computing. In: Oral Presentation, DEWCOM 2017, Opatija, Croatia (2017)
29. Pan, Y., Luo, G.: Cloud computing, fog computing, and dew computing. ZTE Commun. **15**(4), 1–2 (2017)
30. Zhou, Y., Zhang, D., Zhang, Y.: A transparent and user-centric approach to unify resource management and code scheduling of local, edge, and cloud. ZTE Commun. **15**(4), 3–11 (2017)
31. Wang, Y., Skala, K., Rindos, A., Gusev, M., Yang, S., Pan, Y.: Dew computing and transition of internet computing paradigms. ZTE Commun. **15**(4), 30–37 (2017)
32. Goleva, R., Savov, A., Andreev, I., Stainov, R., Achkoski, J., Kletnikov, N., Karafilovski, I.: Cloud, fog, Dew and smart dust platform for environmental analysis. In: The 13th Annual International Conference on Computer Science and Education in Computer Science, Albena, Bulgaria (2017)
33. Gusev, M., Ristov, S., Prodan, R., Dzanko, M., Bilić, I.: Resilient IoT eHealth solutions in case of disasters. In: 9th International Workshop on Resilient Networks Design and Modeling (RNDM) (2017)
34. Mane, T.S., Agrawal, H.: Cloud-fog-dew architecture for refined driving assistance: the complete service computing ecosystem. In: International Conference on Ubiquitous Wireless Broadband (ICUWB), pp. 1–7, Salamanca, Spain (2017)
35. Patel, H., Suthar, K.: A novel approach for securely processing information on dew sites (Dew computing) in collaboration with cloud computing: an approach toward latest research trends on Dew computing. In: 2017 Nirma University International Conference on Engineering (NUiCONE) Ahmedabad, India (2017)
36. Zhou, Y., Zhang, D., Xiong, N.: Post-cloud computing paradigms: a survey and comparison. Tsinghua Sci. Technol. **22**(6), 714–732 (2017)
37. Neagu, G., Ianculescu, M.: Abordarea dew computing ca extensie a arhitecturilor orientate cloud-analiza de oportunitate. Revista Română de Informatică şi Automatică **27**(4) (2017) (in Romanian, abstract in English)

38. Mulay, P., Patel, K., Gauchia, H.G.: Distributed system implementation based on "Ants Feeding Birds" algorithm: electronics transformation via animals and human. In: Kumar, R., Pattnaik, P.K., Pandey, P. (eds.) Detecting and Mitigating Robotic Cyber Security Risks, pp. 51–85 (2017). https://doi.org/10.4018/978-1-5225-2154-9.ch005
39. Risteska Stojkoska, B., Trivodaliev, K., Davcev, D.: Internet of things framework for home care systems. Wirel. Commun. Mob. Comput. **2017**, Article ID 8323646. https://doi.org/10.1155/2017/8323646
40. Klonoff, D.C.: Fog computing and edge computing architectures for processing data from diabetes devices connected to the medical internet of things. J. Diabetes Sci. Technol. **11**(4), 647–652 (2017). https://doi.org/10.1177/1932296817717007
41. Ivanov, N., Lou, J., Yan, Q.: Smart Wi-Fi: universal and secure smart contract-enabled Wi-Fi hotspot. In: International Conference on Security and Privacy in Communication Systems. Springer, Cham, Switzerland, pp. 425–445 (2020)
42. Pustišek, M., Dolenc, D., Kos, A.: LDAF: low-bandwidth distributed applications framework in a use case of blockchain-enabled IoT devices. Sensors **19**, 2337 (2019)
43. Ma, S., Li, H., Yang, W., Li, J., Nepal, S., Bertino, E.: Certified copy? Understanding security risks of Wi-Fi hotspot based android data clone services. In: Proceedings of the Annual Computer Security Applications Conference, pp. 320–331, Austin, TX, USA (2020)
44. Casado-Vara, R., Novais, P., Gil, A.B., Prieto, J., Corchado, J.M.: Distributed continuous-time fault estimation control for multiple devices in IoT networks. IEEE Access **7**, 11972–11984 (2019)
45. Babun, L., Denney, K., Celik, Z.B., McDaniel, P., Uluagac, A.S.: A survey on IoT platforms: communication, security, and privacy perspectives. Comput. Netw. **192**, 108040 (2021)
46. Boursianis, A.D., Papadopoulou, M.S., Diamantoulakis, P., Liopa-Tsakalidi, A., Barouchas, P., Salahas, G., Karagiannidis, G., Wan, S., Goudos, S.K.: Internet of things (IoT) and agricultural unmanned aerial vehicles (UAVs) in smart farming: a comprehensive review. Internet Things **18**, 100187 (2022)
47. Lounis, K., Zulkernine, M.: Attacks and defenses in short-range wireless technologies for IoT. IEEE Access **8**, 88892–88932 (2020)
48. Balcerzak, A.P., Nica, E., Rogalska, E., Poliak, M., Klieštik, T., Sabie, O.M.: Blockchain technology and smart contracts in decentralized governance systems. Adm. Sci. **12**, 96 (2022)
49. Prometheus.: Prometheus/node_exporter: exporter for machine metrics. GitHub (n.d.). https://github.com/prometheus/node_exporter. Accessed 9 Nov 2022
50. Banerjee, P.S., Karmakar, A., Dhara, M., Ganguly, K., Sarkar, S.: A novel method for predicting bradycardia and atrial fibrillation using fuzzy logic and arduino supported IoT sensors. Med. Novel Technol. Dev. **10**, 100058 (2021)
51. Ganguly, K., Karmakar, A., Banerjee, P.S.: ValveCare: a fuzzy based intelligent model for predicting heart diseases using arduino based IoT infrastructure. In: International Conference on Computational Intelligence in Communications and Business Analytics, pp. 229–242. Springer, Cham (2021)
52. Karmakar, A., Ganguly, K., Ghosh, P., Mazumder, A., Banerjee, P.S., De, D.: FemmeBand: a novel IoT application of smart security band implemented using electromyographic sensors based on wireless body area networks. Innov. Syst. Softw. Eng. 1–19 (2022)
53. Karmakar, A., Banerjee, P.S., De, D., Bandyopadhyay, S., Ghosh, P.: MedGini: Gini index based sustainable health monitoring system using dew computing. Med. Novel Technol. Dev. 100145 (2022)
54. Hirsch, M., Mateos, C., Rodriguez, J.M., Zunino, A.: DewSim: a trace-driven toolkit for simulating mobile device clusters in Dew computing environments. Softw.: Pract. Exp. **50**(5), 688–718 (2020)

Security and Privacy Aspects of Authorized and Secure Communications in Dew-Assisted IoT Systems

Mrityunjay Singh and Dheerendra Mishra

1 Introduction

Technology innovations have brought user-friendly portable devices, which have changed the accessibility of data patterns. Now, consumers are also opting for lightweight devices. But, the key challenge is to handle storage in these devices. On the other hand, cloud service emerged as an alternative solution to remotely store digital information. It has emerged as an important, user-friendly, and efficient remote service proving tool [1]. Cloud infrastructure has also emerged great facilitator for IoT services, where collected data from IoT devices can be stored and processed with the help of cloud [2]. But, many loopholes have been realized to ensure real-time communication, and to address this, mechanisms of fog, edge, and Dew were introduced [3–5].

Fog computing, Edge computing, and Dew computing emerged from Cloud computing. Basically, these things exist as layers between the cloud and the user. The top layer is the cloud layer then the fog layer then the edge layer. In Dew computing, we introduce one more layer after the edge layer which people call it Dew layer. Cloud computing came into existence because of many practical reasons. One of them is the decentralization of centralized servers. Because of this decentralization, the load on the centralized server decreases greatly. Many works have been done in the area of Cloud, Fog, and Edge computing, on the other hand Dew computing has much lesser attention. Researchers have designed a Dew computing framework for medical systems, scheduling tasks by using machine learning techniques, etc.

M. Singh
Department of CSE, SRM University AP, Guntur 522240, Andhra Pradesh, India

D. Mishra (✉)
Department of Mathematics, Maulana Azad National Institute of Technology Bhopal, Bhopal 462003, India
e-mail: dheerendra.m@gmail.com

D. De and S. Roy (eds.), *Dew Computing*, Internet of Things,
https://doi.org/10.1007/978-981-99-4590-0_4

The key objective of computing paradigms of fog, edge, and Dew was introduced to maintain data availability near the end user. The introduction of this paradigm will certainly present a great level for data analysis, data processing, and geographical coverage [5]. However, the paradigm of fog computing and cloud computing have a dependency on Internet availability. Moreover, it is being observed that local device access and share data at regular intervals with cloud via Internet connectivity, however, the disconnectivity of Internet may hamper all cloud services. This was the time when Dew paradigms of computing (DPC) emerged. The architecture of Dew fully complements similar paradigms of computing such as cloud, fog, and edge [6, 7]. The cloud-fog-edge-Dew computing has the potential to be utilized for many different applications [8, 9], where various level of hierarchy is involved.

DPC executes its operations near lower-end devices with the objective of creating, modifying, accessing, and deleting data with less coactivity of Internet [10]. The key role of Dew is to store the data on the system where the Dew server is installed and provide it when the connection of the internet is low or there is no internet connection. Thus, Dew node services remain quick as it inhabits near end user device [11]. DPC step-downs latency for transmission and improves the attribute of services [12]. Thus, the design of security frameworks without trusted third-party involvement in session creation is required in this computing platform [13–16].

2 Architecture

To improve the performance of computing, researchers have designed the cloud computing architecture. The whole architecture looks like the following Figs. 1 and 2.

Fig. 1 Overall architecture

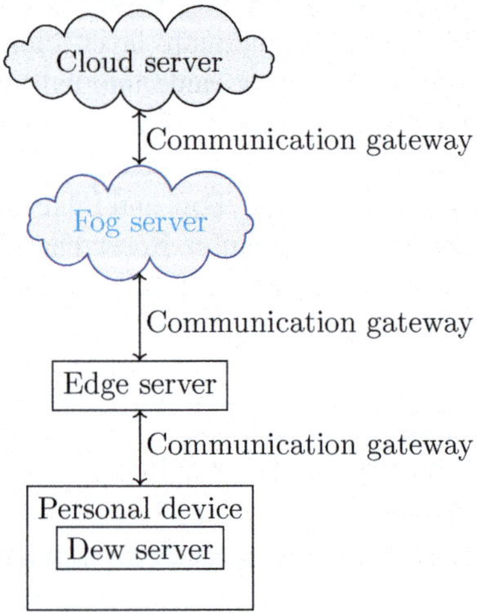

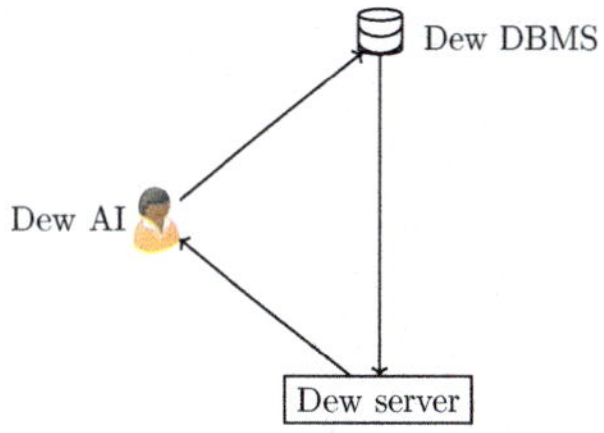

Fig. 2 Dew server architecture

2.1 *Cloud Computing*

Chelappa introduced cloud computing in [17]. After that, it is being used in various domains like document processing, mailing services, video streaming, etc. It suffers from problems like reliability, security, customization, service management, low latency, etc. Thus, it is required to have a low-latency service platform.

2.2 *Fog Computing*

Wang et al. introduced the paradigm of Fog computing in their paper [18]. Fog computing does close computing toward the client side. It is closer than cloud computing. It has certain advantages, e.g., low latency, fast service, and low bandwidth, etc. and disadvantages, e.g., security issues, standardization, portability, etc. with respect to cloud computing.

2.3 *Edge Computing*

Shi introduced edge computing in his paper [19]. Edge computing provides better data processing in comparison to fog computing at the edge of the network. Many uses of edge computing have been proposed in the series of works [20–23]. It suffers from problems with battery life, standardization, service, naming, and programmability.

2.4 *Dew Computing*

Wang et al. gave an idea of Dew computing in their paper [24]. Independence, collaboration, synchronization, scalability, data collection, reorganization, and accessibility are the words that have been included in the definition of Dew computing. Further, we elaborate on these words.

- **Data collection**: It is the first thing that happens in Dew server. A connection gets established between Dew server and cloud server, data from cloud to Dew gets transferred.
- **Synchronization**: Connection between cloud server and Dew server is not always consistent. There must be synchronization between the data which is present on cloud server and on Dew server. Dew computing achieves this.
- **Scalability**: Cloud servers can be scaled easily mean to say multiple cloud servers can be added and where each server deals with separate workstations. In Dew computing paradigm also scalability can be achieved easily.
- **Reorganization**: Sometimes we need to reorganize our collected data. There is a non-consistent connection between cloud server and Dew server which forces synchronization. As we get data on Dew server, we need to reorganize it again and again.
- **Availability**: No matter if there is a connection or not, anyone can access the Dew computing data at any time.
- **Transparency**: The same data is present on cloud as well as on the workstation.

There are various categories of Dew. Some categories are elaborated here.

- **Web in Dew** Cloud server saves the original WWW and the workstation saves a local copy of WWW.
- **Storage in Dew** The data which is present on the workstation will be available on cloud. Any other user can also access the data present on cloud. So, it improves collaboration among users.
- **Software in Dew** Main configuration of the software gets stored on cloud. But a user can create his own account and install that software on his own PC.
- **Database in Dew** Same database must be stored on cloud as well as on Dew server. Out of that one will be called as main or administrator and the other we can call a back-up of the main database.
- **Infrastructure in Dew** The local device can have a duplicate Dew virtual machine instance in the cloud, and its instance gets maintained. The data for each application, system-related setting/data or any other setting/data related to local devices may get stored on cloud.

3 Security Attributes of Authorization and Secure Communication

The remote user authentication protocols for Dew-assisted environment may confront various attacks such as stolen smart card attacks, password guessing attacks, replay attacks, man-in-the-middle attacks, user/server-impersonation attacks, and insider attack [25, 26]. These attacks are based on the following assumptions:

- The stored data in the smart card can be compromised [27–30].
- All the exchanged messages between user and server transmitting via public channel can be compromised. An adversary may be able to modify, delete and resend all the messages. An adversary can also change the route of any message to any other communication party [26, 31].
- An adversary may behave like a legal user or an outside user [26, 32].

In order to make a remote user authentication scheme efficient and secure, the scheme must also resist the following attacks even if an adversary has the above capability:

3.1 DDoS (Distributed Denial of Service) Attack

A Dew node can be compromised by flooding the Dew node with multiple connection requests. Figure 3 shows that an attacker victimized the targeted server with the help of compromised nodes.

3.2 MITM (Men in the Middle Attack)

In an MITM attack, a third party establishes a connection with two communicating parties in such a way that the communicating parties cannot be able to identify that the third party is receiving their exchange messages. If the parties communicating are not properly authenticated, a Dew server may be vulnerable to an MITM attack.

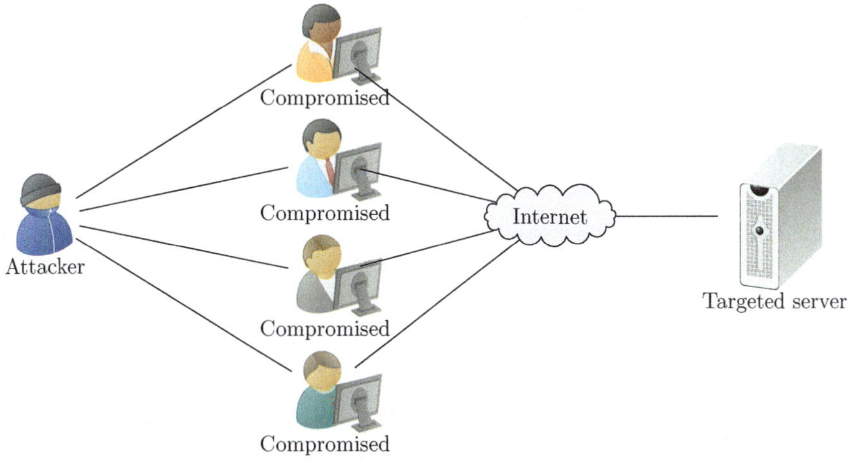

Fig. 3 DDoS attack

3.3 Malicious Dew Node Attack

If an intruder has made a Dew node malicious then the malicious Dew node can get the data from other nodes and cloud. To avoid malicious nodes, we must need to build a good trust management system.

3.4 Fault Tolerance

Even if a Dew node gets crashed, the network should function properly such that the other Dew nodes can still communicate.

3.5 Insider Attack

It is possible that an insider can get the secret credentials of a user because the user uses the same credentials on different accounts in order to remove the burden of remembering many credentials. Therefore, as the user submit his credentials to the server, the insider can get the user credentials and hence, can get access to the user account.

3.6 Stolen Smart Card Attack

An adversary may steal a smart card and that smart card for login on to the server. By doing this, it is possible that the adversary can predict the data stored on the stolen smart card.

3.7 Online Password Guessing Attack

Remembering the different passwords of the different websites is a cumbersome job for any user. So, the users have the right to get the facility of the "forgot password" link. But in the case of authentication using smart card, whenever we use the "forgot password" link we have to generate a fresh smart card for the user. Again this is another difficult job. So, the user tries to select a password that has been picked from a small finite set such that it can be remembered. Since the password has been chosen from a small set, it can be guessed using a dictionary attack. To verify the guess, an adversary may use the stored data of the smart card that has been stolen and the intercepted messages.

3.8 Offline Password Guessing Attack

An attacker will guess a password and perform the login attempt using the guessed password. This procedure will be repeated until the attacker does not get the valid login message.

3.9 Replay Attack

In this attack, an attacker will try to login by using formerly exchanged login messages as the adversary may re-record old communications and can use them wherever possible.

3.10 Known Session Key Attack

An attacker will try to compute the confirmed session key by using the old established session keys.

3.11 Forward Secrecy

In this attack, an attacker computes the right session key by using the compromised secret key of the user.

3.12 Known Session Specific Temporary Information Attack

In this attack, an attacker tries to compute the right session key by using the short-term secret values that are compromised.

3.13 Impersonation Attack

In this attack, an attacker behaves like a legitimate user or server such that the other party believes that he is communicating with the right user or server. It has two categories, namely, user-impersonation and server-impersonation. In the first case, an adversary may masquerade as a registered user in Dew system, and in another case it may masquerade as a server in order to get the session key and other information. Replay and man-in-the-middle attacks can be launched to achieve this attack.

4 Literature Survey

The work done in the area of Dew computing is as follows. There is some general discussion on Dew computing present in [33, 34]. The third international conference on Dew computing named DEWCOM 2018 [35] has eight accepted papers on the various areas related to Dew computing.

- **Medical related work**: Manocha et al. [36] designed Dew computing-based health-meteorological factor analysis for early guess of bronchial asthma. They have proposed Dew-cloud-based cyber-physical systems which detect bronchial asthma at its early stage. They have used an adaptive neuro fuzzy-based system (ANFS) to implement his system and claimed the 94% accuracy in his system. They have also deployed his system in some schools and have done the experimental analysis.
  Afaq et al. [37] designed a Dew computing-based cognitive intelligence-inspired smart environment for diarrhoea prediction. A conceptual framework of diarrhoea prediction has been given by Afaq et al. in Fig. 4.
  They have designed Dew-based cyber physical for detecting diarrhoea and asthma. Many metrological factors affect the health conditions of every individual. Unknowingly or knowingly, because of metrological conditions ever body health is dete-

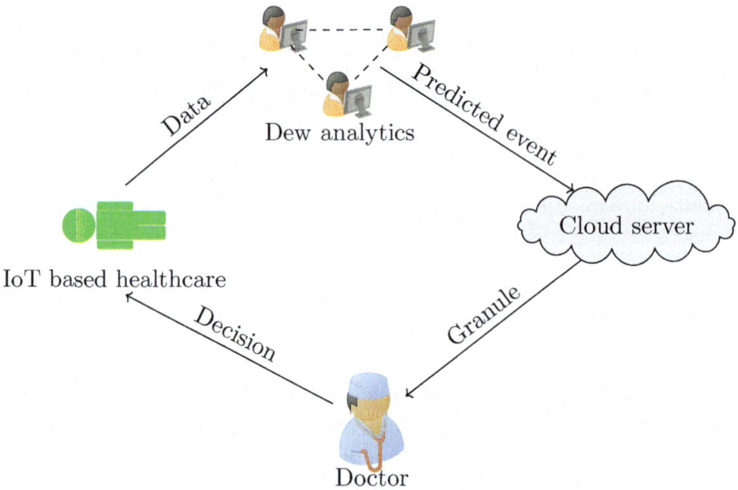

Fig. 4 Conceptual framework for health monitoring

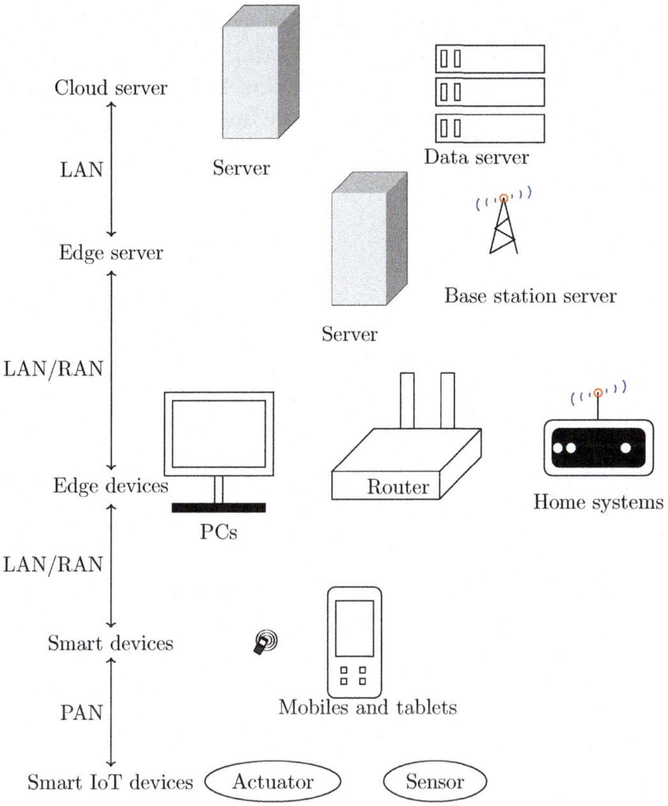

Fig. 5 Gushev architecture of CPS and IoT

riorating. These Dew-based cyber-physical systems help us to monitor our health conditions.

- **System related work**: Gushev [38] designed Dew computing model for CPS, abbreviated as cyber-physical system and internet of things. He has given the following architecture to include Dew computing with cyber-physical system and IoT devices Fig. 5.
  This system can be used by many applications like health monitoring systems, environment management systems, autonomous systems, etc. We achieve energy efficiency, autonomy, independence, collaboration, interoperability, elasticity, and resilience by using Dew computing-based CPS or IoT systems. Javadzadeh et al. [39] designed a mathematical architecture for the arranging of real-time applications in the Internet of things using Dew-based systems. They have given a mixed integer non-linear programming-based mathematical architecture for real-time application in cloud-fog Dew architecture. They have also constructed a more general scheduling algorithm that can be scaled based on non-dominated sorting genetic algorithm II. Their scheme reduces power consumption and traffic load.

Ro et al. [40] designed an IoMT (internet of music things) in Dew computing paradigm. Their system is based on the gathering of information from the crowd. Their crowd-sourcing framework is given in Fig. 6. They have designed a crowd-sourcing-based Dew music framework and the Internet of music framework. After

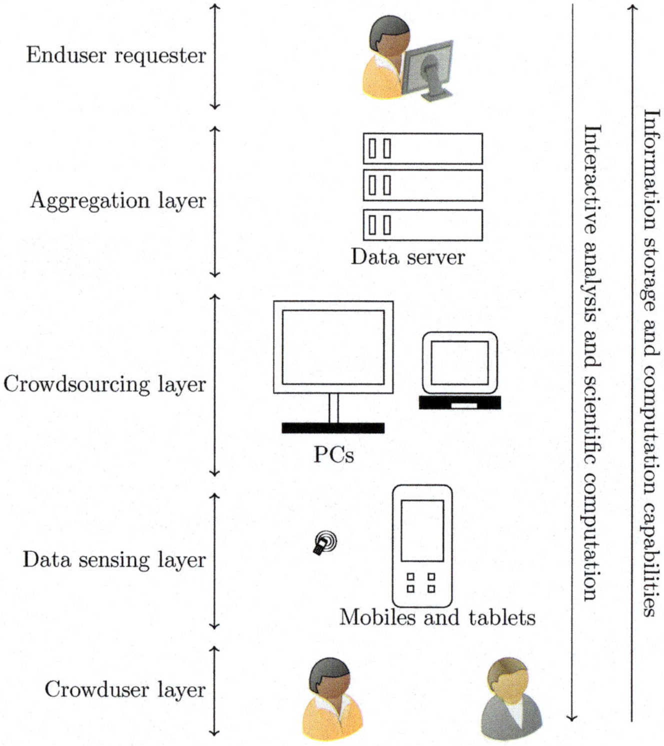

Fig. 6 Crowd sourcing framework

Fig. 7 False alarm reduction protocol using cloud

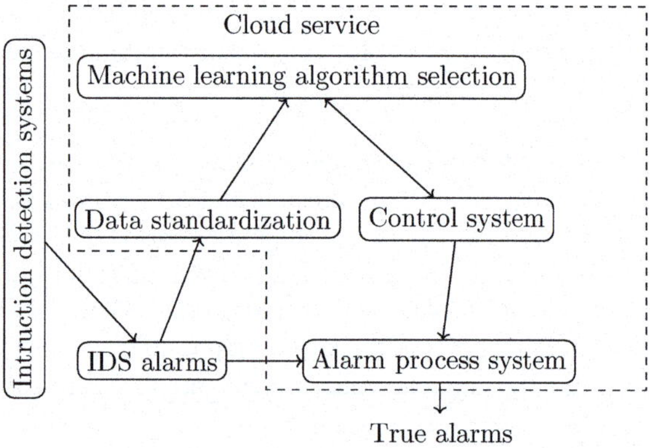

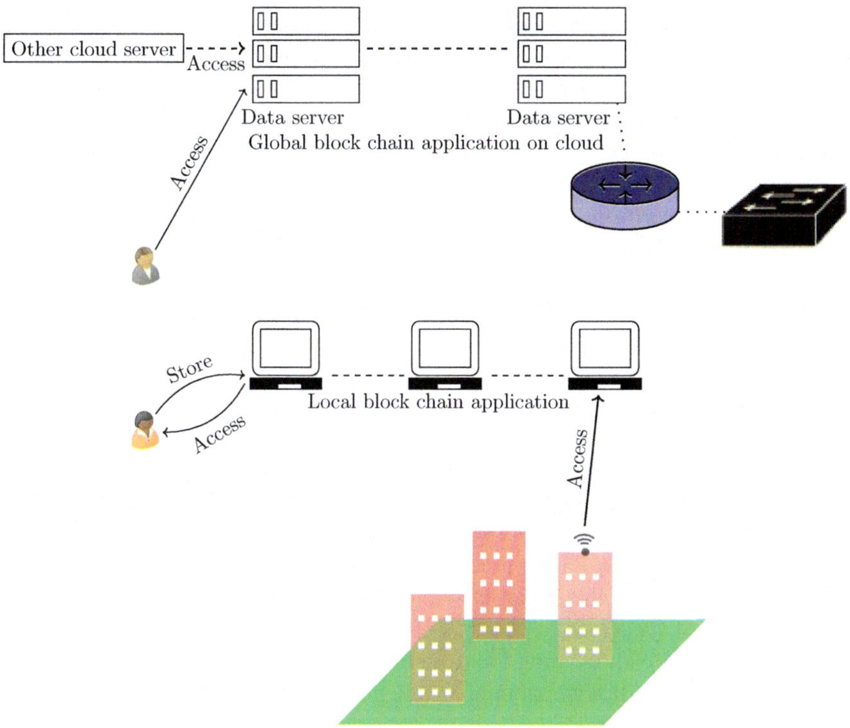

Fig. 8 Blockchain (local block chain; consists of three PCs form Dew server, global block chain; consists of data servers on cloud) based cloud-Dew model

that, they performed data-driven experiments which measure the system efficiency and also formulated transmission, service time, and latency analysis of the proposed system. Finally, an energy-efficient model has been proposed by them. Singh et al. [41] designed a Dew computing as a service for intelligent intrusion detection in the edge-of-things ecosystem. Their system reduces the false alarm ratio. The authors also say that their system gives better efficiency and minimum communication delay. Their system reduces workload on cloud and improves latency. In their system there is a chance of a reduction in classification accuracy. The high-level architecture of their intrusion detection system is given in Fig. 7.

- **Security related work**: Rana et al. [42] constructed a scheme that establishes an authenticated key between Dew server and sensor. Further, it was improved by Ma Yuqian et al. [43] in the paper named cryptanalysis and enhancement of an authenticated key agreement scheme for Dew-assisted Internet of things systems. Data security is a primary concern in cloud computing and protecting data blockchain is a key technology in Today's world. Hati et al. [44] proposed a blockchain network-based Dew cloud modeling for distributed and decentralized smart cities. The proposed four-tier architecture consists of physical nodes, Dew,

edge, and cloud. Local blockchain application has been used for Dew layer and global blockchain has been used for cloud layer. The proposed architecture reduces latency and energy utilization which are the challenges of Dew computing systems. Their model is given in Fig. 8.
Braeken [45] proposed protocols for Dew-based Internet of Things systems that establish a key which is authentic. Their scheme is secure in Yao Dolev attack model and outperforms the existing schemes from a security and performance perspective.

- **Security and authorization related work in cloud, fog and edge computing**: Secure and authorized communication is a major challenge in Dew computing. As Dew server remains near to the user and timely update is required with cloud servers via public channels, secure and authorized communication is really needed to address this. In recent years, many developments have been seen, which try to address different issues.
Due to the failure of the cloud computing model in the handling of a large amount of data, Bonomi et al. [4] gave the idea of fog computing. Stojmenovic and Wen [46] examined the loopholes of fog-based systems. Concurrently some other computing paradigms also started to appear that improve the overall functionality of communication with the cloud. Dew-based systems were one of the computing paradigms that was given by Wang [47] as the cloud-Dew architecture; that is an addition to the already existing client-server architecture. The Dew server performs the following activities which improve the performance of Dew computing.

1. It provides services to the client when the network is not present.
2. It regularly maintains the Dew server database when the connection with the cloud reopens.

Butun et al. [48] talked about the cloud-centric, authentication facility for Internet of things. Concurrently, Wang [24] wrote an article that defines DC formally as software to work separately and jointly with the cloud to do numerous activities such as opening websites without connection to the Internet. This happens by using Dew servers. During the same time, Rindos and Wang [49] gave DC as a reciprocal chunk of the cloud; starting a new kind of computing support known as DC. Ristov et al. [12] executed the cloud, fog, and Dew model for the services that can be extended horizontally. In order to increase the effectiveness of the old load-balancing approaches, they did a case study on improved L3B technology. Sojat and Skala [50] also described the significance and value of Dew computing in control and services. This paper added Dew computing to the earlier ladder of cloud, fog, edge, and grid computing. Dolui and Datta [51] defined the architecture of fog-edge systems. Simultaneously, Gusev [52] extended the Dew computing model for Internet of things streaming gadgets by constructing, processing, and the message transmission among the parties, near Internet of things appliances. Srinivas et al. [53] introduced an authentication protocol that is which has been designed specifically for cloud-based systems, but their protocol is impassable for systems that cannot allow latency in the area of Internet of things healthcare. In [54] Guan et al. discussed the security and privacy challenges in fog-based systems.

Using cloud, fog, and Dew computing model, Gordienko et al. [9] gave a modifiable system that can be modified based on the need and it extends the training environment for non-prominent personalized elderly care. They have also analyzed a new area in which Dew computing can be applied. Mane and Agrawal [55] in their research work proposed the application of Dew model for driving support. Pan et al. [11] gave a summary of Fog, Edge, and Dew Computing. They discussed different models, e.g., fog-cloud, Dew-fog, and Dew-cloud, where Dew adds an extra hierarchy of communication with the Internet of things gadgets. These models came recently. A Dew server can be connected to any network and it saves the energy involved in the communication, and also it is cost-effective because of the closeness of Dew-based systems with the Internet of things devices.

The analysis of Dew-based model was presented in [34], and the Dew-based systems were described as an independent part of cloud computing in Patel et al. [34]. Wazid et al. [56] provide a protocol for fog computing that establishes a secure session key. In their model, the user establishes the connection between his mobile gadget with an Internet of things gadget that utilizes the fog and cloud servers. Botta et al. [57] also constructed a cloud, fog, and Dew-based model for robotics that can be used for the advanced applications.

In the Dew-based systems, the processing and saving are done locally, and then modification happens on a cloud server. By using Dew computing we can reduce the dependency on inter-network, this concept was proposed by Ray [58]. Ray [58] paper described many applications, e.g., playing games online on a phone or system, analyzing images in real time, wearable telehealthcare, etc. By using Dew, Longo et al. [59] reduced the energy consumption of mobile devices. A forecast on the battery availability is done from the actions of the users that have been done in the past. The acceptable scheduling procedures are outlined for long-term availability.

- **Learning related work**: Sanabria et al. [60] solved task arranging problems in Dew-based systems via deep reinforcement learning. In this paper, they propose Dew RL (a family of reinforcement learning which uses deep neural networks to design rules) to distribute jobs in Dew computing environment. They have also developed an interface that helps to connect the Open AI Gym with Dew computing simulator. Future research direction is to come up more efficient RL agent than PPO (Proximal Policy Optimization). Another area of research is to come up with an evaluation strategy that judges the effectiveness of a Dew scheduler. They have scheduled jobs in the fixed Dew environment but in general, this case will not hold always. Chakraborty et al. [61] designed Dew computing-based micro service execution in mobile edge using Q-learning. They propose a Dew computing-based micro service execution scheme based on reinforcement learning which reduces delay, optimizes cost, and gives better responses in real-time. In terms of computation and relocation cost their scheme performs better than FWS (fair-weighted affine-based scheme) and SDGA (smart pre-deployment based on a genetic algorithm) scheme.

- **Smart devices related work**: Hirsch et al. [62] designed a job execution protocol for Dew-based system with technically enhanced smartphones. They tried for industry applications, to improve the efficiency of the job scheduling in Dew computing-based systems. In order to give help in building smart cities, their heuristics improve citizen participation by allowing Dew-based systems.
  Longo et al. [59] integrated mobile devices for Dew-based systems by designing a model for forecasting energy availability in each hour. By using feature extraction and machine learning classifiers they have designed Dew computing-based model which predicts the battery availability in mobile devices. They have performed experiments on their model and achieved a reduction in the MSE metric for 18 users out of 23 users. In their approach, many events get missed by the device analyzer. This is a direction where this work can be extended. It is also possible to use other machine learning models, e.g., neural network based to get more accuracy in the work. Time series analysis can be added to derive more information from the model. It is also possible to integrate Dew-based scheduler given in [63] with battery predictor.
  In [64] Garrocho et al. evaluated, analyzed, and discussed smart watches for the Dew base systems. The outcome of their work showed that smart watches could extend local device services through carrying out activities, collaboration with dispersed cloud computing, and providing support to mitigate the –ve effect of the large data. Authors expect that it can be used with cloud and fog computing. Some tools can be developed which help to improve the efficiency of workload distribution. By using these tool limitation of smartwatches in Dew computing can be evaluated.

5 Security Challenges

Power management, data security, utilization of processors, and data management are the challenges that are persisting with Dew computing. In this article, we will concentrate on the security side of Dew computing. Security in a Dew system can be categorized into two parts communication security and architectural security. We describe some of these issues as follows:

1. The stored code and data must access to the authentic user only. Otherwise, we have to compromise with the integrity of code and data. Even if the integrity violates, our system must be able to detect it.
2. The communication protocol may be attacked. It is also possible that an insecure system may interact with Dew-based system. So, we need to ensure the secure implementation of the communication protocols.
3. In order to access the shared cloud resources individually by using access control protocols we have to share the identities of multiple organizations.

4. An authenticated connection between cloud and Dew-based systems must be established. Otherwise, there is a chance of breaking the confidentiality of information.
5. Differential denial of service attack and advanced persistent attack is also possible in Dew-based systems.
6. Regular back-up of data is required on the cloud side, otherwise in case of any disaster, we will not be able to recover the data.

5.1 Web Application and API Security

The following are the principal suggestions by the cloud security alliance abbreviated as CSA [65] with respect to cloud applications and APIs that will be extended to Dew base systems as well.

1. Based on the need of Dew system development, we need to define privacy and security pre-conditions. These pre-conditions must also be defined based on its effectiveness and possibility of implementation.
2. For the Dew-based systems, the parameters associated with risks and attacks must be explored and incorporated into security requirements. This architecture must be built and maintained on a continuous basis.
3. The security of SDLC (software development life cycle) and software model must be constructed and if there is a need of modification then it has to be modified.
4. In each software, some components of it can be reused. This repeated use of the component may breach the security.
5. An important testing named penetration testing for web applications must be carried out on a regular basis.
6. Another important testing named manual testing should be done on a regular basis. This ensures the security of each session in any web application.

5.2 Access Control and Identity Management

It means restricted access to physical resources to a legitimate user. The user who is accessing the resources must possess the right credentials before he accesses the resource. The access control prevents side-channel attacks in the virtual machine system. The cloud and Dew systems need mutual access control. Security attributes, confidentiality, accountability, and integrity must be maintained in any access control system. If we want that the Dew computing-based systems should be adopted by different bodies then prior authorisation to use the resources and the management of identity in the Dew environment is highly needed. We must consider the following things when we try to establish a secure and efficient access control system (Fig. 9).

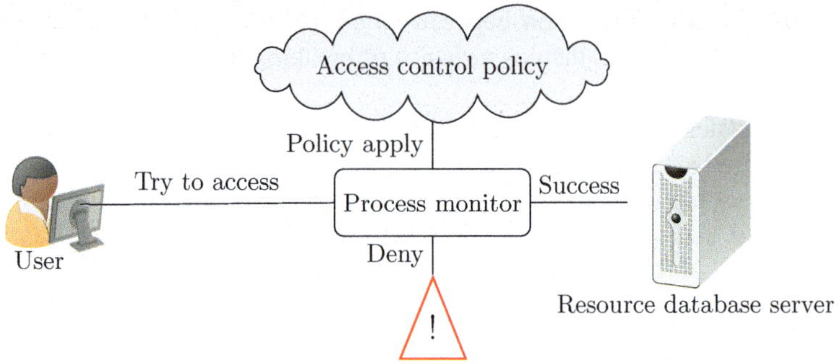

Fig. 9 Access control mechanism

1. Computation cost: If the computation cost of the access control scheme will be high then it will reduce the efficiency of the system, which means say, getting time of a resource by a legitimate user will be high.
2. Communication cost: This cost is measured in terms of bits per session. If the number of bits being transferred in the communication channel will be high then there will be huge traffic, which results in high latency.
3. Data aggregation: There are various users geographically distributed. We need to accumulate their data such that our Dew system face less latency.
4. Privacy: There is a need for user data protection because in Dew-based system data will travel from one domain to a completely different domain, and there is a possibility of data leakage.
5. Scalability: As the number of users increases or decreases, our system must be able to adopt these changes faster.
6. Context awareness: It is possible that Dew system works for different systems like health management, traffic management, etc. Based on the context Dew system must function properly.
7. Network availability: If the network will not be available for the Dew system then all communication will stop.
8. Resource constraints: With limited resources, our system must work efficiently.

It is extremely important to monitor the user's credentials and control illegitimate access to the data. In Dew-based system if we manage identity, proper access control, and quality services then we may be able to correctly answer whether the data has been modified by the adversary, and also we may be able to answer whether it comes from a legitimate user. If we do not ensure strong identity management and strong prior authorization to use resources then many security problems can arise in Dew-based systems. Due to weak identity management and weak access control system in Dew-based systems, the problems, e.g., DoS, weak identity reset procedure, inadequate authorization verification, cross-domain authentication, deficient logging and controlling probabilities, and XML wrapping attacks may arise.

5.3 *Data or Storage Security*

For data/storage security the following recommendations have been given in [65].

1. The used key must be managed by either the trusted cryptographic service or organization/user.
2. The key management practices and encryption decryption scheme get modified from time to time. We must use the latest and verified key management and encryption schemes.
3. If there is a possibility of the use of "off the shelf technology," then it must be used.
4. An individual or a group must maintain the key scope.
5. Updation in encryption algorithms happens from time to time. We have to use the latest algorithms.

In Dew computing environment, implementing these recommendations will be a challenging task.

5.4 *Trust Management*

The security credentials of an entity in the system must follow the established security policies. If the security credential of an entity in the system does not follow the established security policies then there is a need for management which we call trust management. Gao et al. [66] categorized the trust computation scheme into five scopes for the Internet of things. These scopes can be extended for Dew computing systems.

1. Aggregation: Collecting information from others.
2. Formation: By using collected information and using his own observation design the rules for trust formation.
3. Updation: Trust values may update in the future. We should have a system that tracks the changes and based on those changes modifies the trust values.
4. Propagation: It deals with how other entities in the system will compute the trust value.
5. Composition: It defines the number of trust properties that help in computing the trust values.

The attacks like self-promotion, bad mounting, on-off, opportunistic serving, and ballot stuffing can be launched in the trust management systems. We need to build a system that deals with these attacks.

5.5 *Privacy Issues*

Since the Dew computing systems are distributed networks the data will come from various resources and some data are private in nature, so, we need to ensure the privacy of these private data. We have to put privacy on the various types of information, e.g., user, network, data, location, and usage. The preservation of privacy happens by using the techniques of homomorphic encryption, differential privacy, identity obfuscation, trusted third party, and data partitioning.

5.6 *Authentication*

Different authentication mechanism has been proposed by the researchers [42]. Public key infrastructure and biometric-based and password-based authentication mechanisms are being used to achieve authentication. Sometimes researchers also combine the above techniques. But note that when we combine the above techniques then it increases the computation cost of the authentication process and hence we have to compromise with the efficiency or latency of our scheme. For cloud, fog, and edge computing many authentication schemes based on the concept of elliptic curve Diffie Hellman (ECDH) and bilinear pairing have been proposed. Very few authentication schemes are present in case of Dew computing.

5.7 *Virtualization*

By using virtualization we can use the same resources for multiple customers. Several virtual machines, abbreviated as VM, can map to the same physical resources by using a polling mechanism. It makes another security concern in Dew-based systems.

1. VM share image: A user can create his VM image or can use the already created VM image. A user can also share the created VM images which make this functionality prone to attacks.
2. VM isolation: If VMs will not be isolated then it makes the system prone to a data breach or cross-VM attacks.
3. VM escape: VMM is a software that manages all other VMs. It helps to put the request of hardware by a particular VM. If a VM goes away from the control of VMM then it is possible that an attacker gets control of VMM or other VMs.
4. VM migration: Sometimes there is a need for the transfer of VM from one physical device to another device without shutting down the VM. During migration, it is possible that data gets leaked into the network.
5. VM rollback: Sometimes there is a need for VM put in the previous state because there are issues that have been raised during the processing of the VM between two states. This rollback may raise security concerns because if we go to the

previous states then it is possible that the errors that are present in the previous state may raise again.
6. Hypervisor issue: VMM play the role of a hypervisor. VMM controls the working of all the VMs and it provides resources to a VM if a VM put a request for the resources. So any compromised VMM will be a threat to the security of the whole system.

The virtualization may create security issues. To avoid the security issues, CSA [65] suggests the following guidelines.

1. The implementers should secure each virtualized operating system in each of the guest virtual machines.
2. The defined securities parameters must be incorporated for the virtualized operating system.
3. The security technology designed by other parties must be used to reduce the dependency on CSP.
4. The virtual machine at rest should be encrypted.
5. Security weakness estimation devices must work for the virtualized environment.
6. The virtual machine images at rest must be added with the newest fix as early as required. Further, the safeguarding procedure should be in place until the virtual machines are added.
7. The security devices which can handle virtualization-related weaknesses must be applied and used in the cloud-based systems.

5.8 Blockchain Technology

As its name implies that it is a chain of linked blocks where each block is the hash output of the previous block, timestamp, and transaction data. This technology uses peer-to-peer networks for communication [67]. Since it is like a distributed ledger in public use to validate a new transaction block it has to use consensus protocol among the distributed nodes. It possesses various features like immutable, hard to attack, decentralized system, can be made private or public, encryption with validation, and many other features. This technology has been used in the cloud, fog, and edge computing systems. It can be introduced in the Dew computing systems. In Dew computing systems blockchain technology improves security, highly encrypted data hence difficult to attack, supports decentralization, and is immutable.

6 Conclusions

This chapter discusses the security aspects of Dew computing-based systems. In this chapter, we have explored the privacy, security, authentication, and access control aspects of Dew computing systems. The analysis suggests that as Dew has the potential to execute its operation at low latency and with low Internet connectivity, the mechanism of security should be designed without including a legitimate third party to establish a secure session. Moreover, security should also examine and possible solutions should also be explored to ensure data security in the quantum era.

References

1. Guo, Y., Mi, Z., Yang, Y., Ma, H., Obaidat, M.S.: Efficient network resource preallocation on demand in multitenant cloud systems. IEEE Syst. J. **13**(4), 4027–4038 (2019)
2. Sharma, B., Obaidat, M.S.: Comparative analysis of IoT based products, technology and integration of IoT with cloud computing. IET Netw. **9**(2), 43–47 (2019)
3. Vora, J., Kaneriya, S., Tanwar, S., Tyagi, S., Kumar, N., Obaidat, M.: Tilaa: tactile internet-based ambient assistant living in fog environment. Fut. Gener. Comput. Syst. **98**, 635–649 (2019)
4. Bonomi, F., Milito, R., Zhu, J., Addepalli, S.: Fog computing and its role in the internet of things. In: Proceedings of the First Edition of the MCC Workshop on Mobile Cloud Computing, pp. 13–16. ACM (2012)
5. Pradeepa, M.A.M., Gomathi, B.: Towards fog computing based cloud sensor integration for internet of things. Int. J. Comput. Sci. Eng. Commun. **5**(6), 1761–1773 (2017)
6. Naik, B., Obaidat, M.S., Nayak, J., Pelusi, D., Vijayakumar, P., Islam, S.H.: Intelligent secure ecosystem based on metaheuristic and functional link neural network for edge of things. IEEE Trans. Ind. Inform. **16**(3), 1947–1956 (2020)
7. Garg, S., Singh, A., Kaur, K., Aujla, G.S., Batra, S., Kumar, N., Obaidat, M.S.: Edge computing-based security framework for big data analytics in Vanets. IEEE Netw. **33**(2), 72–81 (2019)
8. Šojat, Z., Skala, K.: Views on the role and importance of dew computing in the service and control technology. In: 39th International Convention on Information and Communication Technology, Electronics and Microelectronics (MIPRO)vol. 2016, pp. 164–168. IEEE (2016)
9. Gordienko, Y., Stirenko, S., Alienin, O., Skala, K., Sojat, Z., Rojbi, A., Benito, J.L., González, E.A., Lushchyk, U., Sajn, L.: Augmented coaching ecosystem for non-obtrusive adaptive personalized elderly care on the basis of cloud-fog-dew computing paradigm. In: 40th International Convention on Information and Communication Technology, Electronics and Microelectronics (MIPRO), vol. 2017, pp. 359–364. IEEE (2017)
10. Ray, P.P.: An introduction to dew computing: definition, concept and implications. IEEE Access **6**, 723–737 (2017)
11. Pan, Y., Thulasiraman, P., Wang, Y.: Overview of cloudlet, fog computing, edge computing, and dew computing. In: Proceedings of The 3rd International Workshop on Dew Computing, pp. 20–23 (2018)
12. Ristov, S., Cvetkov, K., Gusev, M.: Implementation of a horizontal scalable balancer for dew computing services. Scalable Comput.: Pract. Exp. **17**(2), 79–90 (2016)
13. Tao, D., Ma, P., Obaidat, M.S.: Anonymous identity authentication mechanism for hybrid architecture in mobile crowd sensing networks. Int. J. Commun. Syst. **32**(14), e4099 (2019)
14. Gupta, D.S., Islam, S.H., Obaidat, M.S.: A secure identity-based three-party authenticated key agreement protocol using bilinear pairings. In: International Conference on Innovative Data Communication Technologies and Application, pp. 1–11. Springer (2019)
15. Wu, T.-Y., Lee, Z., Obaidat, M.S., Kumari, S., Kumar, S., Chen, C.-M.: An authenticated key exchange protocol for multi-server architecture in 5g networks. IEEE Access **8**, 28096–28108 (2020)
16. Meshram, C., Obaidat, M.S., Lee, C.-C., Meshram, S.G.: An efficient key authentication procedure for IND-CCA2 secure Paillier-based cryptosystem. Soft Comput. 1–7 (2020)
17. Chellappa, R.: Intermediaries in cloud-computing: a new computing paradigm. In: Informs Annual Meeting, pp. 26–29, Dallas (1997)

18. Wang, Y., Uehara, T., Sasaki, R., Fog computing: issues and challenges in security and forensics. In: IEEE 39th Annual Computer Software and Applications Conference, vol. 3, pp. 53–59. IEEE (2015)
19. Shi, W., Cao, J., Zhang, Q., Li, Y., Xu, L.: Edge computing: vision and challenges. IEEE Internet Things J. **3**(5), 637–646 (2016)
20. Rudenko, A., Reiher, P., Popek, G.J., Kuenning, G.H.: Saving portable computer battery power through remote process execution. ACM SIGMOBILE Mob. Comput. Commun. Rev. **2**(1), 19–26 (1998)
21. Chun, B.-G., Ihm, S., Maniatis, P., Naik, M., Patti, A.: Clonecloud: elastic execution between mobile device and cloud. In: Proceedings of the Sixth Conference on Computer Systems, pp. 301–314 (2011)
22. Kumar, K., Lu, Y.-H.: Cloud computing for mobile users: can offloading computation save energy? Computer **43**(4), 51–56 (2010)
23. Kosta, S., Aucinas, A., Hui, P., Mortier, R., Zhang, X.: Thinkair: dynamic resource allocation and parallel execution in the cloud for mobile code offloading. In: Proceedings IEEE Infocom, vol. 2012, pp. 945–953. IEEE (2012)
24. Wang, Y.: Definition and categorization of dew computing. Open J. Cloud Comput. (OJCC) **3**(1), 1–7 (2016)
25. Jaspher, G., Katherine, W., Kirubakaran, E., Prakash, P.: Smart card based remote user authentication schemes—Survey. In: 3rd International Conference on Computing Communication & Networking Technologies (ICCCNT), pp. 1–5. IEEE (2012)
26. Boyd, C., Mathuria, A.: Protocols for Authentication and Key Establishment. Springer (2003)
27. Brier, E., Clavier, C., Olivier, F.: Correlation power analysis with a leakage model. In: Cryptographic Hardware and Embedded Systems-CHES'04, pp. 16–29. Springer (2004)
28. Eisenbarth, T., Kasper, T., Moradi, A., Paar, C., Salmasizadeh, Shalmani, M., M.T.M.: On the power of power analysis in the real world: A complete break of the Keeloq code hopping scheme. In: Advances in Cryptology (CRYPTO-2008), pp. 203–220. Springer (2008)
29. Kocher, P., Jaffe, J., Jun, B.: Differential power analysis. In: Advances in Cryptology (CRYPTO99), pp. 388–397. Springer (1999)
30. Messerges, T.S., Dabbish, E.A., Sloan, R.H.: Examining smart-card security under the threat of power analysis attacks. IEEE Trans. Comput. **51**(5), 541–552 (2002)
31. Xu, J., Zhu, W.-T., Feng, D.-G.: An improved smart card based password authentication scheme with provable security. Comput. Stand. Interfaces **31**(4), 723–728 (2009)
32. Yang, C.-C., Yang, H.-W., Wang, R.-C.: Cryptanalysis of security enhancement for the timestamp-based password authentication scheme using smart cards. IEEE Trans. Consum. Electron. **50**(2), 578–579 (2004)
33. Fisher, D., Gloutnikov, S., Xi, Y., Khan, S.: Viability of dew computing for multilayered networks. https://gloutnikovcom/papers/viability.of.dew.computing.for.multilayered.networks.pdf
34. Patel, H.M., Chaudhari, R.R., Prajapati, K.R., Patel, A.A.: The interdependent part of cloud computing: dew computing. In: Intelligent Communication and Computational Technologies, pp. 345–355. Springer (2018)
35. Wang, Y., Skala, K.: The 3rd international workshop on dew computing. In: Proceedings of the 28th Annual International Conference on Computer Science and Software Engineering, pp. 357–358 (2018)
36. Manocha, A., Bhatia, M., Kumar, G.: Dew computing-inspired health-meteorological factor analysis for early prediction of bronchial asthma. J. Netw. Comput. Appl. **179**, 102995 (2021)
37. Afaq, Y., Manocha, A.: Dew computing-assisted cognitive intelligence-inspired smart environment for diarrhea prediction. Computing 1–30 (2022)
38. Gushev, M.: Dew computing architecture for cyber-physical systems and IoT. Internet Things **11**, 100186 (2020)
39. Javadzadeh, G., Rahmani, A.M., Kamarposhti, M.S.: Mathematical model for the scheduling of real-time applications in IoT using dew computing. J. Supercomput. **78**(5), 7464–7488 (2022)
40. Roy, S., Sarkar, D., De, D.: Dewmusic: crowdsourcing-based internet of music things in dew computing paradigm. J. Ambient Intell. Humaniz. Comput. **12**(2), 2103–2119 (2021)

41.` Singh, P., Kaur, A., Aujla, G.S., Batth, R.S., Kanhere, S.: Daas: dew computing as a service for intelligent intrusion detection in edge-of-things ecosystem. IEEE Internet Things J. **8**(16), 12569–12577 (2020)
42. Rana, S., Obaidat, M.S., Mishra, D., Mishra, A., Rao, Y.S.: Efficient design of an authenticated key agreement protocol for dew-assisted IoT systems. J. Supercomput. **78**(3), 3696–3714 (2022)
43. Ma, Y., Ma, Y., Cheng, Q.: Cryptanalysis and enhancement of an authenticated key agreement protocol for dew-assisted IoT systems. Secur. Commun. Netw. **2022** (2022)
44. Hati, S., De, D., Mukherjee, A.: Dewbcity: blockchain network-based dew-cloud modeling for distributed and decentralized smart cities. J. Supercomput. **78**(6), 8977–8997 (2022)
45. Braeken, A.: Authenticated key agreement protocols for dew-assisted IoT systems. J. Supercomput. 1–21 (2022)
46. Stojmenovic, I., Wen, S.: The fog computing paradigm: scenarios and security issues. In: 2014 Federated Conference on Computer Science and Information Systems, pp. 1–8. IEEE (2014)
47. Wang, Y.: Cloud-dew architecture. Int. J. Cloud Comput. **4**(3), 199–210 (2015)
48. Butun, I., Erol-Kantarci, M., Kantarci, B., Song, H.: Cloud-centric multi-level authentication as a service for secure public safety device networks. IEEE Commun. Mag. **54**(4), 47–53 (2016)
49. Rindos, A., Wang, Y.: Dew computing: the complementary piece of cloud computing. In: IEEE International Conferences on Big Data and Cloud Computing (BDCloud), Social Computing and Networking (SocialCom), Sustainable Computing and Communications (SustainCom) (BDCloud-SocialCom-SustainCom), vol. 2016, pp. 15–20. IEEE (2016)
50. Sojaat, Z., Skalaa, K., The dawn of dew: dew computing for advanced living environment. In: 40th International Convention on Information and Communication Technology, Electronics and Microelectronics (MIPRO), vol. 2017, 347–352. IEEE (2017)
51. Dolui, K., Datta, S.K.: Comparison of edge computing implementations: fog computing, cloudlet and mobile edge computing. In: Global Internet of Things Summit (GIoTS), vol. 2017, pp. 1–6. IEEE (2017)
52. Gusev, M.: A dew computing solution for IoT streaming devices. In: 40th International Convention on Information and Communication Technology, Electronics and Microelectronics (MIPRO), vol. 2017, pp. 387–392. IEEE (2017)
53. Srinivas, J., Das, A.K., Kumar, N., Rodrigues, J.: Cloud centric authentication for wearable healthcare monitoring system. IEEE Trans. Depend. Secure Comput. (2018)
54. Guan, Y., Shao, J., Wei, G., Xie, M.: Data security and privacy in fog computing. IEEE Netw. (99), 1–6 (2018)
55. Mane, T.S., Agrawal, H.: Cloud-fog-dew architecture for refined driving assistance: the complete service computing ecosystem. In: 2017 IEEE 17th International Conference on Ubiquitous Wireless Broadband (ICUWB), pp. 1–7. IEEE (2017)
56. Wazid, M., Das, A.K., Kumar, N., Vasilakos, A.V.: Design of secure key management and user authentication scheme for fog computing services. Fut. Gener. Comput. Syst. **91**, 475–492 (2019)
57. Botta, A., Gallo, L., Ventre, G.: Cloud, fog, and dew robotics: architectures for next generation applications. In: 2019 7th IEEE International Conference on Mobile Cloud Computing, Services, and Engineering (MobileCloud), pp. 16–23. IEEE (2019)
58. Ray, P.P.: Minimizing dependency on internetwork: is dew computing a solution? Trans. Emerg. Telecommun. Technol. **30**(1), e3496 (2019)
59. Longo, M., Hirsch, M., Mateos, C., Zunino, A.: Towards integrating mobile devices into dew computing: a model for hour-wise prediction of energy availability. Information **10**(3), 86 (2019)
60. Sanabria, P., Tapia, T.F., Toro Icarte, R., Neyem, A.: Solving task scheduling problems in dew computing via deep reinforcement learning. Appl. Sci. **12**(14), 7137 (2022)
61. Chakraborty, S., De, D., Mazumdar, K.: Dome: dew computing based microservice execution in mobile edge using q-learning. Appl. Intell. 1–20 (2022)
62. Hirsch, M., Mateos, C., Zunino, A., Majchrzak, T.A., Grønli, T.-M., Kaindl, H.: A task execution scheme for dew computing with state-of-the-art smartphones. Electronics **10**(16), 2006 (2021)

63. Hirsch, M., Rodríguez, J.M., Mateos, C., Zunino, A.: A two-phase energy-aware scheduling approach for CPU-intensive jobs in mobile grids. J. Grid Comput. **15**(1), 55–80 (2017)
64. Garrocho, C.T.B., Oliveira, R.A.R.: Counting time in drops: views on the role and importance of smartwatches in dew computing. Wirel. Netw. **26**(5), 3139–3157 (2020)
65. Alliance, C.: Security guidance for critical areas of focus in cloud computing v3. 0, Cloud Secur. Alliance **15**, 1–176 (2011)
66. Guo, J., Chen, R., Tsai, J.J.: A survey of trust computation models for service management in internet of things systems. Comput. Commun. **97**, 1–14 (2017)
67. Chhikara, D., Rana, S., Mishra, A., Mishra, D.: Blockchain-driven authorized data access mechanism for digital healthcare. J. Syst. Archit. **131**, 102714 (2022)

Platforms and Services

Implementation of Dew-Inspired Matrix-Mesh Communication Protocol

Minhajur Rahman and Yingwei Wang

1 Introduction

The main intent of the Matrix-Mesh messaging protocol is to allow the messenger application to adapt to changes in connectivity, passing messages under both normal and disastrous conditions [1]:

1. The internet is working perfectly. Individuals may connect to each other and to servers.
2. The internet is working well enough that individuals may communicate with each other but the messenger app server is not available.
3. The internet is unable to be used by individuals to connect to each other. However Local Area Networks, (LANs) such as local WiFi, are still able to be used for connections.
4. Neither the internet nor LANs are able to be used, and connectivity is only possible through battery-operated mobile devices. Small-scale connectivity is possible through a Personal Area Network (PAN) or a Near-Fear-Communication (NFC) connection.

The proposed messaging protocol is designed in such a way as to allow applications to extend the protocol easily, so as to provide some level of communications support at all of the above-mentioned network connection levels, thereby maximizing utilization. Keeping ease of use and widespread adoption in mind, the protocol was kept as simple and lightweight as possible [1]. The proposed Matrix-Mesh protocol is further discussed and explained in [1]. From this point on, this document will

M. Rahman (✉) · Y. Wang
School of Mathematical and Computational Sciences, University of Prince Edward Island, 550 University Ave, Charlottetown PE C1A 4P3, Canada
e-mail: mrahman2@upei.ca

Y. Wang
e-mail: ywang@upei.ca

D. De and S. Roy (eds.), *Dew Computing*, Internet of Things,
https://doi.org/10.1007/978-981-99-4590-0_5

focus mainly on the proof-of-concept implementation, and the relevant architectural features of the protocol, in order to demonstrate the deployability of this protocol targeting the mobile platform. For expediency, the Android platform was chosen, although the general principles involved can be applied to other platforms, such as iOS.

2 Motivation

Communication plays a very important role in our day-to-day life. People connect to each other in different ways in order to connect with each other and share information. In the modern era, we have given ourselves many different tools to help us make these connections, from instant messaging to video calls. The internet has been crucial to the development of this new era of connection, and as such it is no surprise that most modern communication technologies rely on internet connections. This means that without internet connectivity, communication systems that are currently relied on become unavailable. Worse still, many of these technologies rely on centralized cloud-based services, which can be brought down by a few accidental lines of code. Thus, in times of disasters or turmoil, it is unreasonable to expect these services to operate continuously, or in some cases, to operate at all. Additionally, as infrastructure is not equally well distributed across the globe, there are many places where an internet connection is far from guaranteed, even under optimal conditions. Further, as climate change continues to increase both the frequency and severity of extreme weather events, disruptions will only become more likely. Therefore, an instant messaging system was proposed to provide alternative means for vital communications in such disastrous conditions. The goal of this project is to further explore the details of this new instant messaging system and prove the deployability by implementing a proof-of-concept application.

3 Matrix-Mesh Architecture

As mentioned above, the Matrix-Mesh protocol was designed to be implemented on top of other existing messaging protocols, in particular Matrix [6]. Matrix-Mesh is a hybrid architecture, in order to draw from the strengths of both the client-server and peer-to-peer architectures, while leveraging the advantages of cloud-dew architecture in order to add functionality in as lightweight a way as possible [1–3].

In the above discussions, the client-server architecture was determined to be very efficient. This would be where central servers manage groups/accounts, relay messages, and control access to content, allowing for comparatively simple client applications. This architecture's largest weakness, however, was the single point of failure that this server becomes. As a result, if it becomes unavailable, the rest of the system

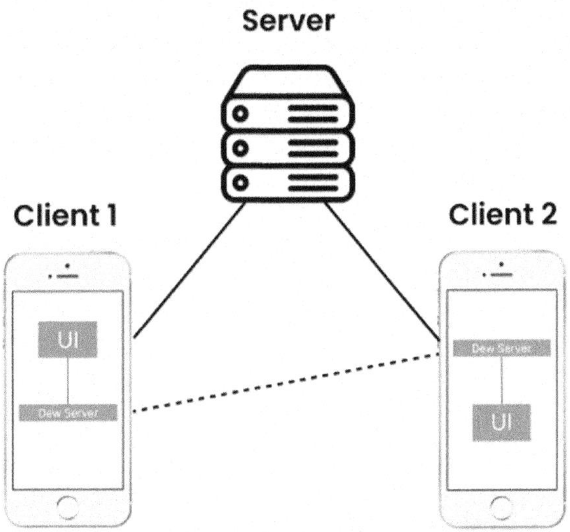

Fig. 1 The proposed matrix-mesh architecture, where the dashed line shows the connections between clients when the server is not able to be reached

is completely unable to function. P2P's decentralization alleviates this weakness, at the cost of increasing the complexity and requisite duties of each participant.

Matrix-Mesh, considering these strengths and weaknesses, will attempt to meet its goals by hybridizing the above structures, to try to gain the advantages of each, while using cloud-dew architecture to maximize performance under difficult circumstances [2]. As such the messaging system will work in client-server mode whenever possible to gain the advantages of that model, however, whenever that is not possible, the system will transition into P2P mode [1]. In addition, Matrix-Mesh will alter the client application, giving it two components: the dew server and the User Interface, (UI) following the conventions of dew computing, or in the terms of Matrix-Mesh, the communication server or mini-router [1]. The dew server includes the functionality of both a client and a server, acting as a client when it serves the requests of the UI, and functioning as either a client or server when interacting with other instances of the app [1]. Overall, the Matrix-Mesh application architecture is inspired by cloud-dew architecture to maintain functionality when the server is unavailable [2, 5]. Figure 1 visualizes the relationships in Matrix-Mesh:

4 The Dew Server and UI

The UI may only operate when required by the user, however, the dew server will function continuously to maintain consistency. The dew server deals primarily with data management and transmission, it does not necessarily incorporate encryption, nor does it deal with user interaction or data display, which fall under the purview of the UI portion of the client. Under normal conditions, dew servers do not need to keep connections with other clients, however, in situations where network functionality is an issue, a dew server should maintain such connections to support its own function and to provide services to other clients. This is vitally important, as the users of these other clients should support the functions of other clients in the event of a disaster. As a result, even when operating as a client-server, the server must collect account information and create connections with other clients. When operating as P2P, emergency conditions will be assumed, as a result neither advanced services nor a complete suite of administrative options need to be provided, which will help to keep the client lightweight [1]. Importantly, the UI does not communicate with external servers or clients, but only with the dew server. This is because, so long as the dew server receives a message, from the point of view of user interacting with the UI, that message is sent and it is the responsibility of the dew server to take care of any additional details [1].

5 Design Considerations and Inspirations

Based on the above, Matrix-Mesh can be looked at as largely an extension of the Matrix protocol, which is Matrix with additional features [1]. As such, an existing open-source implementation of a Matrix messaging client was chosen as the base from which to extend. For this proof-of-concept app, Bluetooth was chosen as the auxiliary link, as it is one of the most used wireless technologies, and as such is now available on most mobile devices.

Functionally, the design goal for the extension is to provide an additional connection, using Bluetooth, to allow an alternate means of connection in situations when the more typical http connection with a home server is not possible. During such a situation, such as a natural disaster, this Bluetooth link will allow peer-to-peer communication to continue, albeit at short distances, as the Bluetooth specification typically specifies a range of 10 m. However, communication between users that are not in close physical proximity will still be possible, as message replication across users will allow messages to eventually reach other devices. In Matrix, users will be members of the same room if they are chatting, with the encryption tied to the room. Further, as messages for a room are replicated across all participants, allowing messages from one participant to propagate to others. Figure 2 depicts the discussed situation, and Fig. 3 depicts the resulting architecture.

6 Technological Consideration and Choices

As this proof-of-concept will be attempting to continue to make use of all Matrix features and maintain full compatibility, the choice was made to extend an existing Matrix client, as such, the choice of client is important. There are already several different existing messenger clients for matrix, most of which are open-source, and

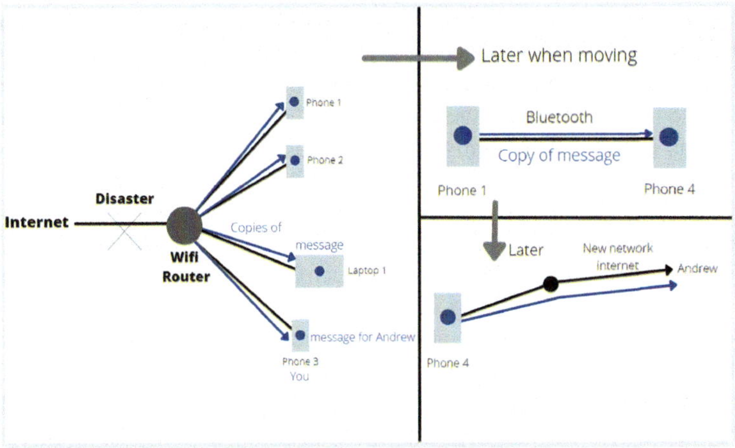

Fig. 2 Visualization of message propagation under disastrous conditions in matrix-mesh. In the depicted situation, some disaster has cut access to the internet. However, messages can still be passed via the auxiliary connection provided by Bluetooth, when a user is in range. Additionally, these messages can be further propagated by devices, either through new contact or by one of the devices achieving internet access

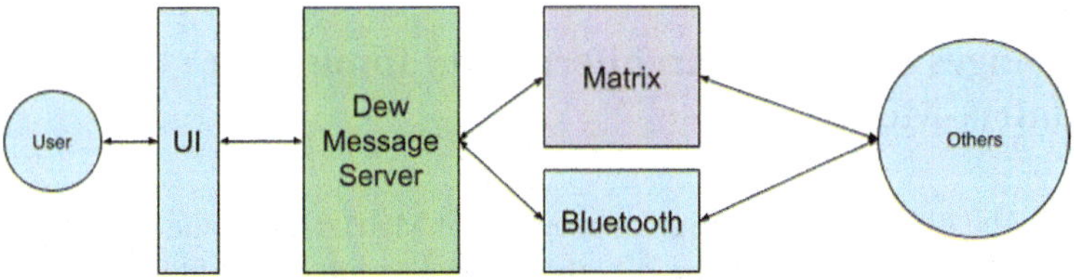

Fig. 3 The resulting matrix-mesh internal architecture

so could be chosen as a foundation, these include: Element, Ditto Chat, NeoChat, and FluffyChat, among others [6]. Though it was chosen to focus on Android as a platform, for more expedient development, the long-term ideal would be to be platform agnostic, and so two cross-platform clients were chosen for final consideration, these being FluffyChat, and Ditto Chat [7, 8].

While both clients were explored as possibilities, ultimately FluffyChat was chosen. Both clients were developed using cross-platform tools to create a code base compatible across multiple platforms [14]. Ditto Chat, for example, was developed in React Native, which is backed by Facebook, with the intention of facilitating development of cross-platform applications that are able to make use of native features when needed [8, 9]. Comparatively, FluffyChat is a Flutter app, written in Dart, where Flutter is a platform backed by Google with comparable goals to React Native [7, 10]. In the end FluffyChat was chosen for a combination of reasons, not least of which being that it seemed to have the most active upkeep of the two, suggesting it would be the most up-to-date and the best maintained. Additionally, Flutter was deemed to be a more convenient and overall better framework when compared to React Native, while also subjectively seeming the more popular of the two platforms. One anecdotal indication of this relative popularity would be to compare the responses to their respective git repositories on GitHub, where FluffyChat had 138k stars, versus 102k for React Native [11, 12]. Additionally, Flutter includes desktop systems in its possible development targets, where React Native focuses only on Android and iOS [9, 10]. As such, FluffyChat was decided to be the best option, thus determining that the proof-of-concept would be a Flutter app as well [7].

Flutter is, fundamentally, a framework which enables cross-platform app development. More precisely, Flutter is an open-source framework developed by Google for building natively compiled multi-platform applications from a single code codebase [10]. Basically, it is a portable and comprehensive software development toolkit, which is based on the Dart programming language [13]. Flutter provides its own widgets, which are drawn by its own high-performance rendering engine, making them both fast and easily customizable. Furthermore, Flutter apps do not need a bridge to interact with native components, something that typically causes performance issues, allowing Flutter app performance to be equivalent to that of native, real-time applications [10].

7 Changes in Matrix-Architecture to Implement Matrix-Mesh

Based on the discussion and original proposal of Matrix-Mesh, the protocol will use the full features of Matrix and modify the layered architecture of the Matrix in order to implement the Matrix-Mesh protocol. In this section, we will be discussing how the architecture of Matrix has been altered in order to implement Matrix-Mesh.

The Matrix protocol has already been established, and in the design, room was made anticipating future development, though the Matrix proposal tries to limit new development to a layered protocol model [17]. That is, the design prefers new features be built on layers of shared abstractions, avoiding tight vertical coupling in the larger stack [17]. The motivation behind this preference is to ensure that extensions are able to evolve quickly, by building on existing layers and swapping out old features without interfering with the rest of the stack, and are thus also unlikely to interact with upgrades to the entire ecosystem [17]. As such, developers extending Matrix are asked to implement features at the highest layer of abstraction possible, so that there is no necessity for changes to the Matrix API [17].

However, despite this preference, and the importance Matrix places in backwards compatibility, it is not desirable for this restriction on feature implementation to prevent the evolution of the algorithm. Thus, when necessary, such low-level changes are permissible when higher level alternatives are not possible. Figure 4 visualizes the architecture and internal structure of the Matrix protocol [18]. As described above, this Matrix extension will add an auxiliary link to extend capabilities with regard

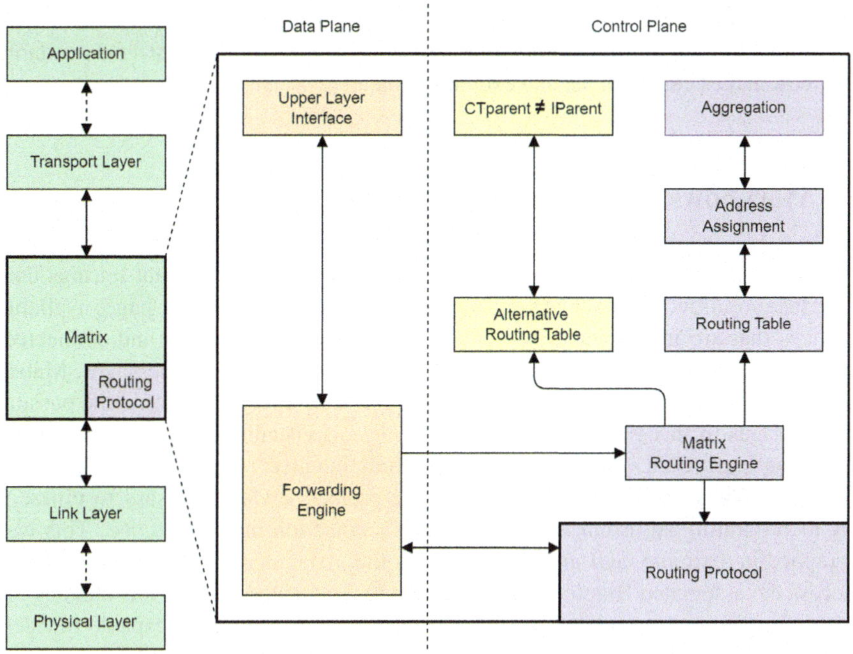

Fig. 4 Matrix protocol architecture

to vital communication, and as such an addition should happen in the link layer. This layer will be in parallel to the existing layer, and as such some changes will have to be made to the Matrix layer itself and cannot be relegated to a higher layer. Further, as the addition of this link will alter function, at least during conditions where Bluetooth is the primary link, there will necessarily be associated changes to behavior and routing in this Matrix layer to handle these circumstances. However, as the goal is not to alter the functionality of Matrix, in terms of functionality offered by the Matrix layer, it is not necessarily required that changes propagate further and necessitate changes to higher layers.

8 Bluetooth

One weakness with the choice of Flutter, and cross-platform frameworks more generally, is that support for low-level access and functionality is not often well incorporated. Comprehensive Bluetooth support is not currently a core feature of Flutter and the Dart built-in libraries. There are, however, a number of open-source libraries that attempt to bring this functionality to the platform. Most of these libraries were tried, however the most comprehensive of the main libraries, Flutter Bluetooth Serial, was chosen. Such choice was in part due to the options being more popular, more GitHub interaction, and more demonstrable uses of the package in other applications [19]. It is through the use of this package that the proof-of-concept app scans for available devices, makes connections, and exchanges data.

9 Workflows of the Integrated System

A Bluetooth devices and settings menu was added to the FluffyChat settings user interface to allow for management of device connections. On this page, available devices that are in range can be chosen. Once devices are paired and connected, the user need only return to the standard chat interface to send messages. Matrix already supports encryption, allowing messages to be encrypted should the passing of messages in this way be a concern. Figure 5a–f visualize the workflows of the modified FluffyChat, with added Bluetooth functionality:

After adding the Bluetooth library to the project, it was necessary to utilize it by incorporating an instance of the library's connection managing class. This was wrapped in a simple GUI and Fig. 5a shows the associated menu entry in order to access the associated Bluetooth management features. Once active, this handles the searching for, pairing with, and connections to nearby Bluetooth capable devices. After connection has taken place, users are able to begin to pass messages, including ones that are encrypted, as these continue to be supported. As a result of the changes including the addition of a new type of communication link, changes at the link layer are unavoidable, however, a wholesale redesign is not necessary, and the additions should have limited ability to meaningfully interact with alterations at higher levels, thus maintaining the encapsulation desired by the Matrix protocol. Other than issues tied to the particularities of a given auxiliary connection technology, the interaction with the Matrix-Mesh architecture should be identical, meaning that the Bluetooth situation is representative of auxiliary links in general.

10 Future Development

Though the demonstration of Matrix-Mesh is an important milestone, there is still much that can be done to progress this research. The most obvious extension to this current work being the extension of the current technique to other forms of auxiliary links, such as local area network sharing, WiFi tethering, and NFC. Additionally, as this is just a proof-of-concept, little effort was put into user friendliness, nor it is a coherent and consistent addition to the messaging and graphical portions of the application. Thus, moving beyond the proof-of-concept stage is a natural continuation of this project.

Looking beyond the immediate application, the original Matrix-Mesh proposal mentioned another protocol for enhanced encryption, called the Tox protocol, which was not examined here [1]. Tox is based on P2P architecture, and it provides end-to-

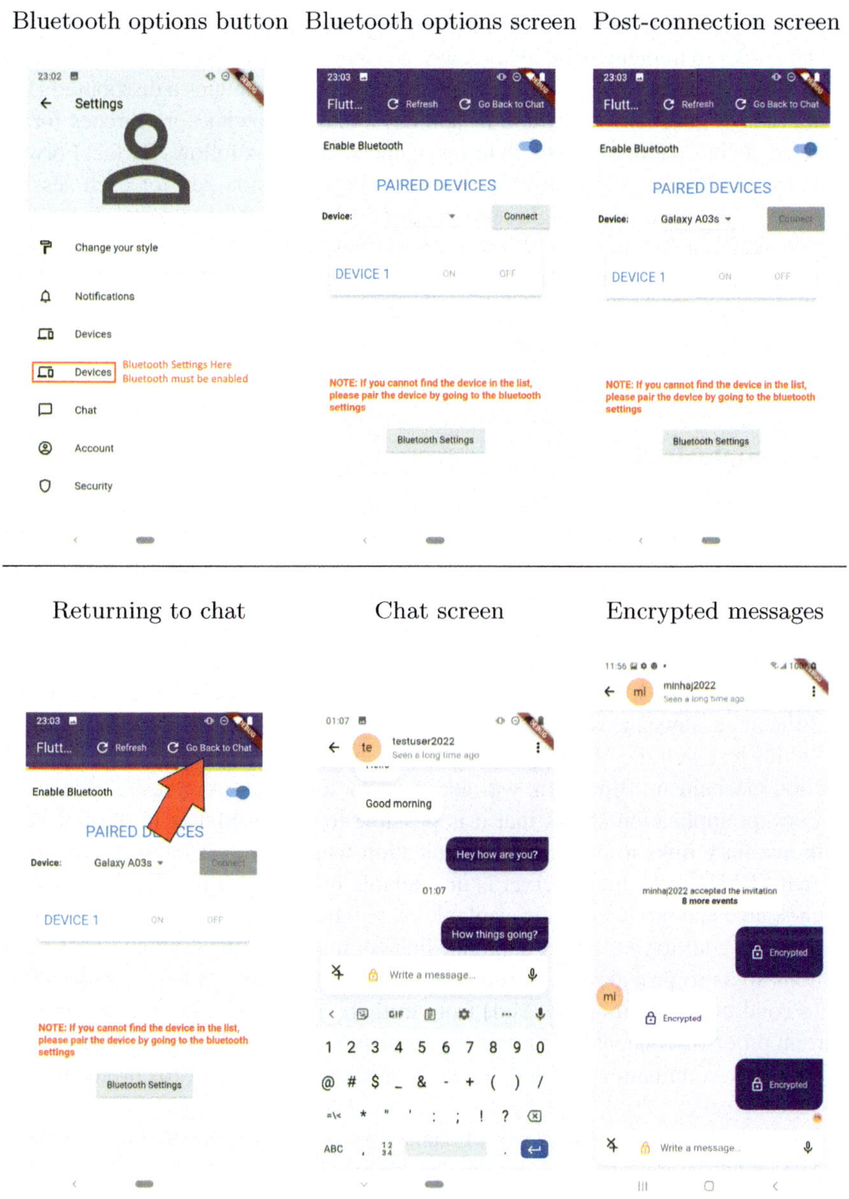

Fig. 5 **a–f** In these figures, the additions to the user interface are demonstrated, along with an example of the chat interface and the use of encryption

end encryption, with the intention of providing secure communication to everyone [15]. In order to implement a P2P messaging protocol such as Tox, each of the clients in the network must know each other's IP address. Tox maintains a distributed Hash table, (DHT) in order to facilitate IP address lookup, allowing for searches for the targeted client's address [1, 16]. In terms of encryption, Tox follows Perfect Forward Secrecy (PFS) encryption which provides a unique session key for each session. That way compromising a single session key does not affect the security of any other data. In addition, Tox uses the Elliptic Curve Diffie-Hellman (ECDH) key agreement scheme to ensure the secure distribution of keys among the clients [1, 16]. Given the features of Tox, extending with Tox, as suggested by Dr. Wang, would be a more ambitious, but worthwhile research effort.

11 Conclusions

Matrix-Mesh is an instant messaging protocol inspired by cloud-dew architecture, and dew computing principles to provide vital communication under disastrous conditions. It works in a standard mode consistent with the Matrix protocol and its client-server-based architecture when under normal operating conditions. However, when that is not possible, it switches to alternate mechanisms and more closely approaches a P2P architecture. As a result, it has the potential to be far more robust in difficult circumstances than the original Matrix protocol on which it is based.

In this research the Matrix protocol was extended, and a parallel, auxiliary connection mechanism, Bluetooth, was added in the link layer. As a result, this proof-of-concept application shows that it is possible to update Matrix in the link layer with auxiliary links to provide communication when a typical internet connection is unavailable, or the home server is unavailable for any reason. This application is open-source and has been made available on GitHub [20]. Future development may include the addition of further auxiliary links to maximize the available connection options so as to provide the most consistent communication possible under disastrous conditions. Additionally, aside from further integration and development in the current proof-of-concept code in order to transition toward a complete user-friendly application, examination of similar integration with regards to the Tox protocol might be investigated.

Given the increasing importance of connectivity, the rising threat of more frequent and extreme natural disasters because of global warming, and the effects on any possible future conflict, increasing the robustness of messaging applications is very important. To better coordinate responses to such disastrous conditions, it is vital for communication applications to incorporate more dew computing principles to better handle disruptions of connectivity, as well as to reduce dependence on single points of failure. This proof-of-concept application demonstrates the feasibility of this goal, and the potential benefits should similar extensions be made to existing messaging systems.

References

1. Wang, Y.: A disaster-resilient messaging protocol based on dew computing. In: 2020 43rd International Convention on Information, Communication and Electronic Technology (MIPRO), pp. 1922–1926, Opatija, Croatia (2020). https://doi.org/10.23919/MIPRO48935.2020.9245286
2. Wang, Y.: Cloud-dew architecture. Int. J. Cloud Comput. **4**(3), 199–210 (2015)
3. Rahman, M., Wang, Y.: Implementation of dewblock clients on a mobile platform. In: 2021 IEEE 45th Annual Computers, Software, and Applications Conference (COMPSAC), pp. 1810–1813 (2021). https://doi.org/10.1109/COMPSAC51774.2021.00272
4. Wang, Y.: Definition and categorization of dew computing. Open J. Cloud Comput. **3**, 1–7 (2016)
5. Ray, P.P.: An introduction to dew computing: definition, concept and implications. IEEE Access **6**, 723–737 (2018). https://doi.org/10.1109/ACCESS.2017.2775042
6. Matrix.org.: Clients. Matrix.org (2022). https://matrix.org/clients/. Accessed 3 Apr. 2022
7. Famedly/fluffychat.: GitLab (n.d.). https://gitlab.com/famedly/fluffychat. Accessed 3 Apr. 2022
8. Ditto chat/ditto.: GitLab. (n.d.). https://gitlab.com/ditto-chat/ditto. Accessed 3 Apr. 2022
9. React native.: Learn once, write anywhere. React Native RSS. (n.d.). https://reactnative.dev/. Accessed 3 Apr. 2022
10. Build apps for any screen.: Flutter (n.d.). https://flutter.dev/. Accessed 3 Apr. 2022
11. Flutter.: Flutter/flutter: flutter makes it easy and fast to build beautiful apps for mobile and beyond. GitHub (n.d.). https://github.com/flutter/flutter. Accessed 3 Apr. 2022
12. Facebook.: Facebook/react-native: a framework for building native applications using react. GitHub (n.d.). https://github.com/facebook/react-native. Accessed 3 Apr. 2022
13. Dart programming language.: Dart (n.d.). https://dart.dev/. Accessed 3 Apr. 2022
14. Contributor, T.T.: What is cross-platform mobile development?—Definition from whatis.com. SearchMobileComputing (2018). https://www.techtarget.com/searchmobilecomputing/definition/cross-platform-mobile-development. Accessed 3 Apr. 2022
15. A new kind of instant messaging.: Project Tox (n.d.). https://tox.chat/. Accessed 3 Apr. 2022
16. Stoica, I., Morris, R., Karger, D., Kaashoek, M.F., Balakrishnan, H.: Chord: a scalable peer-to-peer lookup service for internet applications. SIGCOMM Comput. Commun. Rev. **31**(4), 149–160 (2001). https://doi.org/10.1145/964723.383071
17. Spec change proposals.: Matrix Specification (n.d.). https://spec.matrix.org/v1.2/proposals/. Accessed 3 Apr. 2022
18. Santos, B.P., Goussevskaia, O., Vieira, L.F., Vieira, M.A., Loureiro, A.A.: Mobile matrix: a multihop address allocation and any-to-any routing in mobile 6LoWPAN. In: Proceedings of the 13th ACM Symposium on QoS and Security for Wireless and Mobile Networks (Q2SWinet '17), pp. 65–72. Association for Computing Machinery, New York, NY, USA (2017). https://doi.org/10.1145/3132114.3132126
19. Flutter_bluetooth_serial: Flutter Package.: Dart packages (2021). https://pub.dev/packages/flutter_bluetooth_serial. Accessed 3 Apr. 2022
20. Rahman, M.: Matrix-mesh-V2022.02.01 (2022). https://github.com/Minhaj9800/Matrix-Mesh-V2022.02.01/tree/dev

Blockchain-Based on Dew Computing for Unreliable Network

Amiya Karmakar, Pritam Ghosh, Karolj Skala, Partha Sarathi Banerjee, and Debashis De

1 Introduction

1.1 Dew Computing

Dew computing is a location-based software-hardware organization paradigm in the Cloud computing environment [1], as shown in Fig. 1. These local computers are independent of Cloud services and can also interact with Cloud services. Cloud computing has several advantages such as universal access and scalability [2]. However, the main problem with the cloud is that it is centralized and always requires an Internet connection. All the resources of the cloud are far away from the user's local computer. So, if the Internet connection is lost, the user cannot use his data. This problem can be solved by Dew computing approaches. In Dew computing, a local computer can provide extensive functions independently of a cloud service [3]. It also works

A. Karmakar · D. De
Department of Computer Science Engineering, Centre of Mobile Cloud Computing, Maulana Abul Kalam Azad University of Technology, Kolkata, West Bengal, India
e-mail: amiyakarmakar.ak@gmail.com

D. De
e-mail: debashis.de@makautwb.ac.in

P. Ghosh
Department of CST, Iswar Chandra Vidyasagar Polytechnic, Jhargram, West Bengal, India
e-mail: ghoshpritam25@gmail.com

K. Skala
Ruđer Bošković Institute, Zagreb, Croatia
e-mail: Karolj.Skala@irb.hr

P. S. Banerjee (✉)
Department of Information Technology, Kalyani Govt Engineering College, Kalyani, West Bengal, India
e-mail: psbanerjee.kgec@gmail.com

D. De and S. Roy (eds.), *Dew Computing*, Internet of Things,
https://doi.org/10.1007/978-981-99-4590-0_6

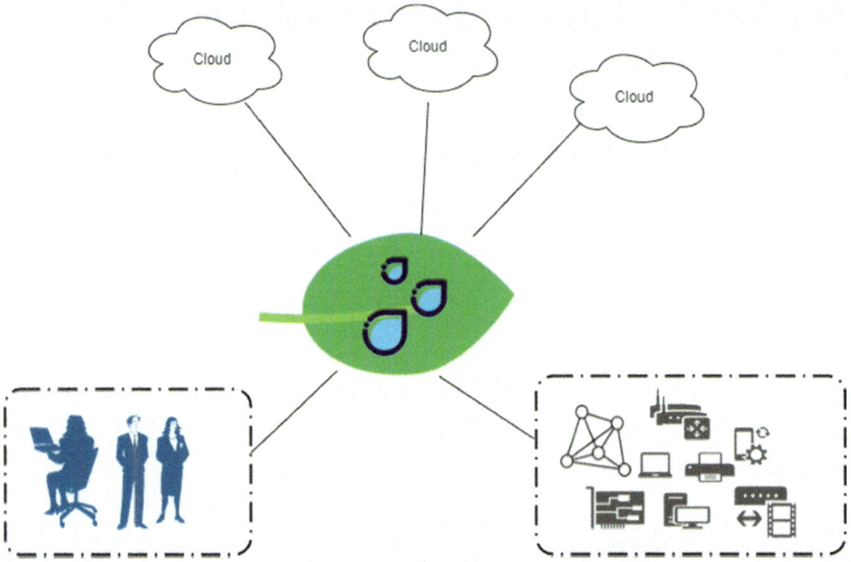

Fig. 1 The architectural structure of the dew computing. (Explanation: Green leaf is the dew layer (on-premises computer/mobile) and dew drops on the leaf represent the application that is running on the dew layer. This dew layer is giving service to the user and it also collaborates with the cloud)

together with Cloud services. The main features of Dew computing are independence and collaboration [4].

Independence:

On-site software can provide all functions without using Cloud services. A stable and continuous Internet connection is also not required for this functionality.

Collaboration: With Dew computing, the application collaborates with the Cloud to synchronize, integrate and correlate data.

Dew computing can be divided into the following categories:

Web in Dew:

In Web in Dew, the on-site computer has a copy of the World Wide Web and provides the full functionality of the Web to the user [5]. It synchronizes with the cloud when the Internet connection is established.

Storage in Dew:

With Storage in Dew, the memory of any local computer is copied to the cloud and automatically synchronized with the cloud.

Database in Dew:

On this basis, the data of the local software is backed up in the cloud or vice versa. In this category, the user must decide which database is the main database.

Software in Dew:

This is the category of Dew architecture where the ownership and setting information of a particular software is not only present in the storage of the user's local computers but also shared with the cloud services. SiD also guarantees that users can download the software again when needed.

Platform in Dew:

Platform in Dew is a category of Dew computing where:

1. A software suite that supports development and operation for specific purposes is installed and run on an on- site computer.
2. The settings and application data of this software suite are dynamically synchronized with a cloud service.

Infrastructure as Dew (IAD):

This is a category of Dew Computing where a local computer is dynamically supported by cloud services.

Blockchain Technology

Blockchain is a secure, immutable special form of distributed decentralized database [6–8]. The main goal of Blockchain is to shop all data in a block format. All blocks are secured with some cryptographic algorithms. Each block contains the following characteristics:

- cryptographic hash of the previous block,
- timestamp,
- transaction dana.

The transaction data of the Blockchain is stored in the Merkle tree format, as shown in Fig. 2.

The Blockchain network is managed by a peer-to-peer network and is based on publicly distributed ledgers. Some protocols must follow each node to communicate with each other and validate the transactions [9].

Since the Blockchain is a distributed database (see Fig. 3), all nodes have a copy of the entire Blockchain data. And this is where the problem lies. How do all nodes agree on a status and how do they verify that they all have the same copy of the Blockchain? In this case, consensus algorithms are needed [5, 10, 11]. There are many consensus algorithms, but among them Proof of Work and Proof of Stakes are very popular. In the Blockchain model, there is no central authority that can validate or authorize a transaction. The Blockchain is secure and immutable because there are strong consensus algorithms.

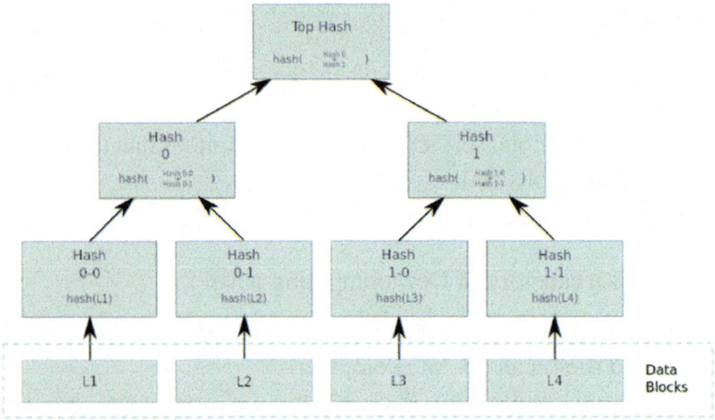

Fig. 2 Merkle tree architecture. (Explanation: L1, L2, L3, L4 represent the transactions. All transactions are converted into hash values: hash(L1), hash(L2), hash(L3), etc. All hash values are converted into the hash value of the root)

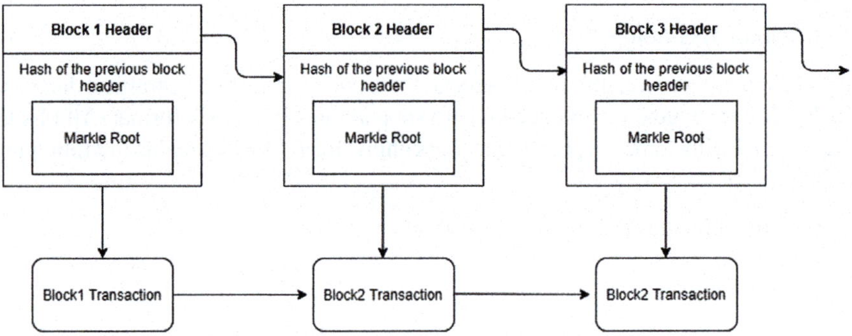

Fig. 3 Architectural structure of the blockchain. (Explanation: block 1 contains a hash of the header of the previous block, the Markle root of the transaction of block 1 and its header. Each block follows the same structure)

2 Literature Study

2.1 Blockchain Algorithms

Proof of work

Pow is a type of consensus algorithm that must solve a computationally intensive puzzle to create a new block in the Blockchain. When a new block appears on the Blockchain network, who adds the block to the Blockchain? Proof of Work solves this problem. According to the Proof of Work algorithm, only the node that has invested the most computing power can add the new block to the Blockchain (Table 1).

Table 1 Limitation of proof of work

Risk	Details
51% attack [12]	If some minor/group of minor has the control more than 50% computing power on the blockchain network can corrupt the blockchain by achieving most of the network
Time	In proof of work, minorities need a lot of time to solve the complex puzzle as they need to check many nonce values
Resource	Minors need a lot of computing power to find the solution to complex mathematical puzzles. For these calculations, minors consume a lot of valuable resources such as money, energy, space and other hardware components. The consumption of these resources can have a big impact on the environment

Table 2 POW versus POS

Measures	Proof of work	Proof of stake
Probability of winning a block	Probability of mining a block depends on the minor's computing power	Probability of confirming a new block depends on the minor's stake
Rewards	Minors receive a reward if they can add the block to the Blockchain	In this algorithm, the validator receives the transaction fees as a reward
Vulnerability	A minor must have control over more than 50% of the blockchain nodes' processing power to add a malicious block	Minors must own more than 50% of the total cryptocurrency to manipulate the system
Resources	Since word proof depends on computing power, a large number of resources is required to mine a block	No resources are required
Tools	Mining tools are required for this algorithm	Since there is no concept of computing power, no mining tools are required

Proof of stake

Proof of stake is an alternative and one of the most popular consensus algorithms [13]. In this algorithm, the block validator or miner must invest a certain amount in the system. In this model, the validator does not have to mine blocks, only confirm them. If he confirms malicious blocks, he loses his stake (Table 2).

2.2 Blockchain Categories and Classification

Blockchain can be classified into the following three types [14–16]

Table 3 Public versus private versus consortium blockchain

Measures	Public	Private	Consortium
Read/write permission	Anyone can read and write in the system	Pre-selected participates from a single organization can read/write	Multiple selected organizations
Speed	Slow	Faster	Faster
Efficiency	Low	High	High
Network	Completely decentralized	Centralized	Partially decentralized
Security	High	Low	Low
Consensus algorithm	Pow, pos etc.	Voting or multi-party consciences algorithms	Voting or multi-party consciences algorithms
Consensus determination	All miners	Organization	The selected set of nodes
Energy consumption	High	Lighter	Lighter
Asset	Native Asser	Any asset	Any asset

Public blockchain

A public Blockchain is open and anyone can be part of it. It is permission-free. Anyone can access this type of Blockchain network by signing up to the platform. Any user of this Blockchain can access all data and mining activities and anyone can contribute to verify the transaction.

Private blockchain

A private Blockchain network is authorized and restrictive. This Blockchain is controlled by a single centralized authority. Compared to the public Blockchain, it is a smaller scale Blockchain.

Consortium blockchain

This type of Blockchain contains the characteristics of both a private and a public Blockchain. Several authorities or organizations control this Blockchain and work together in this decentralized network (Table 3).

2.3 *Restriction of the Blockchain*

There are some challenges in the Blockchain environment [17–19]. Each Blockchain requires a large amount of storage and computing power for the validation process. The block size of each Blockchain environment is fixed. Since the block size is fixed, a constant number of transactions can be performed per second, which increases the delay. When the block size is increased, it takes longer to forward the blocks. To run a proof-of-work algorithm, Minor requires a large amount of energy resources. A large amount of energy is consumed globally for verification in the Blockchain. Energy consumption is a major challenge for the Blockchain [10].

2.4 How Dew Computing Differs from Cloud, Edge, and Fog Computing

Dew computing aims to bring the centralized cloud service closer to the user's light devices (mobile phones, computers, etc.). Users can minimize the use of the Internet through Dew computing as Dew does not require a controversial Internet connection. Edge computing uses a core-edge topology and moves the data, applications and services from the central server to the edge of the network. Fog computing allows heterogeneous devices to communicate and collaborate with each other. In the following table we can see the brief differences between each entity [20–22] (Table 4).

Blockchain in Dew

Cryptocurrencies and a wide range of other applications that require trust between multiple entities have been enabled by Blockchain technology. Some desktop PCs and mobile devices cannot run Blockchain clients because they need to process a significant volume of Blockchain data. The size of Blockchain data is constantly growing, which makes matters worse. This section is about Dewblock, a new type of Blockchain system. In this architecture, a Blockchain client has all the functions of a full Blockchain node without having to manage the Blockchain data. Based on the Cloud Dew architecture and the ideas of Dew computing, Dewblock was developed. The foundation of Blockchain technology is a distributed, secure ledge or Blockchain that resides in each node of the network and enables the establishment of trust between unidentified parties. Blockchain data must be present at every node, and the amount of data grows over time. Therefore, the problem we are trying to solve—that the amount of data on Blockchain clients is too large and constantly growing—is a problem inherent in this technology. The data capacity of Blockchain clients has been reduced in the past, but these clients do not have the status of a full-fledged node. Although these methods make users' lives easier, they cannot serve as the basis for Blockchain networks.

Dewblock introduces a novel method that retains the characteristics of a full-fledged node while minimizing the data size of a client. The key difference is that Blockchain client and Blockchain node are no longer the same in Dewblock. While a client is small and can easily be used on a home computer or mobile device, it works with a remote cloud server to function as a full-fledged node.

Table 4 Cloud versus fog versus edge versus dew computing

Parameters	Cloud computing	Fog computing	Edge computing	Dew computing
Service location	Within the Internet	Within the Internet	In edge network	In edge network
Distance (number of loops)	Multiple loops	Multiple loops	Single loop	No loop
Latency	Very high	High	Low	Negligible
Geo-distribution	Centralized	Semi centralized	Distributed	Highly distributed
Mobility support	Very limited	Limited	Semi supported	Highly supported
Data reroute attacks	Very high probability	High probability	Low probability	Very low probability
Target users	General internet users	General internet users	Semi-mobile users	Purely mobile users
Hardware	Scalable capabilities	Scalable capabilities	Limited capabilities	Very limited capabilities
Internet dependency	Every access time	Every access time	Every access time	Not essential
Client–server connectivity	Yes	Yes	Yes	No
Synchronization feature	Not essential	Not essential	Not essential	Always essential
Green energy compliant	Very low	Very low	Low	High
Delay tolerant	No	No	No	Yes

3 Dew Computing and Blockchain Integration for Unreliable Network

Blockchain technology requires each Blockchain client node to buy the entire Blockchain data starting from the Genesis block. This data grows day by day with the normal operation of the Blockchain network [23]. Since Blockchain is a distributed technology, data plays a crucial role in a Blockchain environment. Data is at the heart of the Blockchain. Each node in the Blockchain must perform a similar task to verify transactions, and it also has to shop a large amount of data. Some mobile phones and computers are not capable of storing these large amounts of data, so these types of devices cannot be used as Blockchain clients. These lightweight devices also have low processing power. To solve this problem, the Dew Block can be introduced. In the Dew Block architecture, the Blockchain client node does not need to shop the entire Blockchain data, but can still perform all the functions of a full Blockchain node

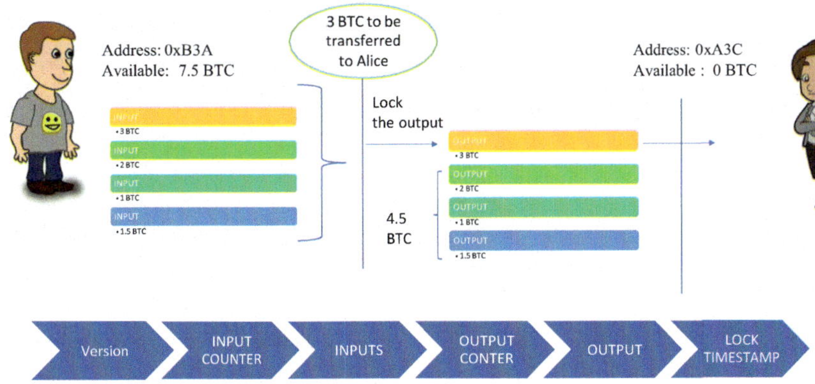

Fig. 4 The operating principle of the UTXO model

[24]. Each Dew node is responsible for maintaining the integrity of the Blockchain, and for deblocking, each user must use a cloud server.

Blockchain can be referred to as state mechanics. There are various models of state maintenance, of which two models of state maintenance are popular

- **UTXO model** [6]: In this model, the user's wallet must keep the list of all unexecuted transactions with the user's address, as shown in Fig. 4. In this model, the total balance is calculated as the sum of unexecuted transactions [25–27].
- **Account model** [28]: This model tracks the balance of each account as a global state

These two models have a different philosophy: UTXO can be described as a history-oriented model, while the account model is reality-oriented. Since the account model is reality-oriented, the Dew block works with the account-oriented model. Although Dew works with the account model, it is not constrained because the UTXO model can also be switched to the account model. In Dew bock, the architecture cloud server contains the main Blockchain, but there is no Blockchain in the client server. The absence of the Blockchain reduces the size of the data. Two types of message channels are used for collaboration between different nodes in Dewblock: Internode channel and Cloud Dew channel. Each system has some disadvantages, and Dewblock is no exception (Fig. 5).

Limitation of dew approach:

- Client nodes are not strong enough to defend themselves against various malicious attacks
- If many client nodes are not full-fledged nodes, the entire Blockchain network can be compromised.

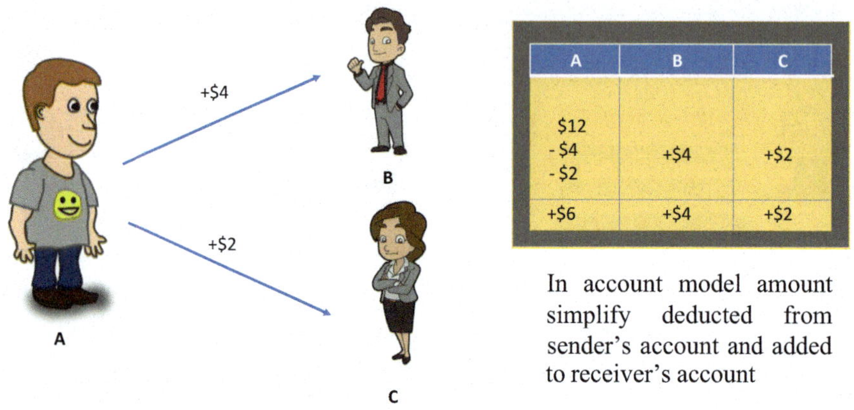

Fig. 5 Operating principle of the account model

4 Decentralized Medical Hardware Ownership a Dewblock Approach

Software in Dew (SiD) can control ownership of hardware, as shown in Fig. 6. SiD not only works with cloud servers but can also work with Blockchain technology [29].

SiD (Software in Dew) talks about the ownership of any software. In the SiD model, ownership of any software resides both on the user's on-site device and in the cloud. Ownership and settings data are stored in the cloud for further use. This feature allows the user to re-download and re-own the software. SiD can transfer ownership of the software, but it is very difficult to gain ownership of the hardware. There is a special software which is the system software of the hardware. Hardware like vehicles, cameras and all other devices are controlled by software. With the help of the system software of the hardware, the ownership of the hardware can be transferred. When the system software is installed on a part of the hardware, the owner of the software can be considered as the owner of the hardware parts. Most importantly, ownership of the system software can be controlled by SiD. If another user wishes to install or modify the system software on this hardware, this will be refused due to conflicts of ownership

Requirement for dew block over cloud block for the unreliable network

Requirements for the Dew block versus the Cloud block for the unreliable network:

In cloud computing, ownership data resides on a single central server and the most important thing is that each cloud server is owned by a specific company [30]. This is where some security concerns come into play. Data privacy, reliability, authenticity, and integrity may be unsatisfactory to the user. But from the user's perspective, both the cloud and the Blockchain do the same thing (Table 5).

Fig. 6 Software in dew

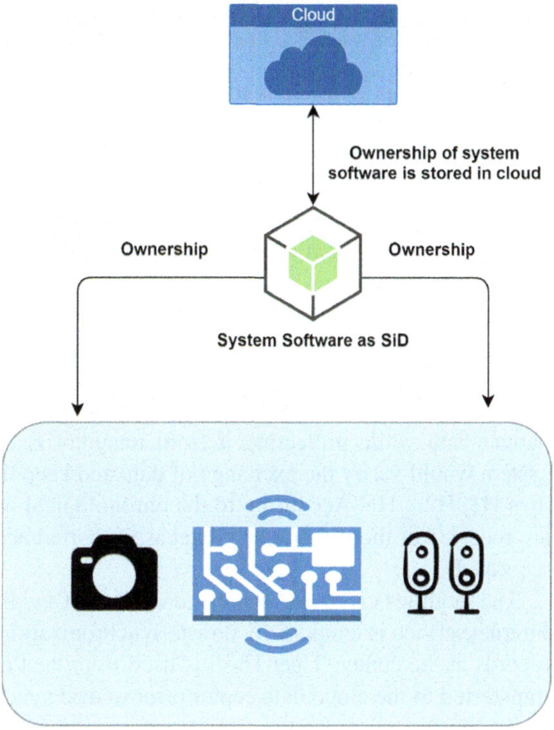

Table 5 Cloud block versus dew block

Measures	Cloud block	Dew blockchain
Network topology	Centralized	Decentralized
Transparency	Less transferred	Transparency
Security	Can be manipulated	Cannot be manipulated

In this proposed method, property-based transactions can take place. The Blockchain is the backbone of this transaction. Previously, the manufacturer had to keep the serial number and verification code of the hardware parts and carry out the entire verification process themselves. To free manufacturers from this burden, function pair verification inspired by asymmetric cryptography is introduced.

5 Dew Computing Based Distributed Healthcare System

For healthcare professionals to make decisions about patients, patient health data should be securely available and accessible to authorized users when needed [31]. When connected, data from Dew systems should be synchronized with data in the cloud. To ensure the integrity of patient data, the cloud server should validate the data from the system Dew. To ensure the integrity of patient data, this study proposes to use a framework that verifies centralized data for distributed healthcare systems in a Dew computing environment [22, 28, 32].

Especially in emergency situations where patients need to be treated outside their primary healthcare facilities, patient health data should be trustworthy and easily accessible to the patients' primary healthcare providers and other healthcare facilities.

Blockchain technology enables the creation, transfer, and exchange of digital patient data while protecting it from unauthorized duplication and forgery. The system would verify the exchange of data and keep the records of the various facilities H1, H2... Hn. According to the chronological order in which medical records are received at the center, each database is sorted according to the date of the new records. B.

The facilities can work with the data in the Dew Nebula environment even if the Internet service is temporarily down. Synchronization then takes place to update the records at the center. Tiger Hash is used to protect and assess the data before it is transferred to the cloud data center prior to data synchronization.

5.1 A System for Continuously Monitoring Health that Uses Dew Computing

The physical device is connected to the Internet via the Internet of Things (IoT). IoT applications are particularly popular in the area of bio-signal monitoring. Our everyday health data is collected and sent via wireless sensor networks (WSNs) to a remote site that uses a cloud computing architecture for further analysis. The current architecture of health monitoring systems has several weaknesses. These include an ever-growing database, lack of user-specific separation of collected data, effective traffic management, symptom-based classification, and feedback to users. The biggest obstacle to implementing sustainable systems is unstable network connectivity and the dangers that come with it. Due to erratic Internet access, there is the problem of data loss in existing eHealth services. Any health-related application needs to be effectively deployed after a comprehensive study of health data. Data loss is a major risk in traditional health monitoring systems. A Dew layer can be installed between the cloud and physical contacts. To solve the problem of uneven data delivery, this Dew layer stores the data briefly [33–35].

Architecture of the Dew-based health monitoring system

In this section, we go into detail about the architecture of Dew in healthcare. The architecture consists of five main components [36–40]: (a) sensor layer, (b) Dew layer, (c) remote processing, (d) physician application layer, and (e) user application

(1) The sensing layers

One of the basic layers of the model is the sensor layer. All physical sensors are located in this layer. After being cached in the exchange, all captured data is connected to the Internet and transmitted to the IoT gateway. Wi-Fi, the remote processing layer, connects the Dew and Blockchain layers.

(2) Dew layer

Storage in the Dew layer is used in this proposal to temporarily shop lower-level data. Given the importance of health data, this model needs to consider data loss. To avoid data loss, a Dew layer is added to the system. The connection to the Internet does not affect this layer. It is directly connected to the sensor layer.

(3) Remote processing

This layer acts as a link between the Dew layer and the application layer. It first sends the data to the application layer. The sensor module then uses the TCP/IP protocol to transmit this data to a remote server. In the proposed model, we use Thingspeak to receive the data remotely. The sensor module first transmits the data to the Blockchain. All the data is then transmitted to the remote server for further calculations.

(4) The application layer for doctors

The data is retrieved by this layer from the remote processing layer. In the backend, there is a Rest API that uses hibernate mappings to communicate with the remote server. In the proposed structure, the relationships between the different tables are created using Hibernate mappings. Our backend uses the data after receiving a request from the remote server, consumes it, and provides a rest API that is then used to populate the frontend. The application layer consists of two sections. This layer populates the data on the doctor's device and is closed to doctors or supervisor nodes.

(5) User application layer

The user node is not far from this layer. Since the sensors are in the lower layer of the architecture, it communicates with the Dew layer, where the data is temporarily stored, and provides a user interface for receiving the result and passing the data to the supervisor node. The data from the Dew layer and the remote processing layer are read by the application layer. Between the doctor application and the user application layer is the remote processing layer. The entire data stream is stored for later analysis in the remote processing layer, which acts as cloud storage. The remote processing layer synchronizes the data from the Dew layer. The doctors receive the data via the doctor application layer, which also allows them to access the recorded data.

6 Conclusions

In this chapter, we have referred to the unified environment of Blockchain and Dew computing. We have described various architectures and models for Blockchain in Dew. We have explained all the basics of Dew Computing and Blockchain. DLT technology in Dew reduces the size of client data, but the client can still use all the features of the Blockchain by using the Blockchain client and Blockchain node architecture. We have also evaluated all related and relevant research papers on Blockchain and Dew Computing. In emerging Smart service applications, Dew computing with DLT technology support will provide a secure and transparent ecosystem.

References

1. Wang, Y.: Definition and categorization of dew computing. Open J. Cloud Comput. (OJCC) **3**(1), 1–7 (2016)
2. Wang, Y.: Cloud-dew architecture. Int. J. Cloud Comput. **4**(3), 199–210 (2015)
3. Chandra, A., Weissman, J., Heintz, B.: Decentralized edge clouds. IEEE Internet Comput. **17**(5), 70–73 (2013)
4. Ray, P.P.: An introduction to dew computing: definition, concept and implications. IEEE Access **6**, 723–737 (2017)
5. Nguyen, G.-T., Kim, K.: A survey about consensus algorithms used in blockchain. J. Inf. Process. Syst. **14**(1), 101–128 (2018)
6. Rindos, A., Wang, Y.: Dew computing: The complementary piece of cloud computing. In: 2016 IEEE International Conferences on Big Data and Cloud Computing (BDCloud), Social Computing and Networking (SocialCom), Sustainable Computing and Communications (SustainCom) (BDCloud-SocialCom- SustainCom), pp. 15–20. IEEE (2016)
7. Nakamoto, S.: Bitcoin: a Peer-to-Peer Electronic Cash System. Manubot (2019)
8. Yaga, D., Mell, P., Roby, N., Scarfone, K.: Blockchain technology overview (2019). arXiv: 1906.11078
9. Karmakar, A., Ghosh, P., Banerjee, P.S., De, D.: Chainsure: agent free insurance system using blockchain for healthcare 4.0
10. Pilkington, M.: Blockchain technology: principles and applications. In: Research Handbook on Digital Transformations. Edward Elgar Publishing (2016)
11. Bach, L.M., Mihaljevic, B., Zagar, M.: Comparative analysis of blockchain consensus algorithms. In: 2018 41st International Convention on Information and Communication Technology, Electronics and Microelectronics (MIPRO), pp. 1545–1550. IEEE (2018)
12. Bamakan, S.M.H., Motavali, A., Bondarti, A.B.: A survey of blockchain consensus algorithms performance evaluation criteria. Expert Syst. Appl. **154**, 113385 (2020)
13. Shanaev, S., Shuraeva, A., Vasenin, M., Kuznetsov, M.: Cryptocurrency value and 51% attacks: evidence from event studies. J. Altern. Invest. **22**(3), 65–77 (2019)
14. Nguyen, C.T., Hoang, D.T., Nguyen, D.N., Niyato, D., Nguyen, H.T., Dutkiewicz, E.: Proof-of-stake consensus mechanisms for future blockchain networks: fundamentals, applications and opportunities. IEEE Access **7**, 85727–85745 (2019)
15. Andreev, R.A., Andreeva, P.A., Krotov, L.N., Krotova, E.L.: Review of blockchain technology: types of blockchain and their application. Intellekt. Sist. Proizv. **16**(1), 11–14 (2018)
16. Guegan, D.: Public blockchain versus private blockhain (2017)
17. Niranjanamurthy, M., Nithya, B.N., Jagannatha, S.: Analysis of blockchain technology: pros, cons and SWOT. Clust. Comput. **22**(6), 14743–14757 (2019)

18. Puthal, D., Malik, N., Mohanty, S.P., Kougianos, E., Das, G.: Everything you wanted to know about the blockchain: its promise, components, processes, and problems. IEEE Consum. Electron. Mag. **7**(4), 6–14 (2018)
19. Chohan, U.W.: The limits to blockchain? Scaling versus. Decentralization (2019)
20. Monrat, A.A., Schelén, O., Andersson, K.: A survey of blockchain from the perspectives of applications, challenges, and opportunities. IEEE Access **7**, 117134–117151 (2019)
21. Nawari, N.O., Ravindran, S.: Blockchain and the built environment: potentials and limitations. J. Build. Eng. **25**, 100832 (2019)
22. Wang, Y., Gusev, M.: Decentralized hardware ownership control: Dew computing with blockchain. Each submission to DEWCOM 2018
23. Skala, K., Davidovic, D., Afgan, E., Sovic, I., Sojat, Z.: Scalable distributed computing hierarchy: cloud, fog and dew computing. Open J. Cloud Comput. (OJCC) **2**(1), 16–24 (2015)
24. Ray, P.P.: Dew computing: a new era of computing implying minimization over internetwork backhaul, vol. 14, pp. 53–57 (2019)
25. Sosu, R.N.A., Babu, C.N., Frimpong, S.A., Essuman, J.: The relevance of blockchain with dew computing: a review. In: 2020 43rd International Convention on Information, Communication and Electronic Technology (MIPRO), pp. 1934–1940. IEEE
26. Wang, Y.: A blockchain system with lightweight full node based on dew computing. Internet Things **11**, 100184 (2020)
27. Chakravarty, M.M., Chapman, J., MacKenzie, K., Melkonian, O., Peyton Jones, M., Wadler, P.: The extended UTXO model. In: International Conference on Financial Cryptography and Data Security, pp. 525–539. Springer, Cham (2020)
28. Park, J.H., Park, J.H.: Blockchain security in cloud computing: use cases, challenges, and solutions. Symmetry **9**(8), 164 (2017)
29. Chakravarty, M.M., Chapman, J., MacKenzie, K., Melkonian, O., Müller, J., Peyton Jones, M., Vinogradova, P., Wadler, P.: Native custom tokens in the extended UTXO model. In: International Symposium on Leveraging Applications of Formal Methods, pp. 89–111. Springer, Cham (2020)
30. Chepurnoy, A., Saxena, A.: Multi-stage contracts in the UTXO model. In: Data Privacy Management, Cryptocurrencies and Blockchain Technology, pp. 244–254. Springer, Cham (2019)
31. Karmakar, A., Ganguly, K., Banerjee, P.S.: HeartHealth: an intelligent model for multi-attribute based heart condition monitoring using fuzzy-TOPSIS method. In: 2021 Devices for Integrated Circuit (DevIC), pp. 1–5. IEEE (2021)
32. Pan, Y., Thulasiraman, P., Wang, Y.:Overview of cloudlet, fog computing, edge computing, and dew computing. In: Proceedings of the 3rd International Workshop on Dew Computing, pp. 20–23 (2018)
33. Buterin, V.: A next-generation smart contract and decentralized application platform. White Pap. **3**(37) (2014)
34. Banerjee, P.S., Karmakar, A., Dhara, M., Ganguly, K., Sarkar, S.: A novel method for predicting bradycardia and atrial fibrillation using fuzzy logic and arduino supported Iot sensors. Med. Novel Technol. Dev. **10**, 100058 (2021)
35. Ganguly, K., Karmakar, A., Banerjee, P.S.: ValveCare: a fuzzy based intelligent model for predicting heart diseases using arduino based IoT infrastructure. In: International Conference on Computational Intelligence in Communications and Business Analytics, pp. 229–242. Springer, Cham (2021)
36. Karmakar, A., Ganguly, K., Ghosh, P., Mazumder, A., Banerjee, P.S., De, D.: FemmeBand: a novel IoT application of smart security band implemented using electromyographic sensors based on wireless body area networks. Innov. Syst. Softw. Eng. 1–19 (2022)
37. Karmakar, A., Banerjee, P.S., De, D., Bandyopadhyay, S., Ghosh, P.: MedGini: Gini index based sustainable health monitoring system using dew computing. Med. Novel Technol. Dev. 100145 (2022)
38. Bhattacharyya, S., Banerjee, P.S., Karmakar, A., De, D. and Rodrigues, J.J.: BCoT: concluding remarks. In: Blockchain Based Internet of Things, pp. 289–293. Springer, Singapore (2022)

39. Karmakar, A., Sengupta, N. and Banerjee, P.S.: I-fresh: an IoT-based system for predicting the freshness of vegetables and flower. In: Proceedings of International Conference on Industrial Instrumentation and Control, pp. 579–587. Springer, Singapore (2022)
40. Karmakar, A., Ganguly, K., Banerjee, P.S.: SafeBand: IoT-based smart security band with instant SOS messaging. In: Proceedings of International Conference on Advanced Computing Applications, pp. 127–140. Springer, Singapore (2022)

DewIDS: Dew Computing for Intrusion Detection System in Edge of Things

Sangita Das, Anwesa Naskar, Rahul Majumder, Debashis De, and Seyed-Sajad Ahmadpour

1 Introduction

At present days, the features of edge devices are upgrading day by day so it is getting difficult for the cloud to match the users' high-performance expectations and demands. Some research works are now focusing on designing more efficient and secure security solutions for EoT devices which include updated edge-based architecture models, firewalls, privacy-preserving protocols, authentication and authorization mechanisms, and more effective intrusion detection and prevention systems. But the research on the security control of edge-based devices is still in progress.

The main goal of the cloud is to provide low utilization of network bandwidth for edge devices. To overcome this problem, the sting computing framework can be used as a complement to the cloud computing system. This technology is known as

S. Das
Department of Computer Science and Engineering, Jadavpur University, 188, Raja Subodh Chandra Mallick Rd, Jadavpur, Kolkata 700032, West Bengal, India
e-mail: sangitadas.jucse@gmail.com

A. Naskar · R. Majumder (✉) · D. De
Department of Computer Science and Engineering, Maulana Abul Kalam Azad University of Technology, Haringhata, Nadia 741249, West Bengal, India
e-mail: rahul.mit10@gmail.com

A. Naskar
e-mail: anwesanaskar@gmail.com

D. De
e-mail: dr.debashis.de@ieee.org

R. Majumder
Advanced Cybersecurity Architect, Honeywell International (Orion Campus), Bangalore, Karnataka, India

S.-S. Ahmadpour
Department of Computer Engineering, Kadir Has University, Istanbul, Turkey

D. De and S. Roy (eds.), *Dew Computing*, Internet of Things,
https://doi.org/10.1007/978-981-99-4590-0_7

Edge of Things computing which enables the processing and movement of data from cloud to edge devices. EoT allows the stock of a large amount of data, registering, and processing the data in real-time [1]. EoT has numerous applications almost in every field, i.e., industrial process, health care, automation, smart environment, etc. for its novel positive aspects. As EoT provides a broad range of services and applications, sometimes it faces **severe** vulnerabilities to exploit. EoT belongs to a heterogeneous environment, so the conventional security method is not enough to control all kinds of attacks. Though the security in EoT has been enhanced by User Authentication (SAML), Authorization, Confidentiality, and Integrity, still there are many issues. An intrusion detection and prevention system are already used in wireless networks but it can be enhanced more in an unstable network connection.

The security mechanism needs to be upgraded to stop unauthorized access and information disclosure to prevent footprinting. Data integrity and confidentiality mechanisms can be used to reduce the threats to the security system of edge devices. Some Cryptography techniques use conventional security mechanism systems that are not capable to access large amounts of data and it is a very time-consuming process. For a malicious user, a short time is enough to access and modify confidential data non-ethically. So, it is important to develop IDS and IPS systems with strong configuration checks to identify and prevent intruders and alert the users. Figure 1 illustrates the IDS and IPS.

Anomaly based IDSs are not enough to prevent the attacks as it has been a series of concerns with false alarm rates. It can be reduced by tracking the incoming traffics and detecting malicious users from unauthorized networks. At a certain threshold, IDS tries to seek out the deviated data from the model which is called anomalies. To get better accuracy IDS should be configured carefully. It is convenient for IDS to

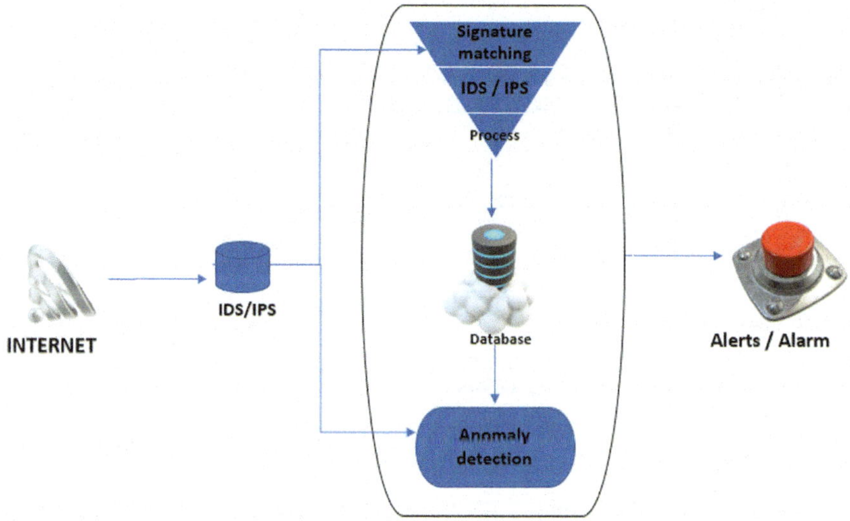

Fig. 1 Block diagram of intrusion detection and prevention system (IDS/IPS)

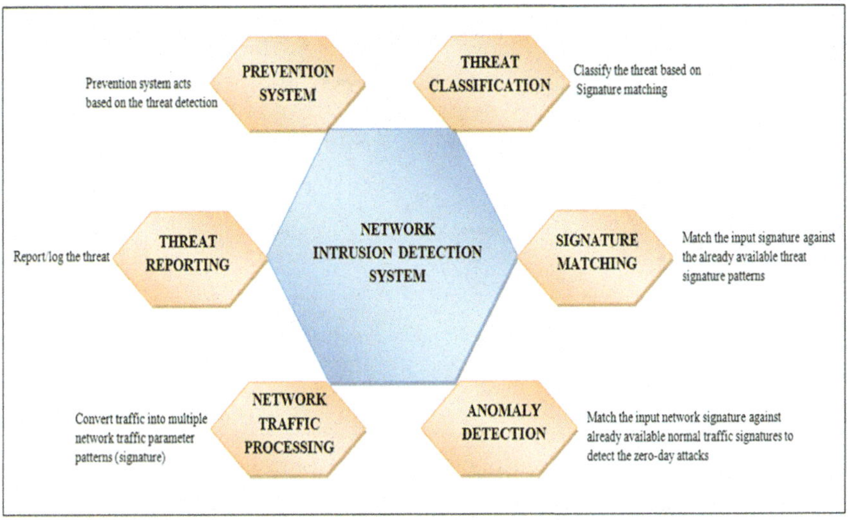

Fig. 2 Characteristics of intrusion detection system

observe the minimum False Positive Rate (FPR) rather than the True Positive Rate (TPR). As the number of false alarms is so much higher in real implementation so False Positive Rate is naturally higher in anomaly-based Intrusion Detection Systems than the other IDSs. To minimize the FPR, an intelligent filter for false alarms can be implemented with some selective machine-learning algorithms. Figure 2 illustrates the characteristics and services provided by an Intrusion Detection System.

In DewIDS, dew computing ensures hazard-free work even without the Internet. So, adding the dew features with the conventional IDS features will help the EoT devices to secure the network from unauthorized attacks from malicious users. Figure 3 represents the characteristics and services provided by an Intrusion Detection System.

1.1 Motivation and Contributions

Nowadays everything is based on a cloud server as it stores a vast amount of information and data using 128-bit AES (Advanced Encryption Standard) encryption to improve data security and confidentiality in the Cloud platform. The AES model uses an SMS mechanism to alert administrator the of the unauthorized access request from intruders. The intrusion Detection and Prevention System of any server is bounded to a certain limit. It may not respond in the absence of an Internet connection or gives false alarms in unstable Internet. So, there is a chance of data loss and information if the system is built with a simple IDS/IPS system without Level 1 cache

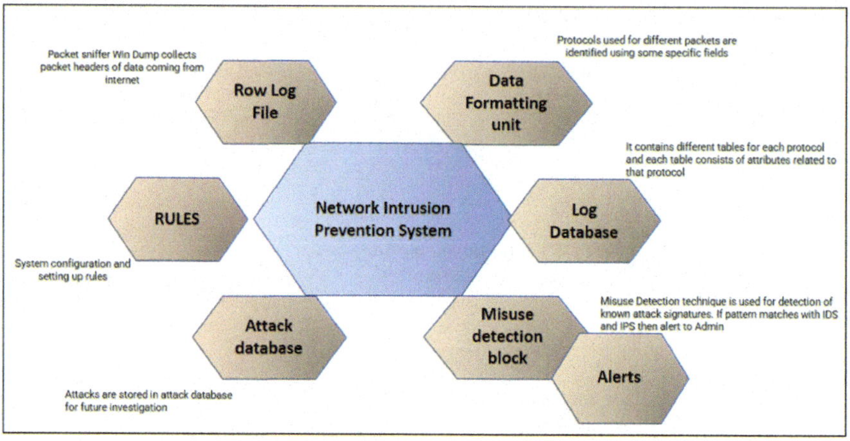

Fig. 3 Characteristics of intrusion prevention system

memory which is small in size, and fastest with limited storage. Dew architecture is a technology that is developed to overcome the drawbacks of traditional systems.

The contributions of this paper are summarized as follows:

(i) DewIDS offers continuous network scanning, monitoring, real-time alert, malicious user detection, and prevention from the server. By these features of dew computing, it can be implemented in between of user and cloud server for reducing the false alarm rate to get the better result.

(ii) In DewIDS, a machine-learning classifier can be used for detecting and classifying the alarm as true or false, and with this system, dew server can be attached so that in the absence of the Internet there will be no issue of detecting the attack or classifying the alarm.

2 Related Works

Many proposals are existed and executed in different methods. In the following subsections, some existing proposals are described.

2.1 Distributed Intrusion Detection System (DIDS)

Distributed IDS is one kind of conventional IDS that consists of multiple IDSs which are interconnected over a large network where a central server monitors all the coming users. With the help of a set of cooperating IDSs, i.e., Collaborative Intrusion Detection Network (CIDN) is possible to enhance the performance of the system for better accuracy.

Wang et al. [2] proposed a hybrid CNN (convolutional neural network) model-based IDS for IoT networks to detect various kinds of attacks. In their proposed model it is appropriate for a broad field of IoT applications. It is authenticated and compared with traditional machine learning and deep learning models and the result of various experiments shows that this model is more responsive to attacks.

Meng et al. [3] used Bayesian inference in the field of healthcare organizations for Medical Smartphone Networks (MSNs) to detect the malicious network and illustrate the efficacy of their approach in real life.

Li et al. [4] designed a malware detection system, i.e., Significant Permission Identification (SigPID), using various machine-learning classification methods. In this work, they develop three levels of trimming methods to filter the most significant permissions that are useful to distinguish between harmless and malicious apps. In addition to the baseline approach and support vector machine (SVM) with SigPID is possible to detect 93.62% and 91.4% of malware from the dataset and unknown samples respectively.

Li et al. [5] designed a special On–Off attack (SOOA) for collaborative intrusion detection networks (CIDNs). In this paper, they investigated the effectiveness of SOOA by responding normally to any one of the nodes and acting abnormally to others. By the experiments, their result shows that the attack is effective for the trust-based computation in CIDNs nodes.

2.2 *Cloud-Based Intrusion Detection System*

Cloud-Based Intrusion Detection System (CBIDS) is helpful for real-time monitoring of intrusion in the era of IoT. Liu et al. [6] designed Comprehensive Transmission (CT) model with the combination of client/server mode and Peer-to-peer (P2P) mode to get stable data transmission. They also added a Two-Phase Resource Sharing Protocol (TPRSP) that has two types of phases one is for pre-filtering and another is for verification.

In the context of smart healthcare systems, the IoT supports medical practitioners to scrutinize insulin levels, body temperature, and heart rate in a real-time environment and warehouses the retrieved data in the Cloud for information dispensation, feature sampling, and illness diagnosis. Health-related information is delivered by the integrated cloud servers for patients' scrupulous health-induced information. Healthcare institutions' confidentiality, integrity, and availability are at stake owing to the insufficient privacy and patch management of the enabled servers. Although IoMT-enabled clinical technologies and processes are vulnerable that are open to exploits and malware. Cyber attackers can attain unauthorized approaches to the IoT-enabled medical peripherals to overindulge the patients, which could have life-captivating consequences on patients' health. Figure 4 demonstrates the proposed security architecture of IoMT.

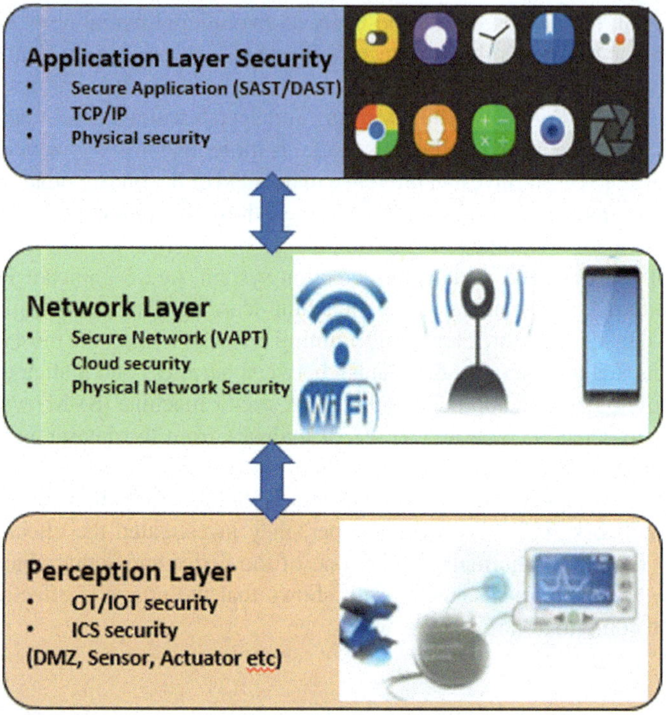

Fig. 4 Proposed security architecture of IoMT

Abusitta et al. [7] proposed a framework of a community that consists of several trustful IDSs distributive. They designed a decentralized algorithm followed by coalitional game theory to increase the accuracy of each detection by combining several cloud-based IDSs.

Maglaras et al. [8] proposed a security-based architecture to control the misrouting of commands and forged commands. This proposed architecture is based on two technologies, i.e., one is Blockchain and another is Software-defined Networks (SDN). In this work, they implemented a Blockchain-based Integrity Checking System (BICS) to control the misrouting of any attack and the combination of K-Nearest Neighbor (KNN) and Random Subspace Learning (RSL) for the prevention of forged commands in industrial IoT systems. The experimental result of this approach gives efficient results in industries.

2.3 DBN-Based IDS

Zhao et al. [9] proposed an IDS which is based on the concept of Probabilistic Neural Network (PNN) and Deep Belief Network (DBN). Low-dimensional data

Table 1 Comparative analysis of intrusion detection and prevention systems

Features IDS / Systems	Continuous Network monitoring	Intrusion rule enforcement	Activity log and insight	Real time alert	Malicious presence detection	Malicious presence blocking	Dew as a service
DaaS [1]	√	√	√	√	√	√	√
Sooa [5]	√	√	√	✗	√	√	✗
Blockchain IDS [8]	√	√	✗	√	√	✗	✗
Deep belief network IDS [11]	√	√	√	√	√	✗	✗
IoT IDS [12]	✗	√	√	✗	√	✗	✗
Soft IDS [13]	√	√	✗	√	√	√	✗
ML IDS [14]	√	√	✗	√	√	✗	✗
PMFA [15]	√	√	✗	√	√	✗	✗

can be formed from the raw data by the non-linear learning process of DBN when taking the necessary attributes of the primary data. They also used Particle Swarm Optimization (PSO) algorithm to get the best result.

Koo and Klabjan [10] proposed the model with the combination of supervised and unsupervised learning based on DBN over the two-phase strategy to get the best and most effective result.

Tian et al. [11] designed an upgraded IDS by advanced Deep Belief Network (DBN) to overcome the problem of overfitting, massive False Positive Rate (FPR), and low accuracy. Here Min–Max Normalization and Probability Mass Function (PMF) are used for the simplification of the dataset. The ultimate experimental result of this work improves the False Positive Rate (FPR) and classification accuracy rate. Table 1 represents the comparative analysis of Intrusion Detection and Prevention systems.

3 Dew Computing-Based Intrusion Detection System (DewIDS)

An Intrusion Detection System (IDS) can also be implemented by dew computing. For reducing the false alarm rate (FPR) increase the classification accuracy rate using the DBN model with dew computing in Distributed Intrusion Detection Systems (DIDSs).

3.1 Dew Computing

Many researchers have discussed the concept of dew computing from different aspects. We surveyed the definitions and descriptions of some of these papers [1, 2, 16–18].

Dew computing mainly has two features according to Wang et al. [2, 16]. Independence and Collaboration. Skala et al. [17] proposed that dew computing is different from the concepts of service and storage or network. They also described that Dew Computing (DC) builds a boundary to compute the applications, low services, and data far from the virtual nodes of the centralized systems. According to Ristov et al. [18] before transferring the process to the cloud server, the maximum resources should be used in dew computing. They described dew computing as the theory of "independence."

Dew computing is nothing but a programming model of peer-to-peer (P2P) communication with universal services. It can be implemented in edge devices, such as laptops, PCs, tablets, and smartphones for analyzing the data with less effort. A few required properties of dew computing are:

(1) Accessibility, (2) Transparency, (3) Re-origination, (4) Scalability, (5) Synchronization, and (6) Rule-based data collection. For services and products, dew computing provides Dew-as-a-Software (DaaS) and Dew-as-a-Infrastructure (DaaI). Unlike cloud computing, Dew Computing (DC) works to edge devices even in an unstable Internet connection [1].

DewIDS is the Intrusion Detection System using Dew Computing in the edge devices for detecting false alarms.

3.2 DewIDS

Dew Computing is used to handle the data with less network load and better response time and processing. Intelligent techniques can be implemented in the Distributed Intrusion Detection System (DIDS) and Prevention system to reduce the false alarm rate. Singh et al. [1] proposed a framework for an Intrusion Detection System using Dew Computing with the combination of three layers, i.e., cloud layer, dew layer, and IDS layer.

CLOUD LAYER: All the computations and processing of data are performed in the cloud layer. Preprocessing of data is done at the dew layer and then the information of the data is sent to the cloud layer.

DEW LAYER: Dew layer is the heart of this architecture. It consists of the application and utility software, and operating system. In this layer, the appropriate algorithm is chosen to minimize the False Alarm Rate (FPR) process. The Dew layer accepts and stores the recent coming data and also uploads the further data to the cloud layer in a certain interval.

IDS LAYER: At this layer, the main task of the intrusion detection process is performed. Various nodes are here to share the data for performing the detection process.

Singh et al. [1] also developed an algorithm based on their proposed framework. The steps of that algorithm are as follows.

1. **Data Normalization**
   (a) Data received by dew server.
   (b) Normalization process extracts the distinctive attribute from the received data.
   (c) Data is represented by an alarm.
2. **Deep Belief Network (DBN) based IDS**
   (a) Dew server sends the alarm to the DBN-based Intrusion Detection System at IDS layer.
   (b) Alarm analysis is performed by DBN-IDS.
3. **Control System**
   (a) DBN-IDS result received by the control system.
   (b) Analyze the received result and perform accordingly.
4. **False alarm detection**
   (a) Dew server receives the DBN-IDS result and the Alarm.
   (b) Reduce the false alarm and produces the true alarm.
   (c) True alarm sends to the edge manager.
5. **True alarm updates the IDS nodes**
   (a) True alarm indicates the true notification.
   (b) Figure 5 illustrates Intrusion Detection System [19–25] using Dew Computing.

3.3 IDS Versus DEW IDS

A basic intrusion detection system detects suspicious activities in a system after the detection prevention system generates an alert to the administrator or authenticated user. Security operations are activated to identify the issues and investigate to solve them. But the problem is here when the connection is lost, cannot detect anything, not for a while. The false alarm rate is also high in this case.

In the DBN model, to reduce the communication delay and workload of the cloud server, Dew computing work in the local cloud environment and collaborate with the public cloud. In dew architecture, the application stays for a limited time even if the connection is lost. Dew environment provides impeccable security service in many

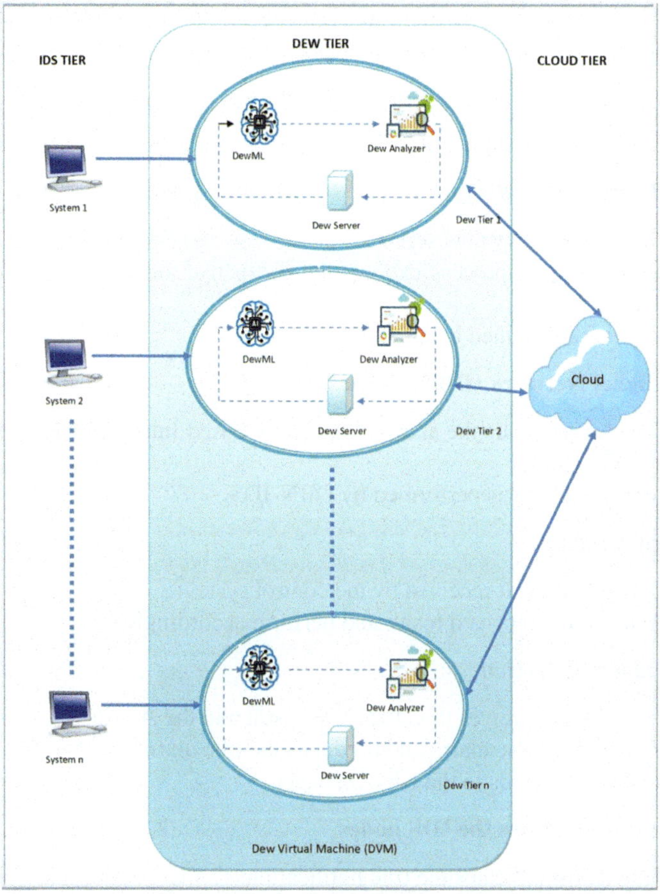

Fig. 5 Intrusion detection system using Dew computing

fields of application. Figure 6 shows the accuracy of DBN-based IDS and Dew IDS. The false alarm rate is less than Basic IDS.

4 Sustainability of DewIDS

This subsection illustrates the sustainability of the proposed DewIDS in Fig. 7.

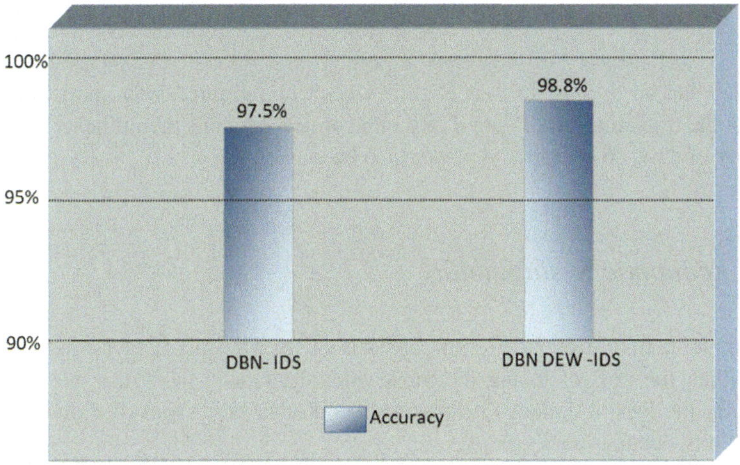

Fig. 6 Accuracy of DBN-based IDS and Dew IDS

Fig. 7 Sustainability of DewIDS

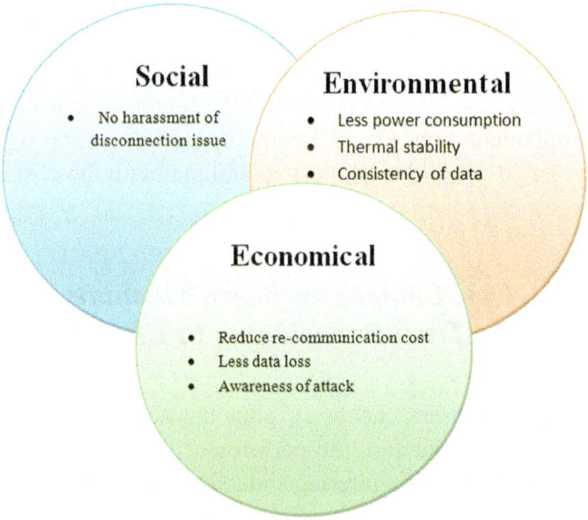

4.1 Social Sustainability

Nowadays, edge devices contain almost every personal and important information. So, the security systems of edge devices must be strong and upgraded. With the implementation of dewIDS, the server can detect malware even in an unstable or without an Internet connection for some time it is possible to alert the edge manager. It can reduce the harassment of the users during a power cut and disconnection issues. It also affects the cultural system of society.

4.2 *Environmental Sustainability*

DewIDS can be used in various edge devices to be secured from malicious users. During the transmission of data if any interruption happens then it never affects the stability of data, so energy and power also be wasted.

4.3 *Economic Sustainability*

The implementation of DewIDS is cost-effective. The initial investment is much better than the cost of losing the most valuable data in the future by traditional IDSs. By the loss of data or information it is a devastating loss of economy to any community occupational groups.

5 Future Research Challenges

DewIDS can work without an Internet connection for a while but not for a long period. As it is a new concept in the IDS field so need more research on this for better implementation. DewIDS is an upgraded version of existing traditional IDSs so, it must be costly. Developers should maintain the cost as much as possible.

5.1 *Dew Computing-Based Machine Learning Approach for Trust Provisioning in EoT*

Malicious users change or alter the actual data which may affect the Quality of Service. To increase the performance, it will be beneficial to implement DewIDS with collaboration filtering and a Bayesian learning-based scheme, i.e., Robust Trust Management [26].

5.2 *Dew Computing for Intrusion Detection System in Heterogeneous Edge of Things*

For Intrusion Detection System (IDS), it is so much difficult and complicated to identify in the heterogeneous network. For increasing the accuracy of detection DBN (Deep Belief Network) based DewIDS model will give a better result [27].

5.3 Dew Game Theory for Intrusion Detection System in Edge of Things

For gaining maximum security in Nash equilibrium with the exact consensus of the game model DewIDS will be beneficial between the attacker and the defender [28].

5.4 Statistical Fingerprint-Based IDS Using Dew Computing

In statistical analysis-based fingerprint IDS [29], for monitoring the flow and detecting the malware DewIDS can be implemented with an appropriate machine-learning model.

5.5 Federated Learning-Based Privacy Preserving Dew Computing in Virtual Tourism

MR/AR/VR and federated learning model-based technology, "FedLens" is used to interact with the scenic beauty of nature, and people are guided virtually at tourist spots. For getting better and more authentic information about unknown places in real time, the privacy of Fedlens can be implemented by the dewIDS model [30].

5.6 Deep Learning-Based IDS Using Dew Computing in Agriculture 4.0

IoT-based smart agriculture consists of several advanced technologies in the modern agricultural system. Many new threats and cyber-attacks like DoS attacks can flood the internal memory and cause the unavailability of services in agriculture 4.0. For safety and security purposes Deep learning-based dewIDS can be implemented [31].

5.7 Artificial Neural Network-Based Fraud Detection in Internet Banking Using Dew Computing

For authenticating the transaction and detecting the intrusion in the banking server DewIDS can be used in a multi-layered, feed-forward artificial neural network [32].

6 Conclusions

Sophisticated and inspired intruders are attacking computers by using modern techniques. The intruders are capable enough to conceal their identities and communications. Therefore, we need advanced intrusion detection and prevention systems based on computer systems to fortify themselves from recent malware. We must have a comprehensive outline of the robustness and weakness of recent IDS research. We have reviewed that DBN and dew computing are used to classify the attacks and reduce the False Alarm Rate. Here, dew performs a major role over the cloud as it reduces computational cost and communication delay. Edge devices exchange a large amount of data, so dew works much more efficiently and gives better accuracy. In future research, hybrid integration of cloud, fog, dew, and EoT can be designed to overcome the limitations of DewIDS. Many other machine-learning algorithms can also be used to improve the classification of attacks with low False Alarm Rates and high alert correlation.

References

1. Singh, P., Kaur, A., Aujla, G.S., Batth, R.S., Kanhere, S.: DaaS: Dew computing as a service for intelligent intrusion detection in edge-of-things ecosystem. IEEE Internet Things J. (2020).
2. Wang, Y.: Definition and categorization of dew computing. Open J. Cloud Comput. (OJCC) **3**(1), 1–7 (2016)
3. Meng, W., Li, W., Xiang, Y., Choo, K.-K.R.: A bayesian inferencebased detection mechanism to defend medical smartphone networks against insider attacks. J. Netw. Comput. Appl. **78**, 162–169 (2017)
4. Li, J., Sun, L., Yan, Q., Li, Z., Srisa-an, W., Ye, H.: Significant permission identification for machine-learning-based android malware detection. IEEE Trans. Ind. Inf. **14**(7), 3216–3225 (2018)
5. Li, W., Meng, W., et al.: Sooa: exploring special on-off attacks on challenge-based collaborative intrusion detection networks. In: International Conference on Green, Pervasive, and Cloud Computing, pp. 402–415. Springer (2017)
6. Liu, Q., Wang, G., Liu, X., Peng, T., Wu, J.: Achieving reliable and secure services in cloud computing environments. Comput. Electr. Eng. **59**, 153–164 (2017)
7. Abusitta, A., Bellaiche, M., Dagenais, M.: A trust-based game theoretical model for cooperative intrusion detection in multi-cloud environments. In: 2018 21st Conference on Innovation in Clouds, Internet and Networks and Workshops (ICIN), pp. 1–8. IEEE (2018)
8. Derhab, A., Guerroumi, M., Gumaei, A., Maglaras, L., Ferrag, M.A., Mukherjee, M., Khan, F.A.: Blockchain and random subspace learning-based IDS for SDN-enabled industrial IoT security. Sensors **19**(14), 3119 (2019)
9. Zhao, G., Zhang, C., Zheng, L.: Intrusion detection using deep belief network and probabilistic neural network. In: 2017 IEEE International Conference on Computational Science and Engineering (CSE) and IEEE International Conference on Embedded and Ubiquitous Computing (EUC), vol. 1, pp. 639–642. IEEE (2017)
10. Koo, J., Klabjan, D.: Improved classification based on deep belief networks (2018). arXiv: 1804.09812
11. Tian, Q., Han, D., Li, K.-C., Liu, X., Duan, L., Castiglione, A.: An intrusion detection approach based on improved deep belief network. Appl. Intell. (2020)

12. Basati, A., Faghih, M.M.: APAE: an IoT intrusion detection system using asymmetric parallel auto-encoder. Neural Comput. Appl. 1–21 (2021)
13. Singh, P., Krishnamoorthy, S., Nayyar, A., Luhach, A.K., Kaur, A.: Soft-computing-based false alarm reduction for hierarchical data of intrusion detection system. Int. J. Distrib. Sens. Netw. **15**(10), 1550147719883132 (2019)
14. Meng, Y., et al.: Adaptive false alarm filter using machine learning in intrusion detection. In: Practical applications of intelligent systems, pp. 573–584. Springer (2011)
15. Li, W., Meng, W., Ip, H.H.S., et al.: Pmfa: toward passive message fingerprint attacks on challenge-based collaborative intrusion detection networks. In: International Conference on Network and System Security, pp. 433–449. Springer (2016)
16. Wang, Y., Meng, W., Li, W., Liu, Z., Liu, Y., Xue, H.: Adaptive machine learning-based alarm reduction via edge computing for distributed intrusion detection systems. Concurr. Comput.: Pract. Exp. **31**(19), e5101 (2019)
17. Skala, K., Davidovic, D., Afgan, E., Sovic, I., Sojat, Z.: Scalable distributed computing hierarchy: cloud, fog and dew computing. Open J. Cloud Comput. (OJCC) **2**(1), 16–24 (2015)
18. Ristov, S., Cvetkov, K., Gusev, M.: Implementation of a horizontal scalable balancer for dew computing services. Scalable Comput.: Pract. Exp. **17**(2), 79–90 (2016)
19. Li, W., Meng, W., et al.: Design of intrusion sensitivity-based trust management model for collaborative intrusion detection networks. In: IFIP International Conference on Trust Management, pp. 61–76. Springer (2014)
20. Vieira, K., Schulter, A., Westphall, C., Westphall, C.: Intrusion detection for grid and cloud computing. It Professional **12**(4), 38–43 (2009)
21. Moustafa, N., Slay, J.: Unsw-nb15: a comprehensive data set for network intrusion detection systems (unsw-nb15 network data set). In: 2015 Military Communications and Information Systems Conference (MilCIS), pp. 1–6. IEEE (2015)
22. Meng, W., Wang, Y., Li, W., Liu, Z., Li, J., Probst, C.W.: Enhancing intelligent alarm reduction for distributed intrusion detection systems via edge computing, In: Australasian Conference on Information Security and Privacy, pp. 759–767. Springer (2018)
23. Sultana, N., Chilamkurti, N., Peng, W., Alhadad, R.: Survey on SDN based network intrusion detection system using machine learning approaches. Peer-to-Peer Netw. Appl. **12**(2), 493–501 (2019)
24. Jose, S., Malathi, D., Reddy, B., Jayaseeli, D.: A survey on anomaly-based host intrusion detection system. J. Phys.: Conf. Ser. **1000**(1), 012049 (IOP Publishing) (2018)
25. Smys, S., Basar, A., Wang, H.: Hybrid intrusion detection system for internet of things (IoT). J. ISMAC **2**(04), 190–199 (2020)
26. Singh, P., Kaur, A., Batth, R.S., Aujla, G.S., Masud, M.: Service versus protection: a Bayesian learning approach for trust provisioning in edge of things environment. IEEE Internet Things J. (2021)
27. Stiawan, D.: Intrusion detection with deep learning on internet of things heterogeneous network. Int. J. Artif. Intell. (IJ-AI) (2021)
28. Wu, H., Wang, W.: A game theory based collaborative security detection method for internet of things systems. IEEE Trans. Inf. Forens. Secur. **13**(6), 1432–1445 (2018)
29. Boero, L., Cello, M., Marchese, M., Mariconti, E., Naqash, T., Zappatore, S.: Statistical fingerprint-based intrusion detection system (SF-IDS). Int. J. Commun Syst **30**(10), e3225 (2017)
30. De, D.: FedLens: federated learning-based privacy-preserving mobile crowdsensing for virtual tourism. Innov. Syst. Softw. Eng. **12**, 1–4 (2022)
31. Ferrag, M.A., Shu, L., Djallel, H., Choo, K.K.: Deep learning-based intrusion detection for distributed denial of service attack in agriculture 4.0. Electronics **10**(11), 1257 (2021)
32. Bignell, K.B.: Authentication in an internet banking environment; Towards developing a strategy for fraud detection. In: International Conference on Internet Surveillance and Protection (ICISP'06), pp. 23–23. IEEE
33. Yu, W., Liang, F., He, X., Hatcher, W.G., Lu, C., Lin, J., Yang, X.: A survey on the edge computing for the internet of things. IEEE Access **6**, 6900–6919 (2017)

34. El-Sayed, H., Sankar, S., Prasad, M., Puthal, D., Gupta, A., Mohanty, M., Lin, C.-T.: Edge of things: the big picture on the integration of edge, IoT and the cloud in a distributed computing environment. IEEE Access **6**, 1706–1717 (2017)
35. Saleem, A., Khan, A., Malik, S.U.R., Pervaiz, H., Malik, H., Alam, M., Jindal, A.: Fesda: fog-enabled secure data aggregation in smart grid IoT network. IEEE Internet Things J. (2019)
36. Ramaki, A.A., Rasoolzadegan, A., Bafghi, A.G.: A systematic mapping study on intrusion alert analysis in intrusion detection systems. ACM Comput. Surv. (CSUR) **51**(3), 1–41 (2018)
37. Hindy, H., Brosset, D., Bayne, E., Seeam, A., Tachtatzis, C., Atkinson, R., Bellekens, X.: A taxonomy and survey of intrusion detection system design techniques, network threats and datasets

Machine Learning-Based Sustainable Dew Computing: Classical to Quantum

Mahua Nandy Pal, Diganta Sengupta, Tien Anh Tran, and Debashis De

1 Introduction

The last decade has witnessed a phenomenal rise in artificial intelligence, penetrating almost all spheres of human life. This incident resulted in the generation of massive scales of data daily, which in turn challenged the storage and processing capabilities of the data. Cloud computing emerged as one of the most promising paradigms to address such challenges, although with a challenge for seamless internet connectivity. As the global geography is yet to observe internet connection at different layers of topologies, seamless internet connectivity itself became another bottleneck. This led to the proposals for fog and edge computing in literature, followed by dew computing. This terminology was coined nearly around 2017 [1]. Dew computing

M. N. Pal (✉)
Department of Computer Science & Engineering, MCKV Institute of Engineering, Howrah 711204, West Bengal, India
e-mail: mahua.nandy@gmail.com

D. Sengupta
Department of Computer Science & Engineering, Meghnad Saha Institute of Technology, Kolkata 700150, West Bengal, India
e-mail: sg.diganta@gmail.com

T. A. Tran
Department of Naval Architecture & Ocean Engineering, Seoul National University, Seoul, South Korea

Department of Marine Engineering, Vietnam Maritime University, Haiphong, Vietnam

T. A. Tran
e-mail: anhtt.mtb@vimaru.edu.vn

D. De
Department of Computer Science & Engineering, Maulana Abul Kalam Azad University of Technology, Nadia 741249, West Bengal, India
e-mail: dr.debashis.de@ieee.org

D. De and S. Roy (eds.), *Dew Computing*, Internet of Things,
https://doi.org/10.1007/978-981-99-4590-0_8

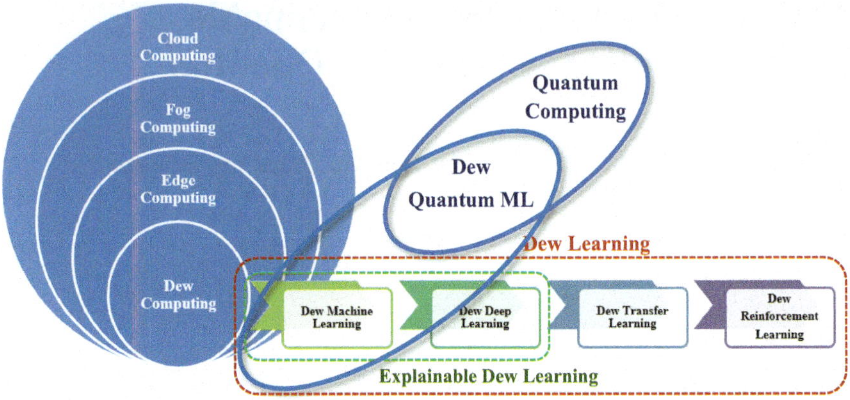

Fig. 1 Four computing paradigms along with dew learning and dew quantum learning prospects

is connected to the cloud through edge and fog computing architectures. Figure 1 presents the hierarchy among the four computing paradigms along with prospects of dew learning.

Further discussions concerning Fig. 1 will help us understand the roles of the four computing paradigms. When a device (Internet of Things device) is present at the edge of the Internet, it is termed edge computing. For example, any device that takes data from the sensors, processes it and sends it to the cloud is an edge device, as it is directly connected to the cloud via the Internet. The massive amount of data transmitted from an edge device to the cloud is filtered by the fog paradigm. Fog computing receives the data from the edge, filters out the redundant data, and then transmits the relevant data only to the cloud, thereby reducing the overhead for cloud computing. As already mentioned, Internet coverage throughout the global geographic topologies is challenging. Hence, we observe cases where the sensors and devices accumulate data but fail to transmit them to the cloud via the edge. In such environments, it becomes imperative to store the data in the device and communicate only when connected to the edge. This challenge is addressed by dew computing, wherein the data from the sensors are stored locally in the device and transmitted to the edge at a later stage. Since the dew devices are required to store data, processor capabilities, memory management, and power management become crucial issues. Few notable proposals are there in the literature which address this issue. Moreover, since the dew devices store the data and later send the data to the edge, processing such data at the dew level has attracted global attention from the research community. Figure 2 presents the growth of research in dew computing from a taxonomical perspective, followed by the current application and prospective domains for dew computing in Fig. 3.

It can be observed from Fig. 2 that the significant attention gained by the dew computing paradigm has been in the last couple of years. Moreover, research in the domain is rising with massive scopes for further study. The glossary for the terms used in Figs. 2 and 3 is presented in Table 1.

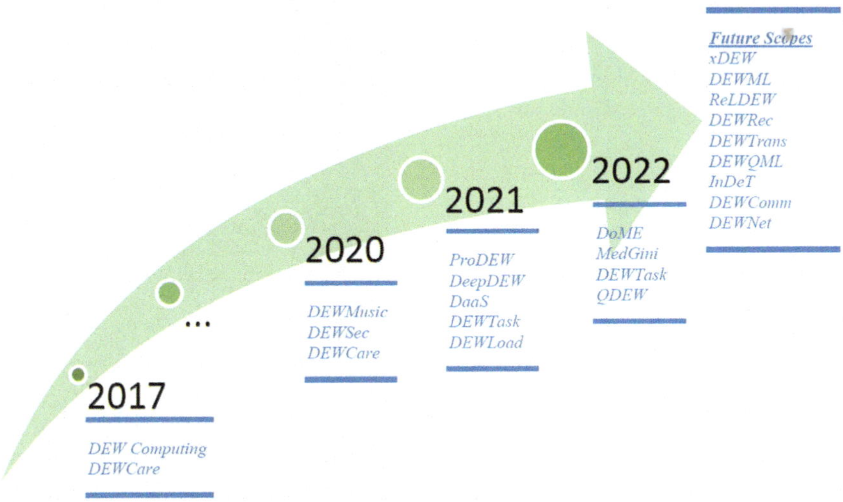

Fig. 2 A taxonomical perspective of growth of research on dew computing

Table 1 also presents some acronyms used to exhibit existing proposals and applications.

As discussed earlier, a few challenges exist in dew computing, such as processor capacities, memory, and power management, as reflected in [1]. In addition to the three challenges mentioned, operating system requirements for the dew devices also demand a certain amount of study, followed by network and communication models. Programming the dew devices and their capacity to integrate with existing programming platforms are potential study domains. The following section presents the current proposals along with the applications.

2 Existing Literature—Dew Learning

We can visualize a few notable existing literature consisting of research insights into dew computing in Fig. 4. The rest of the prior art is discussed in this section further.

Apart from the notable literature presented in Fig. 4, a few more preliminary proposals exist for using dew computing and dew architecture.

The latest literature has reflected the use of a few machine learning models in the Dew paradigm. We discuss a few notable works of literature here and then propose further prospects for using machine learning models.

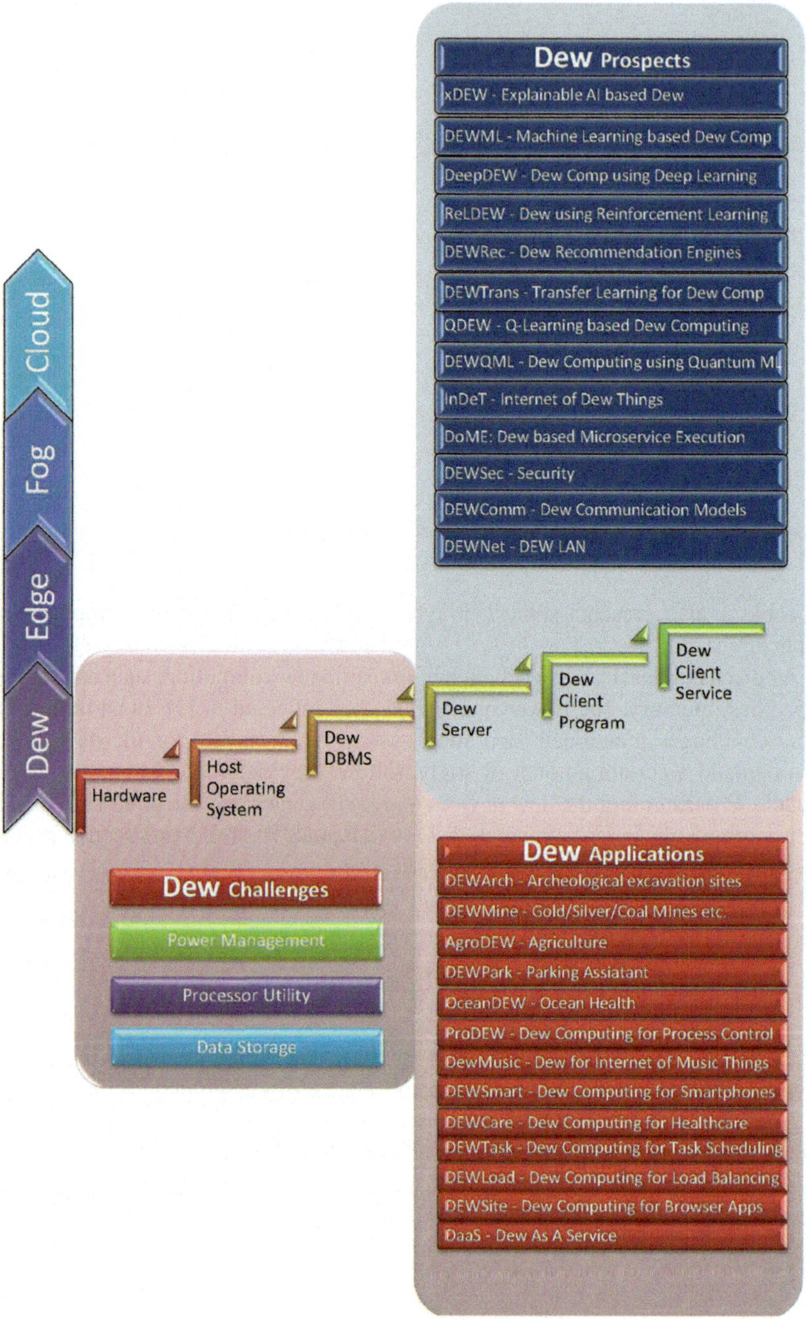

Fig. 3 Overview of dew computing in terms of applications and prospects

Table 1 Glossary of terms related to dew computing

SL	Acronym for dew paradigms	Details for the acronym
1	*DEWCare*	Dew computing for healthcare
2	*DEWMusic*	Dew computing for music crowdsourcing
3	*DEWSec*	Dew computing for security
4	*ProDEW*	Dew computing for process control
5	*DeepDEW*	Deep learning models for Dew computing
6	*Daas*	Dew as a service
7	*DEWTask*	Dew computing for task scheduling
8	*DEWLoad*	Dew computing for load balancing
9	*DoME*	Dew computing-based microservice execution
10	*QDEW*	Q-learning-based dew computing
11	*DEWML*	Machine learning models for dew computing
12	*RelDEW*	Reinforcement learning models for dew computing
13	*DEWRec*	Recommender systems for dew computing
14	*DEWTrans*	Transfer learning models for dew computing
15	*InDeT*	Internet of Dew technology
16	*DEWComm*	Dew communication models
17	*DEWNet*	Local area connectivity for dew computing
18	*DEWArch*	Dew computing for archaeological excavation sites
19	*DEWMine*	Dew computing for mining excavation sites
20	*AgroDEW*	Dew computing for agriculture
21	*DEWPark*	Dew computing for parking assistance
22	*OceanDEW*	Dew computing for ocean health
23	*DEWSmart*	Dew computing for smartphones
24	*DEWSite*	Dew computing for browser applications
25	*xDEW*	Explainable AI for dew computing
26	*QC*	Quantum computing
27	*ML*	Machine learning
28	*QML*	Quantum machine learning
29	*NISQ*	Noisy intermediate scale quantum
30	*VQC*	Variational quantum circuit
31	*PQC*	Parameterized quantum circuit
32	*VQA*	Variational quantum algorithm
33	*DEWQML*	Quantum machine learning for dew computing
34	*QMA*	Quantum minimization algorithm
35	*PSO*	Particle swarm optimization
36	*KNN*	K-nearest neighbour
37	*QPU*	Quantum processing unit
38	*QUBO*	Quadratic unconstrained binary optimization

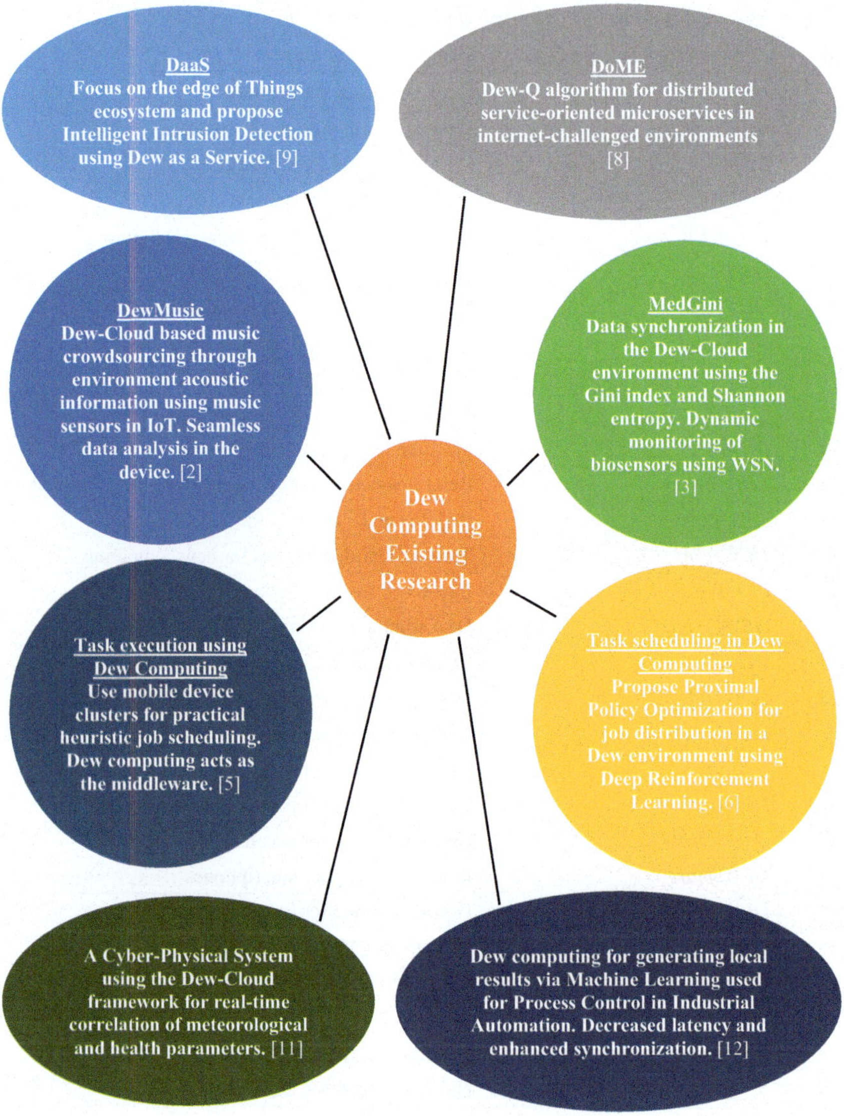

Fig. 4 Existing notable research dimensions in dew computing

2.1 DewMusic

The *DewMusic* architecture in Fig. 2 of [2] comprises the Dew Computing Tier, which fetches input from the physical layer and, after processing, generates results that act as the input for the Fog Computing Layer in music crowdsourcing architecture. The Dew Computing Tier has five elements of processing as follows:

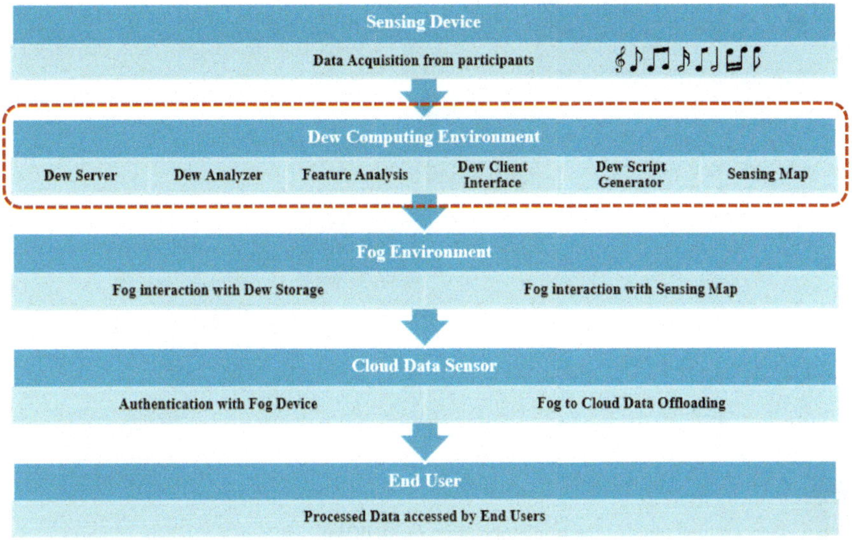

Fig. 5 Activity flow in the dew-cloud framework

1. Data Pre-processing
2. Data Visualization
3. Basic Analytics
4. Filtering/Optimization
5. Data Buffering/Caching.

Each of the five elements proposes enormous scopes for application for machine learning models. Figure 5 presents the activity flow in the dew-cloud framework.

The dotted box in Fig. 5 envelops the dew computing environment. Close observation of the box reveals machine learning deployable prospects. The pre-processing and feature analysis through the dew analyzer are the two most prominent portions of *DEWML* and *DeepDEW*. Scopes for using *RelDEW* and *QDEW* are huge in the dew server as it receives the data from the sensor. *DEWRec* and *DEWTrans* have massive potential in the dew analyzer and the dew client interface. The dew client interface also has potential for the *DEWSite*.

2.2 *MedGini*

MedGini forms a part of *DEWCare*. It presents dew computing for healthcare. Karmakar et al. [3] use dew computing to bridge the gap between the sensors and the cloud. They use the Gini index and Shannon entropy to synchronize when connected to the Internet. The dew layer stores the data fed from the sensors and then synchronizes the data when connected to the internet. Since health data is crucial and loss is

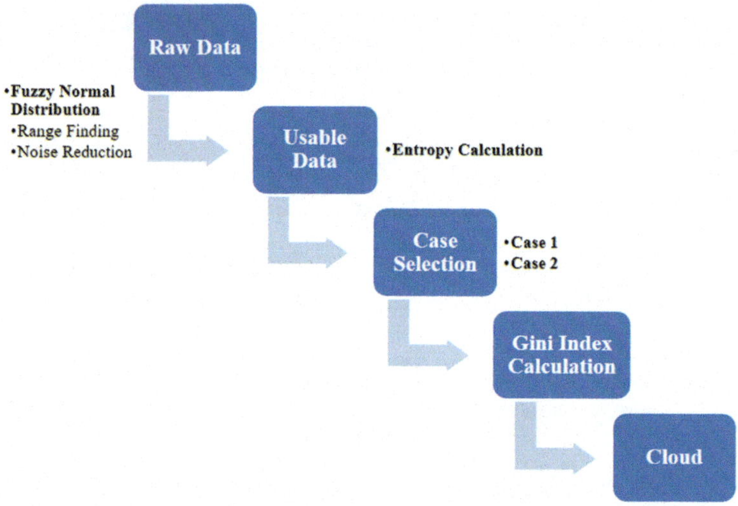

Fig. 6 Dew architecture for the implementation of MedGini

detrimental, the dew layer has been introduced to store the data and transmit the same when Internet connectivity is present. The authors use fuzzy normal distribution and noise reduction in the dew layer for data cleaning. Now, this data cleaning can be done using learning frameworks mentioned in Fig. 1 in the case of dew computing. The architecture for implementing MedGini is presented in Fig. 6. It can be observed that there are four layers of processing prior to the synchronization of the data in the cloud.

It can be observed from Fig. 6 that the generation of usable data can be done using *DewML* and *DeepDew*. *DEWRec* can also be used for the data cleaning part. Fuzzy normal distribution, which has been used for noise reduction, can be done using multiple *DEWML* models and *RelDEW*.

2.3 DEWSec

This deals with the security perspectives of the dew architecture. Moussa et al. [4] propose a cyber-attack detection framework using deep learning for dew-cloud architectures. The cloud-dew architecture for the Connected and Autonomous vehicles proposed by the authors resembles the process flow presented in Fig. 7.

The authors concentrate on the security of the dew databases rather than the cloud infrastructure. The authors focus on four critical factors for the dew database.

1. Cyberattacks on V2X Communications
2. Interfering with Vehicle Sensors and Taking Over Physical Controls
3. Supply Chain and Third-Party Risks

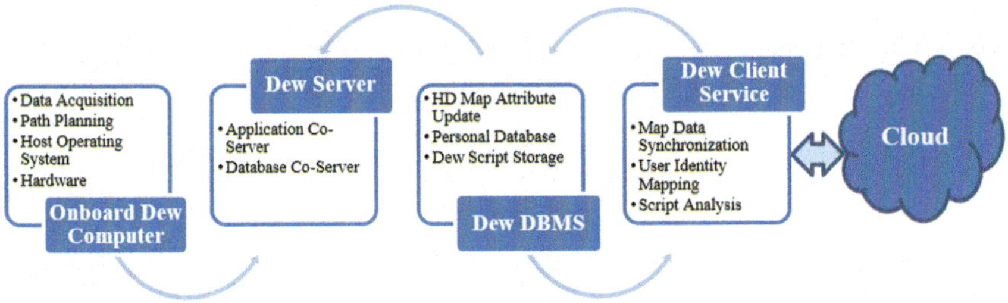

Fig. 7 Cloud-dew architecture for connected and autonomous vehicles

4. Dumpster Diving for Data.

In all these four areas of cyber-attack, options exist for deploying *DEWML*, *DeepDEW*, *RelDEW*, *QDEW*, *DEWQML*, and *DEWRec*.

2.4 *DEWTask*

Task scheduling for dew computing has been another potent area of exploration. Hirsch et al. [5] and Sanabria et al. [6] have addressed this issue. Khalid [7] addressed the deep learning-based offloading strategy for dew computing.

Mobile phones are used as the dew devices, and heuristics are used for load balancing and job assignment for an incoming job to a particular mobile. Hirsch et al. [5] explore the computational capabilities of smartphones for task scheduling in the dew environment. They observe that different smartphones have different computational capabilities depending upon their configurations and, when clustered, can be used for real-time task scheduling. Figure 8 illustrates the mobile cluster performing distributed computing within the context of dew computing.

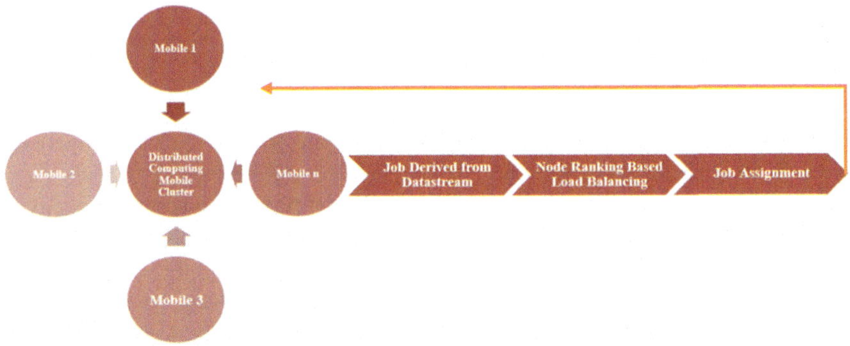

Fig. 8 Distributed computing mobile cluster within the dew context

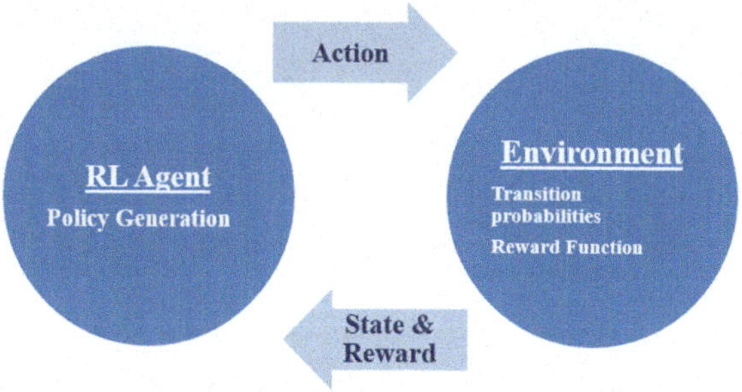

Fig. 9 Training of the edge-dew environment

Device sorting has been done using heuristics in [5]. It can be observed that load balancing methods for mobile devices can be used in the *DEWSmart*, *InDeT*, *DEWComm*, and *DEWNet* options. *DEWTask* itself in this study relates to the *DEWSmart* paradigm which is being proposed.

Sanabria et al. [6] used deep learning frameworks to distribute the jobs among smartphone dew devices. They use the Proximal Policy Optimization technique as an AI agent for load balancing in the simulated dew environment. Again, as in [5], mobile devices are used as intelligent clusters in the dew context. The job assignment is done using the job information coupled with the state of the mobile phone. The training is done as described in Fig. 9.

In the Proximal Policy Optimization technique, the agent receives information about the job and the mobile device's state. The agent decides and schedules the job to a particular mobile device knowing. Once assigned, the agent receives the reward. The authors state one limitation that the device count and presence in the dew cluster remain the same through the job scheduling, i.e., no devices enter or leave the cluster during the evaluation period. The dew environment is fixed. The system designed by the authors conforms to the *RelDEW*, as presented in Table 1. Further, *RelDEW* can be observed in the *DoME* proposal described in the following section. Moreover, the reward system used can be further explored using *QDEW* and *DEWRec*.

2.5 *DoME*

In the study, Chakraborty et al. [8] used *RelDEW* and *QDEW* to execute microservices in the distributed dew-cloud service-oriented architecture. The agent-based *QDEW* offers for real-time microservice execution without any internet connectivity. The authors use a reward-based *RelDEW* environment. The randomly distributed microservice execution is based on service availability in an edge cloud, void of which service migration occurs. The study proposes a two-step approach for the independence and synchronization of dew computing. First, the Dew-Q algorithm proposed by the authors ensures microservice execution at the local server independently, followed by synchronization on resumption of internet connectivity using

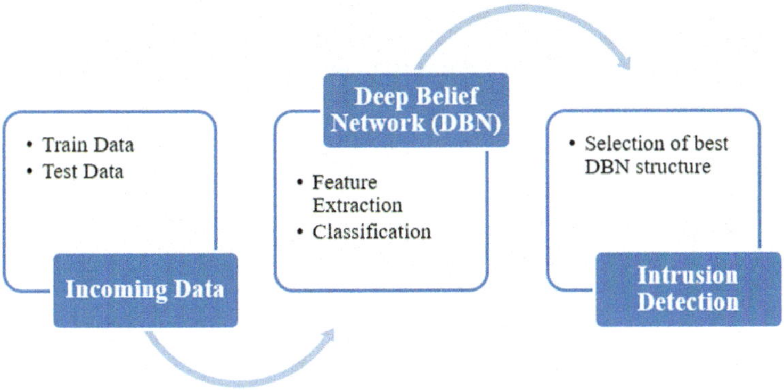

Fig. 10 Deep belief network for intrusion detection system

DEWSite at the 4th edge cloud. The study presents scopes for *DEWSite* as well as *DEWTrans. DEWRec* can help in the node selection and service migration.

2.6 *DaaS*

Singh et al. [9] proposed a Dew-As-A-Service offering intrusion detection in the Edge-of-Things environment. They used artificial intelligence in the dew environment to detect false alarms in the Edge-of-Things environment. The authors use deep learning classifiers to design an alarm filtration mechanism. Deep Belief Network receives the standard warnings from the Intrusion Detection System. The results from the Deep Belief Network are sent to the Control system for analysis by the *DaaS,* which detects the false alarms and eliminates them. Figure 10 describes the Intrusion Detection System based on Deep Belief Network.

A study of *DaaS* exhibits options for the use of *DEWML, DeepDEW*, *ProDEW*, and *DEWSec.*

2.7 *DEWCare*

Gordienko et al. [10] and Manocha et al. [11] have proposed the healthcare use of dew computing. Augmented reality, multimodality of the data, machine learning, and Internet of Things environments have been superimposed to offer elderly care options [10]. The dew layer in their proposal receives the raw sensor data and pre-processes them before sending them to the fog tier. Manocha et al. [11] use dew computing to analyze and monitor the impact of meteorological factors in real time. They use these results to predict the early diagnosis of health, thereby sending relevant alarms

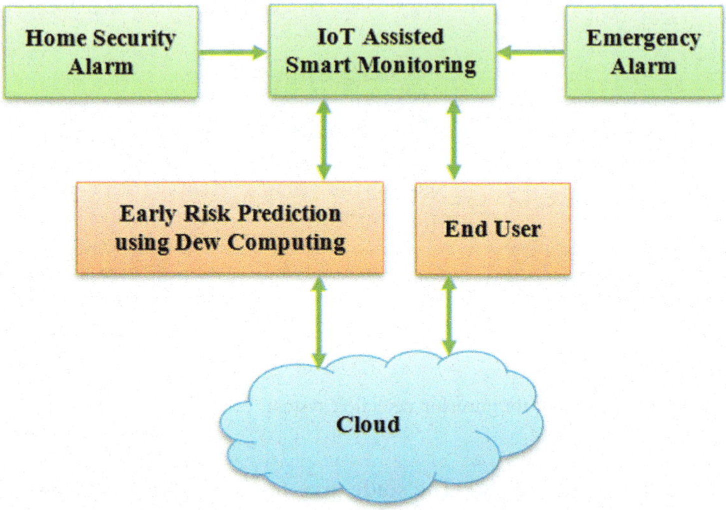

Fig. 11 *DEWCare* framework

based on categories of health. Moreover, they ensure data privacy at the dew layer. Figure 11 shows the framework proposed by the authors.

2.8 XDEW

The rise of dew computing and the rapid advancement calls for Explainable AI-based dew computing. Furthermore, the prospects for using *DEWML*, *DeepDEW*, *RelDEW*, *QDEW*, *DEWRec*, and *DEWTrans* in future research scopes call for the explainability of the technologies. Explainable AI is nascent and stands on the foundation of the four pillars, as presented in Fig. 12. Figure 13 illustrates the pictorial difference between the Artificial Intelligence (AI) models and the Explainable Artificial Intelligence (XAI) models.

As we have observed that the dew paradigm has already enjoyed a few AI frameworks at different layers of the dew environment, Fig. 13 makes it evident that the explainability of the frameworks is a must for enhanced reliability. Both dew computing and explainable AI are new technologies with many research scopes.

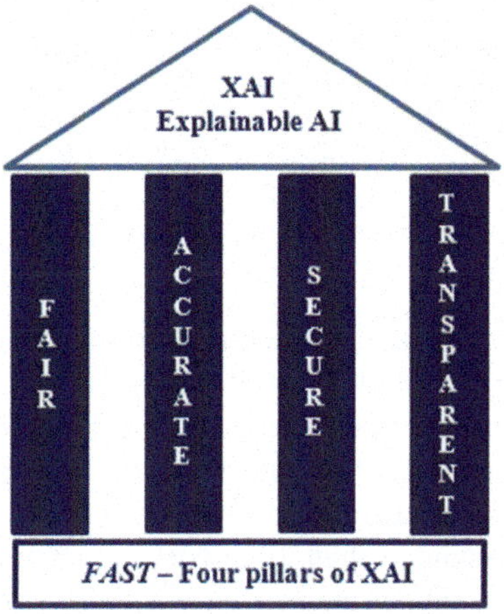

Fig. 12 *FAST—Fast, accurate, secure*, and *transparent* paradigms of sustainable XAI

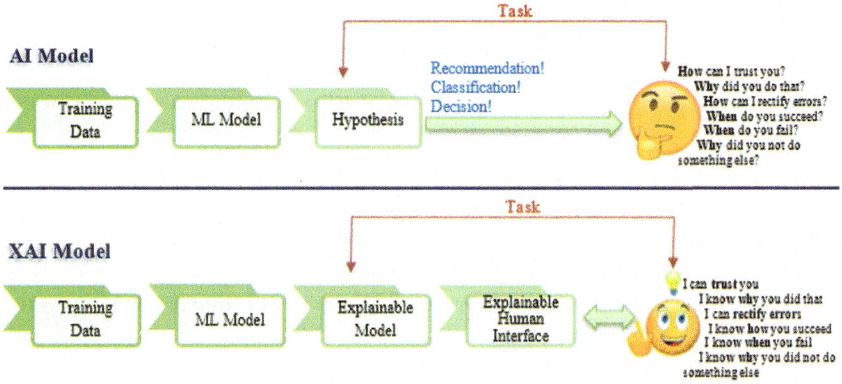

Fig. 13 AI Model versus XAI model

2.9 Dew Quantum Learning

Machine learning (ML) itself is a highly interdisciplinary field. ML algorithms extract generalized knowledge patterns from the collected data to infer some decisions. Recent trends show that the use of traditional machine learning models, such as artificial neural networks, deep neural networks, suffers from the drawback of requirements of the high volume of data for successful network learning. Further, these algorithms pose high computational and time complexity even with sophisticated hardware support like advanced GPU. To overcome these disadvantages, people search for a suitable alternative that may be equally or more efficient in mathematically extracting pattern generalization from data. Researchers are nowadays inquisitive about leveraging the principles of quantum computing and augmenting

Fig. 14 Development phases of QML in the last two decades

them with the principles of classical machine learning. Thus, verifying the applicability and efficiency of quantum machine learning emerges as an exciting research area. Inherent phenomena of quantum computing (QC) are entanglement, superposition, and parallelism. With these quantum mechanical phenomena, Quantum Machine Learning (QML) offers machine learning power with exponential speed up and fewer memory requirements than its classical counterpart.

Machine learning applications are pervasive and inevitable for ubiquitous intelligence. As illustrated in Fig. 14, dew computing promotes independence and collaboration among dew devices with growing computational ability. This section focuses on mapping the implementation architecture required for quantum dew computing within the existing cloud computing framework.

2.9.1 Related Study

Medhi et al. [13] proposed a dew-based distributed architecture for intelligent, low-latency, and reliable healthcare IoT, which utilizes a machine learning model to serve user-specific requirements. Dew computing uses the decision-making power of machine learning. However, the model must not be computation-exhaustive to conform with the lightweight dew framework. Ray et al. [14] modelled the ecosystem with dew-based local computing, especially in e-healthcare.

We incorporate the concept of QML at the dew level to achieve the quantum advantage over classical implementation. Shor solved the factorization and discrete log problems efficiently in the quantum paradigm. Some typical complex issues, like the application of Shor's algorithm that finds out the prime factors of a large number [15] and the application of Grover's algorithm for searching an unstructured database [16], have got an exponentially speedy quantum solution. QPUs with 50–100 qubits can perform tasks that compete with or surpass today's classical computer. Hence, the machine learning research community is keen to explore the intelligent applications of QML. Quantum neural computing was first proposed by Kak [17]. VQC or PQC plays a significant role during the developmental era of QML. Quantum neural network evaluation in NISQ devices is possible because of the development of VQC as QNN [18]. Exhaustive quantum learning implementation needs proper research endeavour. [19–21] are some of the references which deal with VQC evaluation. References [22, 23, 25] have shown performance improvement of quantum learning-based methods over classical methods. A Quantum feed-forward neural network has been trained by Oh et al. [22] with faster optimization and fewer memory requirements. References [22–25] experimented with and evaluated QNN training intricacies in detail.

2.9.2 Evolution: QML in the Last Decade

Machine learning serves various user-specific applications. As user data increases by volume, machine learning faces learning challenges from the algorithms' computational, space, and time complexity. Nowadays, deep learning (DL) performs very well in different user-specific fields like healthcare, speech identification, text recognition, image classification, etc. and has become popular among researchers. DL can extract and learn feature maps of the input without human interference. QML has grown as a prospective research domain for the last several years often it has shown its superiority over its classical counterpart.

2.9.3 Basics of Quantum Computing and Quantum Machine Learning (QML)

In quantum mechanics, the polarization of a photon or spin of the electron decides the mathematical notation of a state. Horizontal polarization of the photon or the up spin of the electron is considered as $|0\rangle$, and vertical polarization of the photon or the down spin of the electron is considered as $|1\rangle$ (Fig. 15). These abstract states are vectors,

Quantum bits or qubits are the superposition states. The superposition state of $|0\rangle$ and $|1\rangle$ is $|\Psi\rangle = \alpha|0\rangle + \beta|1\rangle$, where $|\Psi\rangle$ is the qubit state vector and α, β indicate respective probability amplitudes.

The Quantum gate is the unit of the quantum circuit which takes qubits as the inputs and operates upon the qubits. Quantum gate leverages the property of reversibility due to the unitarity of quantum mechanics, i.e., it does not lose information.

A recent and popular policy to implement QML algorithms is to create and execute variational quantum circuits (VQC) or parameterized quantum circuits (PQC). VQC

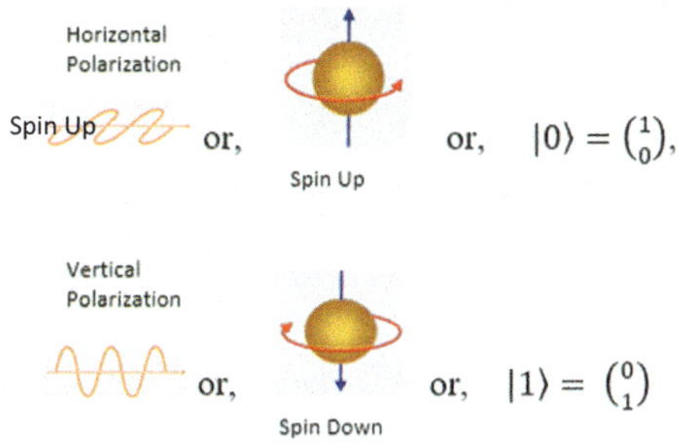

Fig. 15 Quantum notation of states

or PQC comprises a sequence of unitary gates. Quantum gate parameters are optimized, and some cost function is evaluated.

VQCs are executed using variational quantum algorithms (VQA).

There are several specific steps in VQA, such as

1. Data encoding.
2. VQC design—Ansatz.
3. Measurement of output using computational basis.
4. Classifier—Generally first qubit is measured, and based on the probability of labels, $+1$ or -1 (class labels) are decided.
5. Optimizers-classical optimization algorithms are used to compute cost, and ansatz parameters are adjusted to reduce the cost.

Classical data is encoded to quantum data using different quantum encoding algorithms, such as amplitude encoding, basis encoding, angle encoding, QuAM, QRAM, and Qsample encoding. These algorithms embed the data into high-dimensional complex vector space or Hilbert space. Hilbert space is a high-dimensional vector space complete with respect to the norm defined with the inner product of the vector space. Figure 16 represents different quantum encoding algorithms.

Quantum wires carry quantum data or qubits. Now the quantum circuits are highlighted concisely. Quantum circuits comprise quantum gates and their appropriate interconnection through quantum wires. These circuits are unitary. According to the postulates of quantum mechanics, the transformations between quantum states must be unitary. Thus, quantum gate operation, which converts a quantum state to another quantum state, must be reversible. Classical gate operations are not reversible. The only exception is classical NOT gate. Unitary matrices can represent Quantum gates. By definition of unitarity, it exists an inverse operator. Therefore, the quantum gate operation must be reversible. Unitarity preserves the inner product of two arbitrary states. A quantum state $|a\rangle$ is applied with a unitary gate U, and the state transition takes place to the state $|b\rangle$ as shown in Eq. (1).

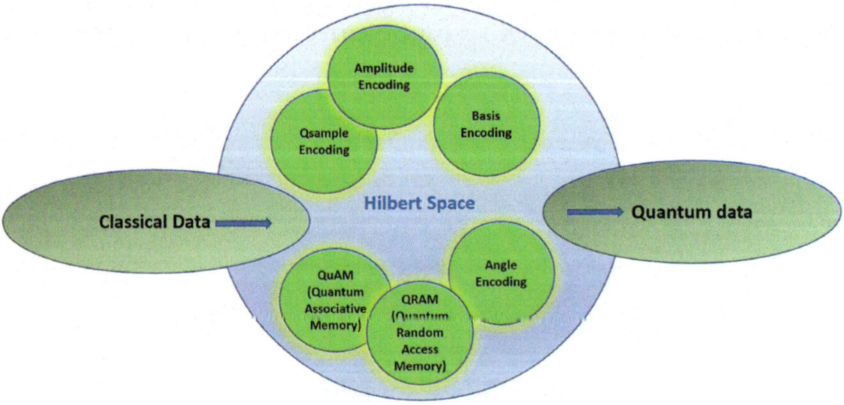

Fig. 16 Quantum encoding algorithms

$$|b\rangle = U|a\rangle \tag{1}$$

It is straightforward to apply the Hermitian conjugate of U to restore the original state, as in Eq. (2).

$$|a\rangle = U^{-1}|b\rangle = U\dagger|b\rangle \tag{2}$$

In other words, any matrix represents a quantum gate if it is unitary.

When qubits are input to the circuit, quantum gates operate on the qubits to produce outputs. These outputs are measured to draw any inference. A block diagram of VQC is present in Fig. 17. Following is a brief discussion regarding Quantum Measurement. When a quantum state is measured or observed, it collapses into one of its basis states. This action converts the qubit to a classical bit and is irreversible. Measurement is usually associated with a z-basis, as in Eq. 3.

$$z = \begin{bmatrix} 1 & 0 \\ 0 & -1 \end{bmatrix} \tag{3}$$

where $z|0\rangle = |0\rangle$ and $z|1\rangle = -|1\rangle$

z-basis is also known as $\{|0\rangle, |1\rangle\}$ basis. During measurement of an arbitrary qubit $(\alpha|0\rangle + \beta|1\rangle)$, probabilities of $|0\rangle$ and $|1\rangle$ are calculated as,

$$P|1\rangle = |\alpha\langle 0|0\rangle + \beta\langle 0|1\rangle|^2 = |\alpha|^2 \tag{5}$$

$$P|0\rangle = |\alpha\langle 1|0\rangle + \beta\langle 1|1\rangle|^2 = |\beta|^2 \tag{6}$$

$$\text{and, } |\alpha|^2 + |\beta|^2 = 1 \tag{7}$$

Quantum circuits are differentiable; hence, the circuit control parameters can be adjusted and trained like neural networks. The parameter updation through multiple training iterations minimizes the error. So, VQC or PQC is considered a QNN. Figure 18 is the block diagram representing QNN training.

Fig. 17 Variational quantum circuit

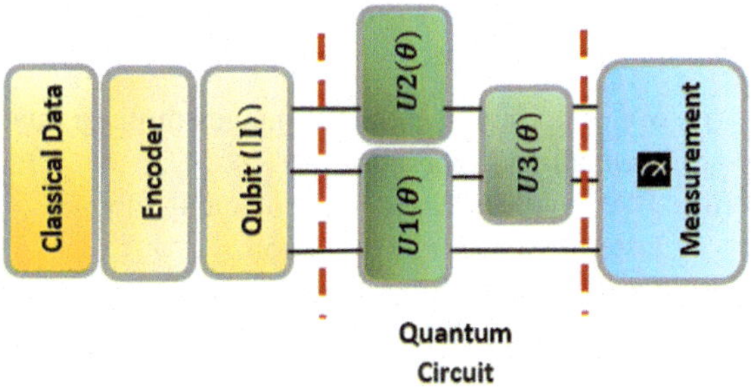

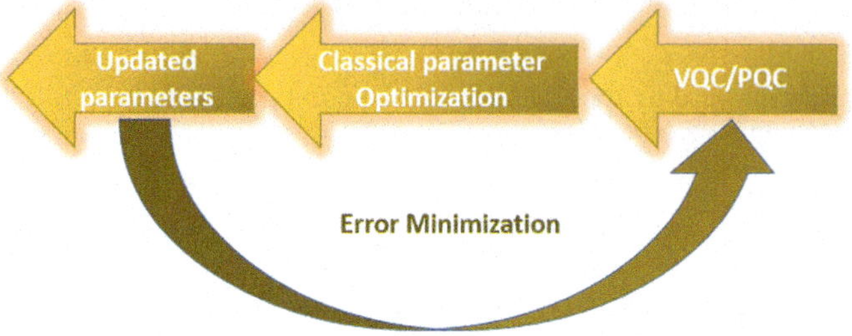

Fig. 18 Hybrid QML architecture: VQC training over circuit parameters through classical optimization

Farhi and Neven [27] used VQC for binary classification. The circuit is created with several quantum gate operations considering the continuous variables. Meinhardt et al. [28] used the same circuit with an efficient data representation. The authors used the Qubit encoding, where n input bits convert to $n + 1$ qubits, and the extra qubit is the readout qubit, denoting the class representation. Several parameter-dependent unitaries apply to the qubits. For error minimization, QNN training updates these parameters. In work [28], Meinhardt et al. used the stochastic gradient descent algorithm for parameter optimization. The authors of [28] used another quantum encoding method—amplitude encoding. In amplitude encoding, n bit string is represented as the superposition of the computational basis states.

2.9.4 QML Benefit in Dew Computing

Dew computing permits edge devices to compute and analyze data swiftly. Dew independence means non-dependency on the availability of continuous internet service. The dew device is capable of providing the service considering internet service interruption. Dew collaboration helps to connect and synchronize data with the server at the appropriate interval. Dew computing can produce local decisions through machine learning algorithms and synchronize the information as and when internet connectivity is available. Quantum machine learning associated with dew computing incorporates the special quantum advantages into dew machine learning, as there is a fundamental difference between quantum computers and classical computers in information processing.

(i) We reap the quantum advantages through quantum properties such as entanglement and superposition.
(ii) The unique characteristic of qubit-superposition makes quantum execution faster than classical execution.
(iii) Quantum computing is advantageous with respect to time complexity and cost of execution. When the classical computation is impossible due to extreme time and computation complexity, quantum computation provides the potential solution.

(iv) For multi-dimension systems and multi-variate statistical analysis, quantum computing is quite efficient. A multi-dimension classical computing system is challenging to implement and execute for its large number of degrees of freedom [26]. Quantum superposition and parallelism are effective in handling these multi-dimension models particularly. Thus, quantum machine learning has the advantage over classical machine learning.

However, currently available NISQ devices cannot handle large numbers of qubits and are not fault tolerant. But the research trend tends to bright future direction of the QML application domain. Dew QML will harvest the benefits of both QC and DEWML.

2.9.5 Coordination of Dew Site, Dew Server, and Cloud Server with Quantum Hardware

The dew computing system represents a real-life metaphor of dew and the cloud. A dew server runs in the local dew devices. It has the same working principle as the cloud server. But the dew server only serves one specific client. A user may have the dew system available on his/her local devices (such as mobile phones); with this device, the user may register to the dew server as well as to the cloud server of the system. Dew servers and cloud servers both use user credentials to identify the users. The client is connected with the remote service provider through the cloud server. When the user data is captured at the dew level, it is stored temporarily in the dew database. Quantum hardware as of now is available to the common users through quantum cloud only. There may be two different approaches for implementing Dew QML, i.e., QML-based dew computing architecture.

In one of the approaches, the dew server may communicate with quantum hardware through quantum cloud services for data analysis and analytical report generation. It also updates and synchronizes with the cloud server database as and when the internet is available. Cloud server segregates the multiple-user huge database userwise and stores it. In contrast, the dew server deals with a single-user small database. In another QML-based dew computing approach, instead of a dew server, the cloud server itself is associated with a quantum cloud. It can access quantum hardware to analyze the data. This interpreted data is synchronized between the dew databases at the user/physical service provider end and the cloud database.

Suppose the dew server disappears like a dew drop due to hardware failure, virus attack, etc. In that case, it can be regenerated as all the data is available in the cloud server in a synchronized way. Both the approaches described above are pictorially represented in Fig. 19.

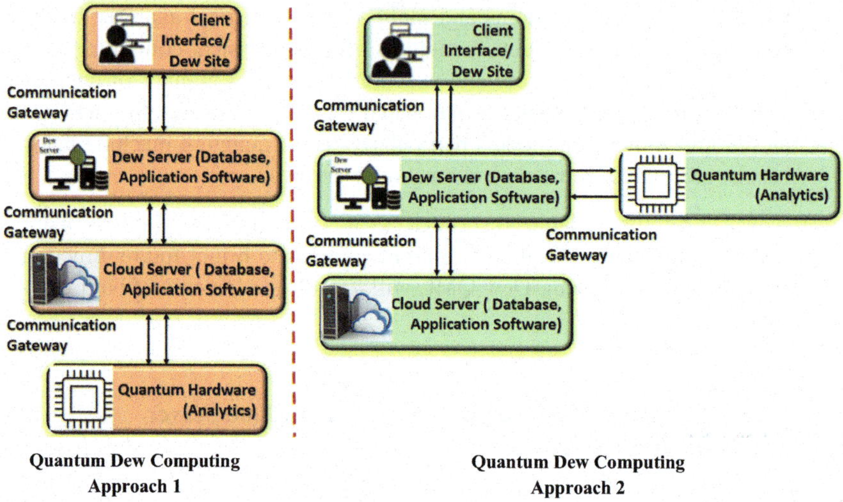

Fig. 19 Quantum dew computing approaches

2.9.6 Implementation Requirement: Quantum Compiler Toolkits

Quantum operations involve noisy qubits. Quantum execution platforms, which are available nowadays, still need to be capable of providing fault tolerance, i.e., extended quantum operations are error-prone. Thus, there is a restriction on the depth of the logical circuit due to the factor mentioned above. These currently available quantum devices are "Noisy Intermediate-Scale Quantum" (NISQ) devices. Quantum circuits are used to implement quantum algorithms. Quantum circuits are implemented with a series of quantum gates connected sequentially. Next to the circuit block is a measurement block, which provides the result of circuit execution with classical post-processing. The Quantum compiler toolkit executes quantum processing. Figure 20 shows some of the quantum compiler toolkits.

IBM Qiskit

IBM Qiskit is an open source software development kit, developed by IBM Research. It supports execution of quantum algorithms and circuits in local quantum simulator and quantum hardware.

ProjectQ

ProjectQ is open-source and allows quantum implementation in Python. The notable feature is that ProjectQ provides programme translation to any backend, including simulators and quantum computers.

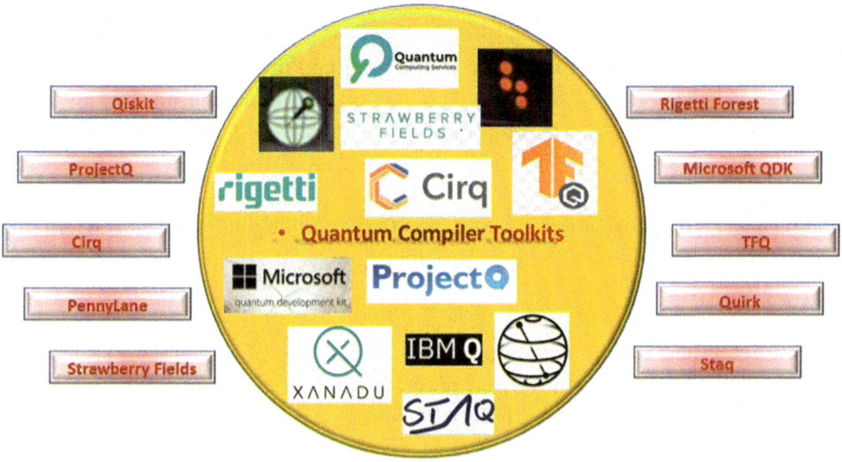

Fig. 20 Some quantum compiler toolkits

Cirq

Cirq is an open-source Python library. It enables the user to execute and optimize quantum circuits in NISQ devices using both simulator and actual quantum processor. Google AI Quantum Team promotes Cirq.

PennyLane and Strawberry Fields from Xanadu

PennyLane is a Python library for QML and quantum–classical training optimization. The important feature of PennyLane is that it provides a cross-platform interface through plug-ins and supports multi-platform project development with IBMQ Qiskit and ProjectQ.

Another product from Xanadu is Strawberry Fields. It is also a Python library for full-stack development of QML and quantum–classical training optimization.

Microsoft Quantum Development Kit

Microsoft provides the Quantum Development Kit with the quantum programming language, Q#, integrated with their Visual Studio code development interface. Simulators run locally, and the actual quantum processors can be accessed through the Microsoft Azure cloud platform.

Rigetti Forest and Cloud Computing Services (QCS)

Rigetti provides the Rigetti Forest suite with a Python library for Quil programmes called pyQuil, Grove, a quantum programmes library and Quantum Virtual machine (QVM), a simulation environment. Access to QVM and QPU is provisioned through QCS.

Quirk

Quirk provides a user-friendly editor with a drag-and-drop facility for the quantum simulator in a web browser. Moreover, it has a good graphical runtime representation of circuit modification and re-execution.

TensorFlow Quantum

Google provided TensorFlow Quantum (TFQ), which is a Python-based QML library. It can be integrated with the Cirq environment.

D-Wave

D-Wave released a quantum computing toolkit that accepts large QUBO (Quadratic Unconstrained Binary Optimization) problems and divides them into smaller sub-QUBOs. These sub-QUBOs are fed to the D-Wave quantum processor.

Staq

Staq is also a quantum computing tool written in C++. It supports openQASM language.

3 QML Based Dew Health Monitoring System: A Case Study

Rapid analysis of health data is very crucial for the successful deployment of an e-health monitoring system. Fast and efficient network training with ever-growing data, efficient network traffic management, efficient object level and symptom level classification of medical data, user-wise segregation of health data, and timely feedback to the patients are the challenges posed to the available cloud-based e-health monitoring system. Dew-based architecture with machine learning model, provisions user-specific services in healthcare decision-making. Also, dew-based e-healthcare provides real-time dew services to the dew client [12].

We include the quantum advantage into the dew-based e-health monitoring system and present a case study, particularly on quantum dew computing for retinal image analysis and decision-making. We focus on the mapping between the implementation framework required for quantum dew computing within existing distributed cloud computing frameworks. A dew layer between the cloud and user interface must help improve the e-health system. Data inconsistency is removed from the actual database stored in the cloud as and when the internet facility is available.

At the client end of QML-based dew health monitoring system, the dew site has two application interfaces-patient interface and doctor interface. Each interface has two databases—the Raw database and the Report database. The Raw database contains the client's own raw medical test data. It is created when data is captured at the patient end. In contrast, the Report database contains the analyzed results received when analytics software uses raw medical data and draws inferences with the help of a pre-trained learning model. Sensor-captured raw medical data of the patient may be transferred to a Cloud server as and when the internet is available. The copy of raw data is not kept or backed up in the dew server.

Considering the example of quantum dew computing for retinal image analysis and decision-making, the patient's dew device such as a mobile phone is equipped with a special purpose fundus camera to capture retinal images. Raw retinal images may be sent to a cloud server, which is responsible for accessing quantum hardware to get high-speed analysis. If required, Raw retinal image data from the Patient dew interface can be directly channelized to quantum hardware through quantum cloud service. This QNN analysis report is synchronized between the Doctor dew interface and the cloud server. After verification is done at the Doctor dew interface, the final analysis report will be synchronized among the Patient dew interface, Doctor dew interface, and cloud server. Ophthalmologist's manual verification is essential before the generation of the final report. This two-layer verification will lead to effective decision-making for medical data. Raw medical data is captured at the Patient dew interface, but they are not preserved in the Patient dew database. From the doctor's dew application interface, as per the requirement, the doctor may fetch both the patient's Raw data and Report data for further manual reference or analysis. Patient dew interface database has no need to store any raw medical data as the patient is concerned only about the analysis of his/her health data.

Other medical dew computing applications using QML use daily health data like heart rate or blood pressure monitoring systems, generating an alert when any discrepancy in behaviour is found. The cloud database is updated accordingly with the availability of internet connections. But it is undesirable to exhaust the system with a huge amount of redundant data that may be generated due to improper data capturing. This noisy data is not of proper use of the medical analysis system. Further, once a particular sample has been collected; it may not be required to collect the successive samples immediately as it provides no new information to the system during a certain period of time until the anatomical system undergoes possible changes. For example, once the retinal image is captured correctly through the fundus camera associated with a dew device, it is not required to capture more samples until a certain period when the anatomical structure of the retina possibly undergoes further changes. Data redundancy exploits the system's storage capability and hinders the system's performance.

The diagrams of Figs. 21 and 22 represent two different approaches of QML-based dew health monitoring system. Figure 21 represents the phenomenon when cloud server is responsible for communicating with quantum cloud. Figure 22 represents the alternative approach where dew server can access quantum cloud directly.

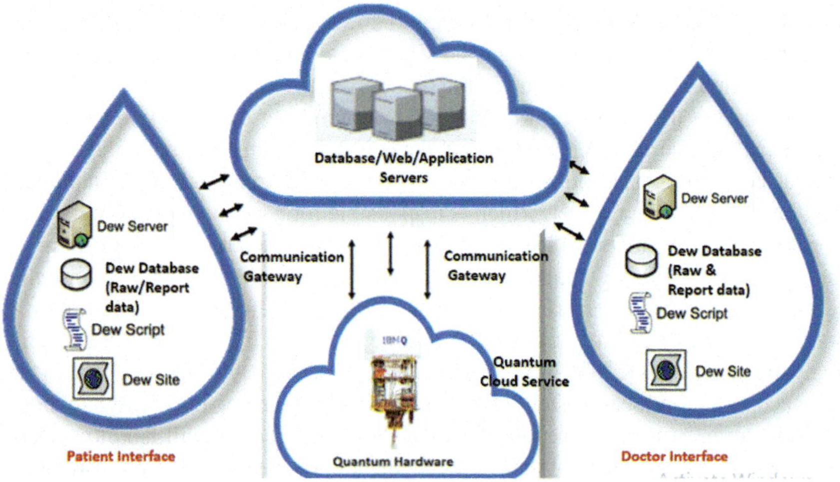

Fig. 21 QML-based dew health monitoring system—cloud server communicates with quantum hardware

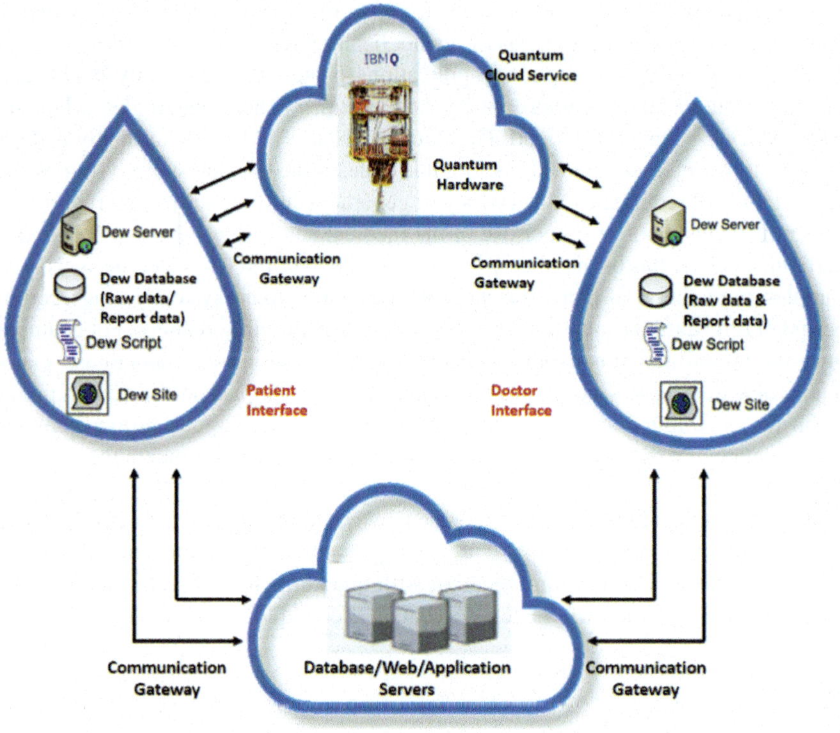

Fig. 22 QML-based dew health monitoring system—Dew server communicates with quantum hardware

4 Potential Challenges

Dew computing is a very recent computing paradigm, hence the scopes for research are enormous. In this regard, technology has attracted multiple challenges as of date. One of the most concerning issues lies in the synchronization part when internet connectivity resumes. In this regard, data sequencing and seamless data transmission demand huge research. Another important issue is the security of the dew server as well as the dew database. Since dew devices are computationally constrained, resource-intensive security protocols fail to succeed in the device level implementation.

Moreover, computationally complex algorithms find less appreciation in the dew device executions. Dew computing is yet to recognize the man in the middle attacks in the data flow between sensors and the dew devices. Learning model applications in the dew environment, such as *DEWML*, *DeepDEW*, *RelDEW*, *DEWRec*, *DEWTrans*, and *xDEW*, are data-intensive models, thereby encouraging high storage dew databases coupled with exponential computational power, which at present is subject to further research explorations. Model validation is usually done in the cloud environment, which can be delivered to the dew device only on synchronization. Device-level model validation is another challenge as of date. The scalability of dew computing is another potential challenge for the utilization of the dew device processor. The operating system on a dew device needs to be further improvised to support dew service-specific applications. As of date, the dew devices communicate with the cloud in their personal protocols void of any standards. Standardization of the dew communication protocols has become imperative with global research being conducted in dew computing.

In addition to these, the following are some specific challenges related to DewQML.

Hybrid System Implementation

VQC/PQC can be implemented for dew level analysis. We may implement quantum circuit, coupled with a CNN model to obtain a quantum convolutional neural network (QCNN) [29]. In QCNN, images may be convoluted to obtain higher-level feature representations, which are reduced to smaller dimension vectors before converting classical information to quantum information and then passing quantum information through the quantum circuit. QKNN is quantum K Nearest Neighbour (KNN) learning algorithm which is implemented with Swap Test to measure the similarity between quantum feature vectors and then Quantum Minimization algorithm (QMA) to find the nearest neighbour using Grover's search [30]. Similarly, QNN, in dew applications, can be coupled with Particle Swarm Optimization (QPSO) [31] to achieve the benefits of both. So, hybrid quantum circuit implementation and performance evaluation is a prospective research direction.

Advanced Circuit Design

Quantum gates are noisy. This limitation constrains the size of the quantum circuit to be small. Different parameterized circuits affect the model's landscapes. QNN can be designed and experimented with multi-qubit gates. An increasing number of qubits and higher circuit depth are heading towards sophisticated circuit design. In that case, a proper gradient estimator is also required for executing the QNN training cycle. This is an open door for further research advancement in this domain.

PQC Gradient Estimation

For gradient-based learning, truncation and round-off error of finite difference method sometimes ill-estimate the gradients. In that case, circuit learning becomes unstable. The probable solution may be analytical gradient estimation. Li et al. [32] provided an efficient way for analytical gradient estimation in the quantum computing paradigm. However, even gradient-based learning becomes unsuitable in some situations, like the noisy experimental environment. In such conditions, gradient-free methods for circuit learning may be suitable. Particle swarm gradient-free optimization [33] and genetic algorithm-based gradient-free optimization [34] may be adopted as potential solutions.

Vanishing Gradient Problem

QNN uses gradient descent optimization algorithm; hence, it is required to avoid vanishing gradient or disappearance of gradient problem during learning iterations. In the quantum learning domain, this vanishing gradient problem is also called the "barren plateau" problem [35, 36]. McClean et al. [35] proved that the probability of a "barren plateau" increases exponentially as the number of qubits increase, and the phenomenon can be prevented in a small-scale circuit by initializing with suitable parameters. According to them, the mean of the circuit parameters is zero, and the variance decreases exponentially with an increasing number of qubits. As proposed by Grant et al. [36], a good choice of remedy to this problem may be to cut the whole circuit into several shallow blocks and initialize the network with some randomly selected parameters. Other parameters are chosen later. In this way, circuit depth can be controlled. However, the problem "barren plateau" is yet to be studied by researchers.

5 Conclusions

Dew Computing is a novel computing paradigm; thereby, scopes of research exist. Machine Learning, Deep Learning, as well as Quantum Machine Learning form one of the most sought-after emerging technologies. In this context, Dew Machine Learning and Dew Quantum Machine Learning have enormous potential in the near future. Most of the geographical topologies are yet to witness seamless internet connectivity to date, thereby generating a huge opportunity for Dew Computing to fill in the gap between internet-connected sites and the sensor deployment sites void of Internet connectivity. Applications which require instant results at the sensor/device end followed by synchronization with the cloud exhibit potential scopes of Dew Computing. In situations constrained by seamless internet connectivity, arresting data loss further involves much importance of Dew Computing. The last decade has observed the emergence of the cloud, fog as well as edge computing and their significance and value to society. In this regard, the research and application gaps provide ample scope for Dew Computing to offer solutions in future.

Dew analyzers in the dew layer may use QML for successful analysis and extraction of knowledge from the user data captured if the quantum hardware is made available to the dew client directly through cloud quantum service. Another alternative approach may be to access the quantum cloud and, subsequently, quantum hardware through the cloud server. Though the quantum computers in the NISQ era do not have any error correction policy [37, 38] or cannot exploit the benefit of high feature dimension or circuit depth [39], the transformation is shaping towards a more competent and fault tolerant quantum computing technology in the near future. Quantum error correction and minimizing the use of quantum hardware by minimizing quantum involvement as far as possible without affecting the actual benefits, are the future direction to reap the benefit of "Quantum supremacy".

References

1. Ray, P.P.: An introduction to dew computing: definition, concept and implications. IEEE Access **6**, 723–737 (2017)
2. Roy, S., Sarkar, D., De, D.: DewMusic: crowdsourcing-based internet of music things in dew computing paradigm. J. Ambient. Intell. Humaniz. Comput. **12**(2), 2103–2119 (2021)
3. Karmakar, A., Banerjee, P.S., De, D., Bandyopadhyay, S., Ghosh, P.: MedGini: Gini index based sustainable health monitoring system using dew computing. Med. Novel Technol. Dev. 100145 (2022)
4. Moussa, M.M., Alazzawi, L.: Cyber attacks detection based on deep learning for cloud-dew computing in automotive IoT applications. In: 2020 IEEE International Conference on Smart Cloud (SmartCloud), pp. 55–61. IEEE (2020)
5. Hirsch, M., Mateos, C., Zunino, A., Majchrzak, T.A., Grønli, T.M., Kaindl, H.: A task execution scheme for dew computing with state-of-the-art smartphones. Electronics **10**(16), 2006 (2021)
6. Sanabria, P., Tapia, T.F., Toro Icarte, R., Neyem, A.: Solving task scheduling problems in Dew computing via deep reinforcement learning. Appl. Sci. **12**(14), 7137 (2022)
7. Khalid, M.N.B.: Deep learning-based dew computing with novel offloading strategy. In: International Conference on Security, Privacy and Anonymity in Computation, Communication and Storage, pp. 444–453. Springer, Cham (2021)
8. Chakraborty, S., De, D., Mazumdar, K.: DoME: Dew computing based microservice execution in mobile edge using Q-learning. Appl. Intell. 1–20 (2022)
9. Singh, P., Kaur, A., Aujla, G.S., Batth, R.S., Kanhere, S.: Daas: Dew computing as a service for intelligent intrusion detection in edge-of-things ecosystem. IEEE Internet Things J. **8**(16), 12569–12577 (2020)
10. Gordienko, Y., Stirenko, S., Alienin, O., Skala, K., Sojat, Z., Rojbi, A., Jervan, G.: Augmented coaching ecosystem for non-obtrusive adaptive personalized elderly care on the basis of cloud-fog-dew computing paradigm. In: 2017 40th International Convention on Information and

Communication Technology, Electronics and Microelectronics (MIPRO), pp. 359–364. IEEE (2017)
11. Manocha, A., Bhatia, M., Kumar, G.: Dew computing-inspired health-meteorological factor analysis for early prediction of bronchial asthma. J. Netw. Comput. Appl. **179**, 102995 (2021)
12. Sverko, M., Tankovic, N., Etinger, D.: Dew computing in industrial automation: applying machine learning for process control. In: 2021 IEEE 45th Annual Computers, Software, and Applications Conference (COMPSAC), pp. 1789–1794. IEEE (2021)
13. Medhi, K., Ahmed, N., Hussain, M.I.: Dew-based offline computing architecture for healthcare IoT. ICT Express **8**(3), 371–378 (2022)
14. Ray, P.P., Dash, D., De, D.: Internet of things-based real-time model study on e-healthcare: device, message service and dew computing. Comput. Netw. **149**, 226–239 (2019)
15. Shor, P.W.: Polynomial-time algorithms for prime factorization and discrete logarithms on a quantum computer. SIAM Rev. **41**(2), 303–332 (1999)
16. Long, G.L.: Grover algorithm with zero theoretical failure rate. Phys. Rev. A **64**(2), 022307 (2001)
17. Kak, S.C.: Quantum neural computing. Adv. Imaging Electron. Phys. **94**, 259–313 (1995)
18. Cerezo, M., Arrasmith, A., Babbush, R., Benjamin, S.C., Endo, S., Fujii, K., et al.: Variational quantum algorithms. Nat. Rev. Phys. **3**(9), 625–644 (2021)
19. Havlíček, V., Córcoles, A.D., Temme, K., Harrow, A.W., Kandala, A., Chow, J.M., Gambetta, J.M.: Supervised learning with quantum-enhanced feature spaces. Nature **567**(7747), 209–212 (2019)
20. Grant, E., Benedetti, M., Cao, S., Hallam, A., Lockhart, J., Stojevic, V., et al.: Hierarchical quantum classifiers. NPJ Quantum Inf. **4**(1), 65 (2018)
21. Chen, S.Y.C., Huang, C.M., Hsing, C.W., Kao, Y.J.: Hybrid quantum-classical classifier based on tensor network and variational quantum circuit (2020). arXiv:2011.14651
22. Oh, S., Choi, J., Kim, J.: A tutorial on quantum convolutional neural networks (QCNN). In: 2020 International Conference on Information and Communication Technology Convergence (ICTC), pp. 236–239. IEEE (2020)
23. Kerenidis, I., Landman, J., Prakash, A.: Quantum algorithms for deep convolutional neural networks (2019). arXiv:1911.01117
24. Beer, K., Bondarenko, D., Farrelly, T., Osborne, T.J., Salzmann, R., Scheiermann, D., Wolf, R.: Training deep quantum neural networks. Nat. Commun. **11**(1), 1–6 (2020)
25. Abbas, A., Sutter, D., Zoufal, C., Lucchi, A., Figalli, A., Woerner, S.: The power of quantum neural networks. Nat. Comput. Sci. **1**(6), 403–409 (2021)
26. Sammut, C., Webb, G.I.: Encyclopedia of Machine Learning and Data Mining. Springer Publishing Company, Incorporated (2017)
27. Farhi, E., Neven, H.: Classification with quantum neural networks on near term processors (2018). arXiv:1802.06002
28. Meinhardt, N., Dekker, B., Neumann, N.M., Phillipson, F.: Implementation of a variational quantum circuit for machine learning with compact data representation. Digit. Welt **4**(1), 95–101 (2020)
29. Hur, T., Kim, L., Park, D.K.: Quantum convolutional neural network for classical data classification. Quantum Mach. Intell. **4**(1), 1–18 (2022)
30. Padha, A., Sahoo, A.: Quantum enhanced machine learning for unobtrusive stress monitoring. In: Proceedings of the 2022 Fourteenth International Conference on Contemporary Computing, pp. 476–483 (2022)
31. Zhou, J., Qin, H., Zhang, Y., Yang, R., Liu, Y., Li, C.: Binary quantum elite particle swarm optimization algorithm for spectrum allocation in cognitive wireless medical sensor network. J. Phys.: Conf. Ser. **1924**(1), 012030 (IOP Publishing)
32. Li, J., Yang, X., Peng, X., Sun, C.P.: Hybrid quantum-classical approach to quantum optimal control. Phys. Rev. Lett. **118**(15), 150503 (2017)
33. Bas, E., Egrioglu, E., Kolemen, E.: Training simple recurrent deep artificial neural network for forecasting using particle swarm optimization. Granul. Comput. **7**(2), 411–420 (2022)

34. Zhang, H., Thompson, J., Gu, M., Jiang, X.D., Cai, H., Liu, P.Y., et al.: Efficient on-chip training of optical neural networks using genetic algorithm. ACS Photon. **8**(6), 1662–1672
35. McClean, J.R., Boixo, S., Smelyanskiy, V.N., Babbush, R., Neven, H.: Barren plateaus in quantum neural network training landscapes. Nat. Commun. **9**(1), 1–6 (2018)
36. Grant, E., Wossnig, L., Ostaszewski, M., Benedetti, M.: An initialization strategy for addressing barren plateaus in parametrized quantum circuits. Quantum **3**, 214 (2019)
37. Leymann, F., Barzen, J.: The bitter truth about gate-based quantum algorithms in the NISQ era. Quantum Sci. Technol. **5**(4), 044007 (2020)
38. Preskill, J.: Quantum computing in the NISQ era and beyond. Quantum **2**, 79 (2018)
39. Sharma, K., Cerezo, M., Cincio, L., Coles, P.J.: Trainability of dissipative perceptron-based quantum neural networks. Phys. Rev. Lett. **128**(18), 180505 (2022)

Applications

Dew Computing-Based Sustainable Internet of Vehicular Things

Sushovan Khatua, Daniele Manerba, Samir Maity, and Debashis De

1 Introduction

Future autonomous and intelligent transportation systems consist of vehicular networks. Nowadays, the development of vehicular network applications such as ineffective data storage, access, retrieval, etc. A promising architecture called vehicular edge computing (VEC) uses roadside equipment to function as edge servers for task offloading and caching. A task-based architecture for content caching in VEC, where three key tasks, predicting the content popularity, adding material to the cache, and retrieving content from the cache, are distinguished. Thus a tasks-based architecture can be more effective using artificial intelligence (AI) approaches like regression, deep Q-learning, etc. A recent study of AI-based vehicular edge computing through cache memory is developed for secure caching, effective sub-channel allocation for content retrieval in vehicle to everything (V2X), and collaborative data sharing by [25]. The present study gives some novel applications of dew computing, particularly in the vehicular network system.

Future traffic control applications are made possible by a number of essential technologies, including the Internet of Vehicles (IoV). Autonomous driving and

S. Khatua (✉) · D. De
Department of Computer Science and Engineering, Maulana Abul Kalam Azad University of Technology, Haringhata, Nadia, West Bengal 741249, India
e-mail: sushovankhatua79@gmail.com

D. De
e-mail: dr.debashis.de@ieee.org

D. Manerba
Università degli Studi di Brescia, 25123 Brescia, Italy
e-mail: daniele.manerba@unibs.it

S. Maity
Department of Materials and Production, Operations, Research Group, Aalborg University, Aalborg 9220, Denmark
e-mail: samirm@mp.aau.dk

D. De and S. Roy (eds.), *Dew Computing*, Internet of Things,
https://doi.org/10.1007/978-981-99-4590-0_9

intelligent traffic congestion control are possible with dependable wireless communications between vehicles, and with roadside infrastructure (RSUs) [61]. An IoV communication architecture can also be used to provide a variety of infotainment and entertainment applications, including multimedia streaming, content sharing by service providers, and commercial adverts. This process opens new challenges for logistic routing, procurement, transportation, etc., which need to design mathematically and depend on cloud-based dew computing, may another novel strategy to solve the above-given research areas (see Sect. 9).

Big data will be produced in vehicular applications and it will need to be shared among many IoV nodes (vehicles and RSUs). The IoV nodes will also need to carry out many activities, such as determining the safety of moving vehicles, calculating traffic routes, downloading content, etc. More computing and storage will be required for all of these new duties. One option for saving urgent and popular data and offloading computationally demanding operations is cloud computing. Another option is introducing quantum computing for fast task-based architecture. Due to the large geographical distance between vehicles and cloud servers, storage and computation may not be as effective due to increased latency (when retrieving content from the cloud cache or sending task offloading requests) and uses up communication resources (for long-distance communication between vehicle and cloud) [10].

In the semi-urban/rural areas Internet is disrupted most of the time.[1] Dew computing overcomes this problem in vehicular networks. Effective caching and RSU storage management are significant obstacles in dynamic vehicle networks. Fast mobility and transient connectivity in the vehicular network make it more difficult to rank the contents in terms of their priority; as a result, storage must be updated with the pertinent contents regularly, and in this case, an optimal resource allocation strategy is required to allow downloading of contents from the cache. It works as fast and in an effective way according to [47]. AI approaches can gather and evaluate many forms of data to assist make better-informed judgments on caching, methods of like matching theory [27] employed for effective content caching in VEC. Data in vehicular networks is spatially and temporally variable. Static algorithms might therefore not perform well in-vehicle networks. Data caching and learning approaches can both help applications run more reliably and quickly by gaining meaningful insights from the data.

To support improved driver assistance systems and self-driving, smart vehicles are developing as a new paradigm that accesses popular content and shares transportation information. Innovative automotive applications for improving driving safety and trip comfort have been made possible by advances in sensing, and AI technologies [2, 63]. The essential for communication, processing, and storage resources is rising as a result of these applications. If the contents are fetched from distant data centers, the low content access latency and different application requirements might not be met. Fortunately, by moving cloud cache servers to edge devices, vehicle edge computing, and networks can provide an efficient framework for reducing the load on

[1] https://economictimes.indiatimes.com/tech/newsletters/ettech-unwrapped/indias-dubious-record-on-internet-shutdowns-set-to-continue/articleshow/94105055.cms?from=mdr.

the backhaul links [62]. The article of [49] studied novel research areas by introducing decentralized adaptive resource-aware communication through dew computing under the caching framework, which can adapt the dynamic network environments based on Deep Reinforcement Learning (DRL). The authors proposed adaptive micro cloud services to the edge of proximity resource-constrained smart communication and Internet of Everything (IoE) devices for cellular subscribers. But this work's limitation is that cellular network disruptions still need more investigations. To fill this lacuna, the present article design some practical applications of dew computing under the caching framework in the remaining Sect. 3.

The currently used mobile edge caching techniques in heterogeneous networks can be broadly categorized into four groups [39]: device-to-device (D2D) communication networks, small-cell base stations (SBSs), macro-cell base stations (MBSs), and mobile devices [9]. These edge caching techniques cannot be used in automotive edge caching networks, though it is acceptable to access all content during the connection duration within the range of SBSs or D2D communication networks based on the assumption of low-speed moving scenarios. These edge caching techniques cannot be used in automotive edge caching networks, However, in the course of content distribution, high-speed mobility vehicles may pass a number of roadside units (RSUs), and the reliability of vehicle-to-vehicle (V2V) communication links may not always be guaranteed. As a result, obtaining the entire content from a single edge caching node may not be possible. To increase the spectrum usage in fog access networks [53], examined social aware edge caching. The study [59] was proposed a caching system based on the pattern of vehicular mobility which made in order to increase the effectiveness of content distribution in the Internet of Vehicles (IoV). The basic goal of this technique is to use moving cars as relay nodes to store, transport, and send popular contents. The aforementioned work, however, does not take into account how content delivery and placement might be optimized together to improve caching performance. These problems call for the creation of a network for vehicle edge caching that combines content placement and delivery. Now present study includes dew computing under caching in vehicular networks.

The present investigations are organized as follows: Sect. 1 a brief introduction is given. Section 2 elaborates on a concise literature survey. In Sect. 3, vehicular cache-based dew computing is described. An application of dew computing under secure IoV through cloud and fog environment is presented in Sect. 4. In Sect. 5 studied the vehicular data under Dew computing. The applications of AI in IoV communications are studied in Sect. 6. In Sect. 7, an ML-based IoV network is described. Section 8 studied cloud-based vehicular localization. Smart procurement planning under cloud-based dew computing is elaborated in Sect. 9. Finally, in Sect. 12, we conclude the paper by discussing important research questions, the relevance of the insights obtained, limitations, and future research.

2 Literature Survey

Dynamic real-time traffic-aware vehicle routing employing V2V communication has several difficulties, like (i) How to get actual time information, (ii) How to route a vehicle dynamically based on in-time information as the vehicle moves on the road, and (iii) How to get a consistent network. Real-time traffic data collection requires effective data packet routing from cars to vehicles. The connectivity between two vehicles in V2V communication is referred to as a link. Vehicle-assisted networks (VANETs) are characterized by highly dynamic topologies with frequent link breakages, network fragmentation, and a high number of packet collisions and interferences. This is due to the highly dynamic nature of vehicle mobility as well as the complex road condition and building blockage. The evaluation of Quality of Service (QoS) in the delivery of arbitrary data packets from a source vehicle to a destination vehicle has thus been the main focus of studies of VANET protocols [26]. However, the process of gathering traffic data typically involves several sources (vehicles on the desired route giving data) and a single destination (the vehicle requesting data), which increases the number of hops and latency. Four V2V protocols are put forth by [32] for the use of beacon messages to detect traffic congestion along routes of interest. To improve the robustness of the data transmission system in the event of a bad network environment [54], suggest dew computing. Dew computing's main goal is to give edge devices the ability to function without the internet according to [43]. Edge devices can still synchronize data with the cloud server when they can access the internet. The authors describe dew computing has two crucial characteristics: independence and collaboration. A dew computing model needs to meet a number of conditions. For instance, it should at the very least include all types of dew computing applications offered as a service, including not being constrained by a particular class of applications. For instance, one straightforward way to model dew computing is to treat each dew computing system as an Internet. It is denoted that to build a network of computers that are linked together using TCP/IP protocols. Taking this strategy into account, each on-premises computer or a collection of such computers is structured as an Internet; websites are made on this Internet and numerous services are available in this separate dew world. The lacuna of this work is that not much investigations are available for disrupted Internet in IoVs. This limitations of the literature motivated to study cache-enabled dew computing. The article [37] studied the edge computing, sometimes referred to as mobile computing, which is the process of performing computation at the edge of a device's network. This implies that a computer is linked to the device's network, processing the data before instantly sending it to the cloud. This device is referred to as a "edge node" or "edge computer." An addition to cloud computing in a disrupted interconnections of communication via Internet, fog computing work between edge and cloud computing systems. Fog computing becomes a layer that exists between the cloud and the edge. Fog nodes receive the massive volumes of data that edge computers send to the cloud and sort through the vital information. Then the fog nodes move the vital data to the cloud to be stored and either delete or keep the unnecessary data on their own for later analysis. Recently [51] is shown that how fog computing transfers critical data rapidly while conserving a lot of space on the cloud. Dew computing (see cf. [33]) can be assumed as a hybrid extension of the existing Cloud-Fog-Edge hierarchy. It provides localized user-concentric support. Users can use dew devices to access Internet data without or with minimal use of inter-network connectivity [58].

Chen et al. [8] investigated that an user-centric data is utilized for effective cache-enabled unmanned aerial vehicle (UAV) deployment to optimize the quality of the users' experience while using the least amount of overall transmit power for the strength of UAVs. The paper of [6] describes the adoption of a single cache-enabled UAV and the design of an online wireless caching scheme by jointly optimizing UAV trajectory, transmission power, and caching content scheduling. Finding the best routes for numerous cars traveling to a set of places is the objective of the Vehicle Routing Problem (VRP) according to [50]. The shortest overall distance is the definition of an optimal route. However, if there are no other restrictions, assigning just one vehicle to visit every place and figuring out its shortest path is the best course of action. In many ways, this issue is similar to the VRP. Minimizing the length of the longest single route taken by all vehicles is a more effective technique to determine optimal routes. Constraints on the vehicles, including (i) Constraints on capacity: Although the vehicles have a maximum carrying capacity, they must pick up items at each location they visit, and (ii) Time windows: Each place must be visited during a predetermined period of time. Now under IoV how VRP can work is a big challenge for the future logistic routing.

Now, the cloud-dew computing architecture, a new type of application can access personal cloud data continuously even without an Internet connection. Wang et al. [56] studied a dynamic real time vehicle routing employing V2V communication has two key difficulties: (1) How to get real-time traffic information, and (2) How to dynamically route a car while it's on the road using real-time traffic information. (3) Real-time traffic data collection requires efficient routing of data packets from V2V. Connectivity between two vehicles is called V2V communication a connection. Because the motion of vehicles is so dynamic, and complex road conditions and blocking of buildings, VANETs are distinguished by their extremely dynamic topologies. The disconnections, network fragmentation, and high numbers packet collisions and interference are regular issues in this situation. Because of this, the research on the VANET protocols (see cf. [26, 28]) mostly concentrated on measuring the QoS throughout the transmission of any data packet between a source and a target vehicle. The number of sources (vehicles on the desired route that submit traffic data) and destinations (the vehicle seeking traffic data) are often required for the collection of traffic data, which increases the number of hops and delays. The recent study [32] developed four V2V protocols to employ beacon messages to monitor traffic congestion along desirable routes. The difficult part is incorporating it into dynamic vehicle routing. Stochastic algorithms that replicate the dynamic routing of social animals have drawn a lot of interest in addition to advancements in conventional algorithms (see cf. [13]) because of their efficacy and resemblance to the dynamic vehicle routing challenge. These algorithms are centralized solutions that

Table 1 Literature survey on dew-based Internet of vehicular things

Scientific contributions	Analyzed features										
	IoT	Comp. & data sensing	Cache computing	IoV	ML methods	Dew computing	Fog computing	VANET	Deep learning	Edge computing	Cloud services
Rindos et al. [44]	✓	×	×	×	×	✓	×	×	×	✓	✓
Patal et al. [40]	✓	×	×	×	×	✓	✓	×	×	✓	✓
Meng et al. [33]	✓	✓	×	✓	×	✓	×	×	×	×	✓
Mane et al. [30]	✓	✓	×	✓	×	✓	✓	✓	×	✓	×
Li et al. [27]	✓	✓	✓	✓	×	✓	✓	✓	×	✓	✓
Javad et al. [25]	✓	✓	✓	✓	✓	×	×	×	✓	✓	✓
Gusev et al. [17]	✓	✓	✓	✓	×	✓	✓	×	✓	✓	×
Hou et al. [24]	✓	×	✓	✓	✓	✓	×	✓	✓	✓	✓
Guo et al. [15]	×	×	✓	×	×	×	×	✓	×	×	×

call for in-depth familiarity with the changing road network. The authors of [14] suggest a powerful Geographic Source Routing (GSR), an enhancement to the GSR protocol for a route recalculation algorithm, where to exchange traffic information. In this work, a system for routing message communications for VANET that substitutes traffic information for real driving is presented. The VANET is organized into several categories (see cf. [52]). Thus the present chapter includes the dew computing using cache under dew-edge-cloud network. A concise survey in tabular form is available in Table 1.

3 Vehicular Cache Based on Dew Computing

The high-definition maps needed for autonomous driving can produce enormous amounts of data transmission, which the current networking architecture is unable to handle. The effects of downloading high-definition maps for vehicles, in particular when the vehicle is in a location with a bad network connection, might substantially impair driving safety due to packet failures and delays. To assure the vehicle's driving safety, a dew computing-based vehicular edge caching architecture is suggested in this article. The suggested technique ensures the vehicle's stability when acquiring the high-definition map in areas with weak network coverage and keeps the high-definition map the vehicles have stored up to date.

The dew computing-based vehicle edge caching architecture in detail using the network overview. The high-definition maps caching model suggested by [27]. In dew computing-based architecture for vehicle edge caching is described according to [27]. When other cars visit an area with a bad network environment, they can share the maps that are cached in their local cache. We categorize the map into two categories. Normal maps are one of the varieties, and high-definition maps are another. The local cache of the car can be used to pre-store a normal map. Vehicle-to-vehicle and vehicle-to-base station (V2V and V2B) connectivity can update high-definition maps.

(i) Regular map: It stores information on the lane and intersection configurations, as well as the network of roads.
(ii) High-definition map: It caches information on certain road conditions, such as the location of hazards, the number of vehicles, and traffic congestion.

The purposes of these two kinds of maps vary. A regular map can be used to plan a route and for navigation. A high-definition map can direct the car to steer clear of hazards and instantly change its pace.

A dew computing-based system for vehicle edge caching strategy may work more appropriately in navigation system. For navigation, every car should cache a standard map, when a vehicle comes across a roadblock, the high-definition map cached in each car needs to be updated as soon as possible. While this is going on, vehicles in places with weak network can use V2V to transmit the most recent high-definition map to other vehicles. The moment the vehicle connects to the Internet, high-definition map data is synchronized in real time with the cloud server. The safety of automobiles can be ensured by these methods.

4 Secure IoV in Cloud, Fog, and Dew Computing

The designing an IoV architecture, security concerns are always a top priority. Some security measures are required for the cloud computing and fog computing connected to the vehicular network in order to build a secure IoV architecture. Vehicle networks can be protected to some extent using conventional security mechanisms like authentication, digital signatures, and encryption. The usual security procedures, however, fall short of the unique security requirements of the IoV due to the frequent network connection handoff caused by the swift movement of the vehicles. In a vehicular network, connections are brief, transient, and include numerous quick handover operations. The security problems in the IoV have some potential remedies. Modern automobiles are intricate electromechanical machines with multiple moving parts. The vehicle has steadily evolved into an intelligent terminal with a wealth of features that go beyond smartphones as the IoV has grown. In order to achieve more advanced driving control capabilities, such as auxiliary driving, autonomous driving, and other functions, the vehicle itself integrates an increasing number of sensors, controllers, and computers. Second, the car will make it easier to connect to the IoTs, communicate with distant data centers, and progressively implement the features of intelligent transportation. The following issues are also of concern with regard to the security of connected vehicles, in addition to the security of information systems and their networks:

Anomaly detection: Finding illegal vehicles and fraud vehicle identification cards using anomaly detection.
Risk assessment: Information security risk assessment for IoV's information systems and intelligent vehicles.
Identification: In normal conditions, license plates and RFID are largely used for vehicle identification, however in abnormal situations, where the vehicle identification information cannot be recognized, many other features are used.
Privacy protection: Safeguard sensitive data such as the identities of drivers and other moving objects.
Safe driving: Technology like automatic driving, auxiliary driving, and others are designed to make cars themselves safer and easier to use.

The distributed and intelligent computing models that are being gradually improved are cloud computing, fog computing, and dew computing according to [45]. The interaction between the very complicated real-world entities and a range of roles, including all forms of motor vehicles, various services, various regulatory bodies, and road network maps, makes up the enormously complex information system known as "car networking." The Vehicular ad hoc network (VANET), IoV, and smart IoV may be gradually built thanks to the benefits of cloud, fog, and dew computing models, and as a result, integrated intelligent traffic will be realized. The following new security challenges have been researched as a result of the gradual application of cloud computing, fog computing, and dew computing to the vehicle network due to its practical necessity.

A mobile cloud computing security problem was proposed by [5]. In the automotive network, cloud computing offers a centralized data processing and computing platform. Data security issues and the leakage of sensitive information are among the security issues associated to outsourcing of computing and data storage. Arwa et al. [3] studied the IoT fog computing security and privacy issues were researched. Fog computing, in contrast to cloud computing, does not entirely rely on the processing resources in the distant data center; as a result, data produced by IoT devices can be kept in the nearby fog nodes, reducing the risk of data loss and the cost of full outsourcing.

Dew computing, which is more suited for distributed intelligent IoV computing platforms, can support smart objects after cloud computing (see cf. [44]). In order to streamline the deployment and service approaches, intelligent computing typically involves high complexity scientific computing tasks, such as image recognition, path planning, automatic driving, driver assistance, voice recognition, and other high level intelligence functions in large information systems. Artificial intelligence algorithms have the ability to manage the vehicle and rely significantly on input data, therefore these algorithms could lead to new security issues by operating with disastrous results due to initial input data flaws.

5 Analyzing Vehicular Data Using Dew Computing

The Intelligent Transportation System (ITS) has been recently developed to offer numerous technologies. In centralized ITS solutions like Google Map, the crowd-sourcing software Waze, and Toyota's Entune dynamic routing, onboard smartphone devices or vehicles transmit regular local traffic data to a central server or cloud and receive regular traffic information on their desired routes. VANET is used by the decentralized ITS solutions to gather traffic data via vehicle-to-infrastructure (V2I) and V2V communication.

Today's vehicles have various technology features, like heated seats, backup cameras, cruise control, keyless entry, GPS, smartphone integration, automatic emergency braking, and others. Thus one can forecast that best on past data using the above features, multi-processor chips will be as common in cars as they are in smartphones. According to this information, drivers use it to plan their routes. Currently, the path-finding algorithm is run on the cloud, and data is provided straight to the car. Roadside equipment with sensors gathers the information needed to estimate the route accurately. Only one path is discovered using conventional path-finding methods according to [12]. The issue that can arise is the interruption of Internet access, which will make it harder to navigate the cloud. The routing algorithm in the suggested solution identifies many paths. Each vehicle has a modest, specific dew computer installed. Graphics processors are affordable today. A graphic processor will be installed on each dew computer to handle the calculations.

Through short-range communication technology, the vehicles communicate with each other in the zone. The moving vehicles communicate about the state of the road's traffic. The ant colony optimization (ACO) algorithm is applied to the road network using this data. There is a scope to design the meta-heuristics like ACO, Genetic Algorithm, etc., to perform better while using machine learning for real-time data. Thus machine learning in meta-heuristics is another paradigm still to be explored in, like ITS. The routing algorithm will be built using the information gathered through V2V and V2I communication channels. ACO is a highly parallelizable algorithm [57]. On a graphic processor, the algorithm is easily parallelizable. The authors suggested that the system, which may have a tiny vehicle, will be used to implement the ACO algorithm on the provided road network. Graphic processing units on the computer will parallelize the algorithm and present effective paths in real time. Thus Dew computing-based vehicular networks have a new research area in ITS under cache which work in a disrupted Internet.

6 Artificial Intelligence for Multimedia IoV Communication

A device that supports data transfer and communication with the devices around it is necessary to adopt IoV in multimedia communications. Any technology, including Personal Area Networks (PAN), the IoT, and Wireless Sensor Networks (WSN), can accomplish this. Scalability and flexibility of data interchange are crucial for IoV since they allow for integrating sensors, vehicles, actuators, machines, etc. While synchronized traffic data transfer in the IoV system network improves vehicular system efficiency, the sensor in intelligent IoV increases the safety of vehicles and traffic systems. However, there is a heavy data transmission risk due to the energy consumption, needed capacity, green buffer awareness, and message exchange through IoVs [1]. Self-driving car AI stimulates a variety of applications with various intelligent advantages. It is accurate and valuable for future directions, especially given the rise in complexity and volume of data that the algorithms will be analyzing. To effectively monitor and manage the demand for intelligent IT technologies, a smart utility is needed to keep up with the growing amount of high-traffic information in IoVs [22]. The development of multimedia is dependent on the IoV system due to the rapid advancement of digital technologies. It requires a portable device to gather significant data to support and direct the specific trend for IoT-based platforms to analyze the transportation sector.

7 Machine Learning in the IoV Network

For prediction issues and intelligent management, machine learning is many models, classifications, clustering, etc. techniques. Reinforcement Learning (RL) will offer directional behavior in IoV applications to support robustness and scalability. In IoV networks, it can provide path selection or route optimization. Throughput maximization and latency minimization can be guaranteed as the operation and maintenance method when ML is used with the Software Defined Network (SDN) in the Internet of Vehicle (IoV). With reliable and improved routing services, ML and SDN will work together to enhance IoV network performance [64].

It guarantees the best routing policy adaption based on sensing and learning from the IoV environment to achieve higher usage. The ML features used in IoV networks is depicted in Fig. 1. SDN and ML offer some special benefits to adopting security solutions in the area of IoV network security. To address security concerns, it will be practical to construct ML software interaction with the SDN data plane to deliver statistical information to the application layer in response to vehicle requests [36]. Automation and connectivity are crucial for self-driving aspects of Cognitive Internet of Vehicles (CIoV) applications, such as automatic driving, which should have enough intelligence to reduce traffic accidents. ML can take over vehicles to provide error-free driving. For security and privacy reasons, CIoV permits cloud-based ML in a transportation system [21].

In addition to network resource allocation and optimization, ML in the CIoV cognition and control layer offers strategic services for various function levels, such as driving habits, health monitoring, and pattern and mood analysis. Deep learning (DL) methods offer intelligent decision-making to assess the crucial, influential collision probability elements and danger of potential accidents in the IoV, which will increase

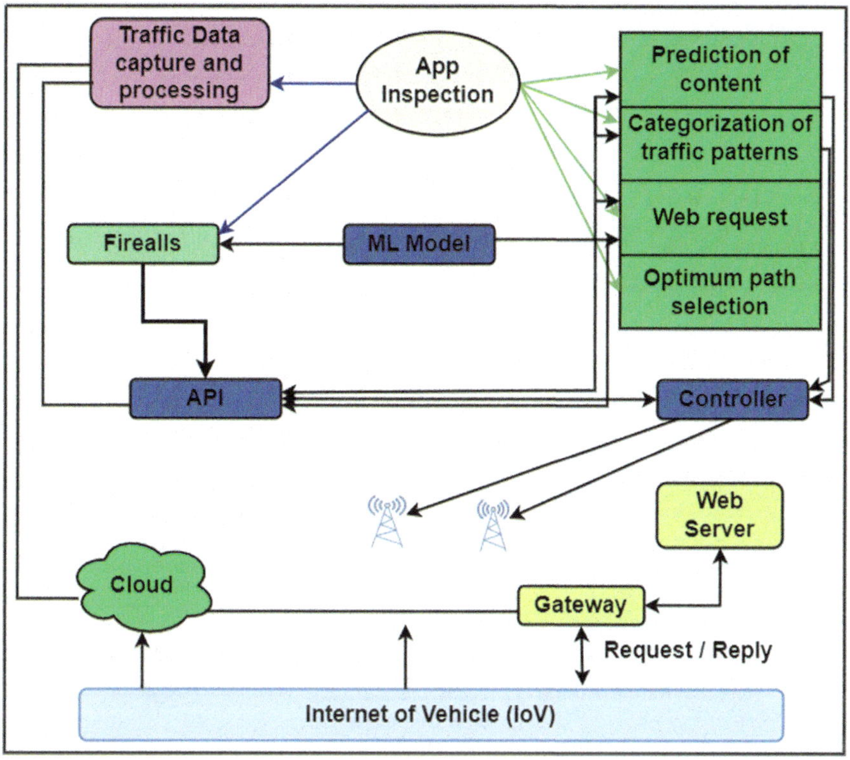

Fig. 1 ML in IoV network

driving safety and efficiency in the IoV transportation system [7]. Different ML methods, such as Support Vector Machines (SVM), Genetic Algorithms (GA), and Neural Networks (NN), can be used to anticipate collisions and accidents.

8 Cloud Based Dew Computing Applications for Vehicular Localization Systems

The dew computing method manages resources like smartphones, tablets, and intelligent sensors that are always connected to the cloud-dew network. As a result, it spans a broad range of technologies, such as extensive data access, distributed peer-to-peer networking, cooperative applications, and data processing environments. It also has implications for data-intensive applications, augmented reality, autonomic self-healing networks, and general-purpose use in residential and commercial settings. In terms of practical applications, that might be helpful in various fields, including

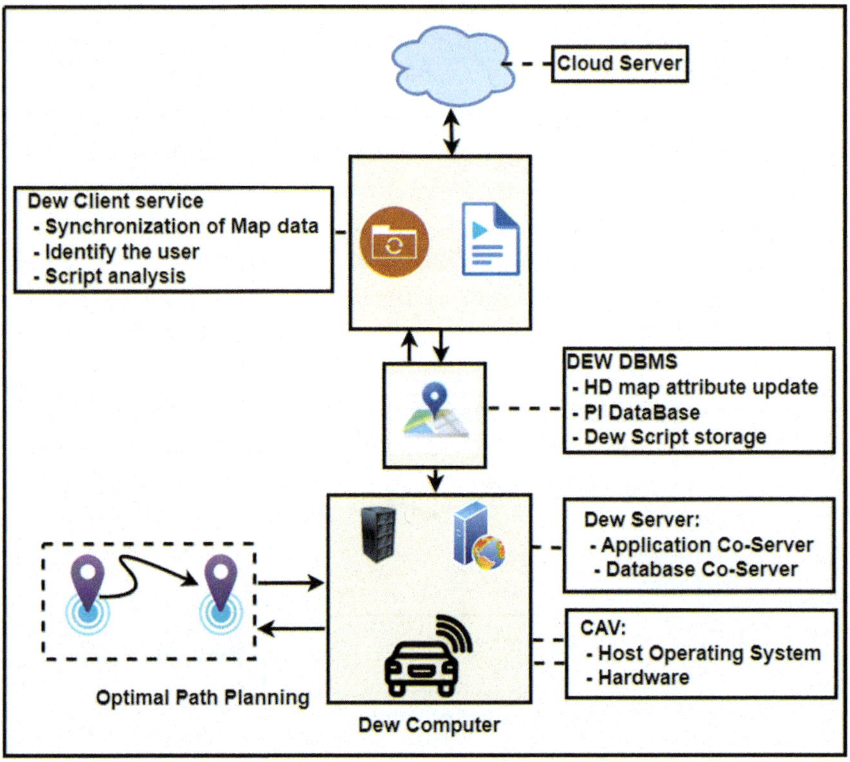

Fig. 2 CAVs' in cloud-dew structure

natural language processing for image recognition and detection, streaming services, financial market activity, and financial market behavior, to mention a few.

As it integrates mobile terminal, localization technology, and communication network, cloud-dew computing in IoT systems is a flexible platform to carry out e-mobility, offering a novel method for location information transmission and consumption. The architecture is given in Fig. 2 where Connected and Autonomous Vehicles (CAVs) are likely to benefit from such technology since path planning requires location data to produce a dependable and safe trajectory. To build the appropriate track, data is sent back and forth in the cloud (see Fig. 2) by positioning the moving dew device on the road. Rapid reaction speeds are one of the essential aspects of CAVs. That might account for the intolerance for service disruptions or latency-related delays. The sensors in CAVs aim to provide an onboard computer with a stream of real-time data as an apparatus for nearby processing. Therefore, it becomes the cloud-dew network design robustness' responsibility to maintain vehicular active safety features through data privacy, which practically demands centralized management. For developers to continuously improve the algorithm, self-driving cars must be connected to a secure Internet to download essential updates and continuously feed data into the cloud. In such an application, the requirement to always be online poses a severe risk to user safety, especially when using mobile devices where connectivity may be spotty. Which we already discussed in previous sections. Dew computing characteristics are helpful in the current circumstance. An autonomous vehicle is required in the scenario shown in Fig. 2, which is dependable on dynamic map information going to the dew server. That is possible with a sustainable amount of knowledge from various cars at the same location uploaded to the cloud, moved to the targeted vehicle while maintaining a safe route is the ultimate goal of planning. This method might be beneficial to adjust the position of the vehicle to any changes in the road environment. The dew client program facilitates several applications, including scalability, rule-based data collecting, raw data synchronization, and identity mapping. The most crucial step may be identity mapping [33]. When a user logs in and out of a website, a Server Identification mapping is restricted as a credential checkpoint regarding the user's identity before granting access to secure material. User identification on the dew site may be based on already-in-use privacy and authorization methods. If numerous users attempt to visit one dew server simultaneously, that becomes necessary. The connection between a user's IDs from various domains and the user ID of the back-end system is known as user identity mapping. It makes it possible to audit a single user's activity across several fields and compares these activities to find anomalies. Thus cloud-based dew computing is a novel research area where different practical problems can be solved in the near future.

9 Sustainable Procurement Planning Under Dew Computing

The traditional food supply industry has recently undergone significant changes due to a market structure paradigm shift. Increased market competition has forced an industry-wise consolidation that instantly transforms traditional online shopping companies such as Amazon into merchants with physical outposts in hundreds of neighborhoods. In developing countries such as India, since the foreign direct investment was 100% allowed for food retail, there has been competition between online e-commerce industries with supermarket chains [23]. Two product characteristics in the food/beverage industry, perishability, and breakability, present additional challenges to the purchasers in decision-making. Thus, one of the research goals in this study is to design efficient and effective procurement and transportation planning for the food retail chain considering source locations spread across various locations under a cloud-dew cache environment. In this scenario, dew computing plays a significant role. Intelligent transportation makes procurement planning less time-consuming and which makes the system sustainable. An intelligent supply chain proposed by [35] first draws attention to the use of dew and cloud-based computing. The traveling purchaser problem (TPP) [42] is a mixing of procurement planning and routing. The present study proposes that dew computing enhances procurement and route design decisions. Also a quantum cloud-dew and quantum-inspired cloud-dew computing in vehicular is given in Fig. 4 for sustainable procurement planning or logistic routing. For breakable and perishable products to be purchased by then, the transportations of the purchased goods have a vital role here. The price and availability are yet to be important since they properly/intelligently do not transport the procured goods. So there are enormous opportunities for implementing/installing intelligent devices/sensors on the roadside and vehicles, in the semi-urban areas, dew computing enabled routing and, according to that purchased plan, making the entire system cost or time minimum. Also, most perishable products are transported through refrigerated vehicles, which produce greenhouse gases (GHG) and may reduce global emissions using proper utilization of dew computing. This becomes sustainable in terms of the environment, economy, and society. Thus through dew computing, sustainable procurement planning can be designed for better decision-making.

10 Some Novel Areas in Intelligent Vehicular Network

Dew Computing: Dew computing is an emerging technology it has a minimal dependency on inter-network. It provides server-less and connection-less architecture. The data are kept on the dew server as a local copy and synced with the master copy at the cloud side using the Internet. In the era of cloud computing, "Dew Computing" is a paradigm for organizing personal computer software. Realizing the full potential of personal computers and cloud services is its aim. In this paradigm, software on a personal computer is set up following the Cloud-dew architecture; cloud services and local computers work together to deliver rich functionality [54].

Between end-user devices, such as smart modules, that translate physical parameters into digital information and vice versa, processing and coordination in the Edge/Fog/Cloud levels, Dew computing adds an extra layer that provides autonomy, independence, and collaborative features [19]. Dew computing is a cloud-based mechanism that solves the major cloud storage-based issues (e.g., Dropbox). Cloud storage stores the user's data in a cloud server through the Internet with a local backup. When a sufficient network is not available, the user can use the uploaded data for a certain time. A fresh paradigm for computing called "Dew Computing" emerged with the broad adoption of cloud computing. The initial study on dew computing began with cloud computing architecture, and the metaphor of dew was developed as a logical extension of the cloud metaphor. Dew computing may be thought of as the physical or ground edge of cloud computing [46].

Vehicular Cloud-Dew Computing: A dew server's final characteristic points to a potential new use: if a website adopts the cloud-dew architecture, it will always be accessible, with or without an Internet connection, since a dew server operates on the user's local computer. Let's say a user keeps their private information, such as photos

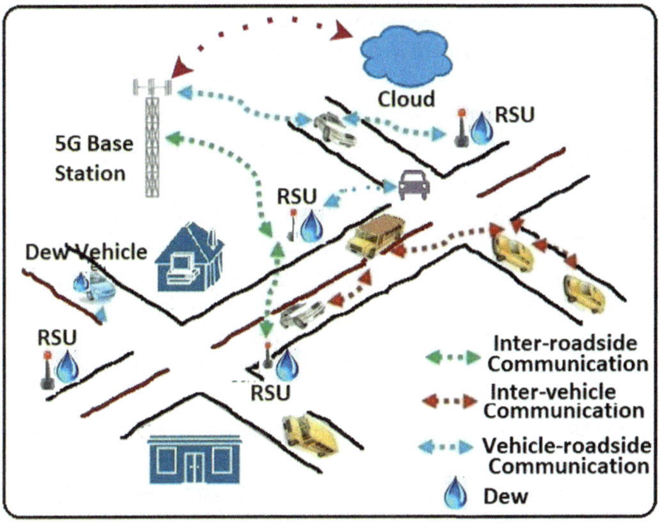

Fig. 3 Cloud-dew architecture in vehicular network

and texts, on a car. Although the data is accessible to the general public, a lack of an Internet connection prevents the user from accessing their own data. The user can locally store a copy of their personal information on any linked device. Saving images and messages in files, however, could be uncomfortable and challenging to manage. Consider a vehicle that uses the cloud-dew architecture. A dew server running on a user's local PC will mirror the website. Duplication differs from copying in some ways.

The idea of "vehicular clouds" is put out to take advantage of underutilized vehicular resources in addition to cloud computing services to reduce the load on centralized cloud data centers and enhance service quality [4]. According to the National Institute of Standards and Technology definition of cloud computing in [31], this architecture enables universal, on-demand access to a shared pool of computer resources (e.g., storage, networks, servers, services, and applications). Cloud computing can be quickly facilitated with few administrative tasks or service provider interactions. Service models based on Platform as a Service (PaaS), Software as a Service (SaaS) [34], and Infrastructure as a Service (IaaS) [43] are taken into consideration in cloud computing. An architecture for cloud-dew in the vehicular network is given in Fig. 3.

In the past two decades, many papers kept their remarkable footsteps in dew computing which are very helpful in building a modern society and providing a more accessible mechanism for IoT devices in IoV when the network is unavailable for a certain time. With the incorporation of smart vehicles and high-speed communications, inter-network access has reached a new level of sophistication. Various cloud domains and fog services are currently engaged to supply a required set of information to its users or customers. On developing countries' roads, vehicles are really aggressive, and lack of network is a common issue. To handle this problem, dew can be proposed to solve this problem as a powerful weapon that can turn vehicle routing issues into a cloud of cosmic dust. One of this model's key qualities is availability. Standard mechanisms can be used to access the services on heterogeneous client platforms like laptops, mobile phones, and tablets. One more characteristic of cloud computing is automation. One can access the necessary computing, a popular choice for individuals and corporations. Consumers don't need to engage with humans to get necessary computer resources like storage and server time. Elasticity is a benefit of cloud computing over other types of computing. In the cloud paradigm, capabilities can be infinitely available, elastically available in any amount at any moment. Another aspect of cloud computing is a metered service that uses a metering capability to regulate and optimize resource utilization automatically.

Vehicular Fog-Dew Computing: Fog computing is a decentralized computing environment where data, storage, and applications are distributed between the data source and the cloud. It is a word for moving some operations and resources to the cloud's edge rather than creating channels for cloud storage and usage. According to [44], fog computing addresses one of the major challenges in cloud computing after collecting vehicular data at specific access points rather than transmitting every piece of information through cloud channels. They studied and stated that this style of the dispersed method increases efficiency while reducing expenses. The investigation [38] invokes more intriguing is that it's one strategy for addressing the burgeoning IoT. As per [16] was proposed, Fog computing is a computing architecture in which nodes receive data from IoT devices in real time. Each node communicates with the others in a flash. Data are sent periodically from the nodes in the cloud, and the cloud monitors them continuously. After analyzing the data cloud, reverse the decisions to all the nodes. Here dew computing plays a significant role in reversing back the consistent information. A cloud-based application analyzes the data that has been received from the various nodes with the goal of providing actionable insight.

Vehicular Edge Computing: To speed up data processing before devices connect to the IoT [29] and transmit the data for use by the smart vehicle, edge computing occurs in intelligent devices precisely where sensors and other instruments are receiving and processing data. Although the edge's physical architecture might be complex, the fundamental concept is that client vehicular devices that link to neighboring edge modules for faster processing and more seamless activities. IoV sensors, a worker's vehicle, smartphone, security cameras, or even Internet-connected vehicles are connected to the edge devices [37]. Edge computing is a distributed information technology architecture in which client data is processed at the periphery of the network, as close to the originating source as possible. If we consider all the vehicles as originating sources, then the data is processed through the installed mechanism of vehicles and sent to the cloud. These processed data originated from vehicles with the insufficient network.

Quantum Vehicular Dew Computing: Very recently, researchers have focused their investigation towards novel areas using quantum physics that may solve complex real-life problems very fast. This novel research direction is termed Quantum Computing (QC). Since quantum technology operates to solve the practical problem potentially very fast compared to other state-of-the-art technologies. Thus present study focuses on two approaches (i) Quantum Computing in Dew for Vehicular Networks and (ii) Quantum-inspired Computing in Dew for Vehicular Networks.

Quantum Computing: The classical quantum computing the information is stored in quantum bits (qubits) [20]. A quantum qubit represents state 1, state 0, or a superposition of both. The quantum state of a qubit can be described as [48]:

$$|\Psi\rangle = \alpha|0\rangle + \beta|1\rangle \tag{1}$$

where $|0\rangle$ and $|1\rangle$ represent the classical bit values 0 and 1, respectively, with α and β complex numbers such that

$$||\alpha||^2 + ||\beta||^2 = 1 \tag{2}$$

α^2 and β^2 are the probability values of the qubit in states 0 and 1, respectively. In classical quantum computing, a quantum register with n qubits can represent 2^n different values. However, when considering the "measure," the superposition is demolished and one single value becomes accessible for use. The exponential growth of the state space with the number of particles recommended a possible exponential speed-up of computation on quantum computers vis-a-vis classical computers. The exponential increase of the state space with particle number suggested a possible exponential speed-up of processing on quantum computers in comparison to classical computers.

(i) **Quantum Computing in Cloud-Dew for Vehicular Network**: The classical computing system faces performance deterioration and quality of service (QoS) degradation problems, especially in IoT-based scenarios under disrupted Internet services in vehicular networks. Thus existing caching strategies need to be enhanced to augment the cache hit ratio and manage the limited storage to accelerate content deliveries. Alternatively, quantum computing (QC) is a prospect of more or less every typical computing problem. Hence, QC-enabled cache-based cloud-dew vehicular networks may solve the above-stated problem. An architecture is given in Figs. 4 and 5. According to Fig. 5, the local cache and neighboring cache may be built through quantum memory in vehicular networks, which may access and operate the large local data very smoothly in a reasonable time. Another research direction may design the algorithm under QC parameters. We have limited here, but this is an undiscovered research field and can perform under cloud-dew computing architecture.

(ii) **Quantum-inspired Computing in Cloud-Dew for Vehicular Network**: Some recent study [11, 41], shows that quantum-inspired approaches have some power

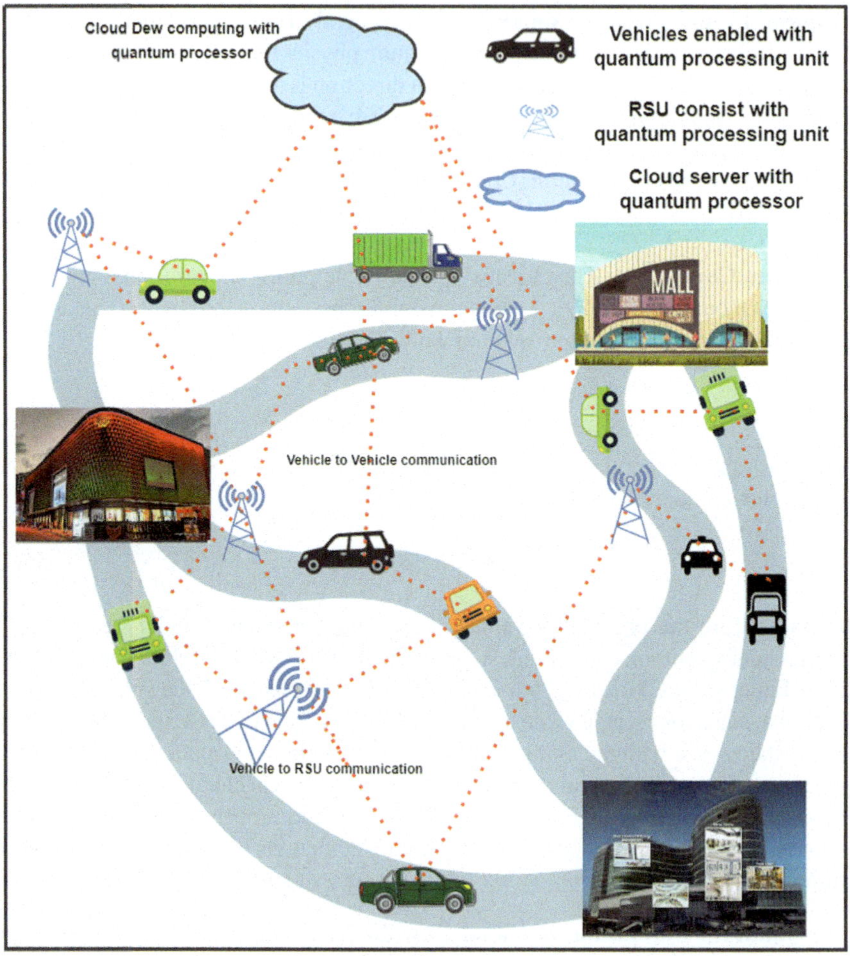

Fig. 4 Vehicular networks under quantum dew computing

to enhance the classical computing systems. Cloud-dew architecture in vehicular networks should perform better than the present one. Since in urban areas have under developing mode so in minor disrupted communications system can be handled through quantum-inspired approaches without massive changes of the devices like RSU, Cloud server, smart vehicles, etc. Dew computing has real challenge to access, control, and process the enormous information in real time. This challenge can be mitigated through quantum-inspired dew computing. A Fig. 5 for both QC and quantum-inspired cloud-dew architecture in vehicular network are shown.

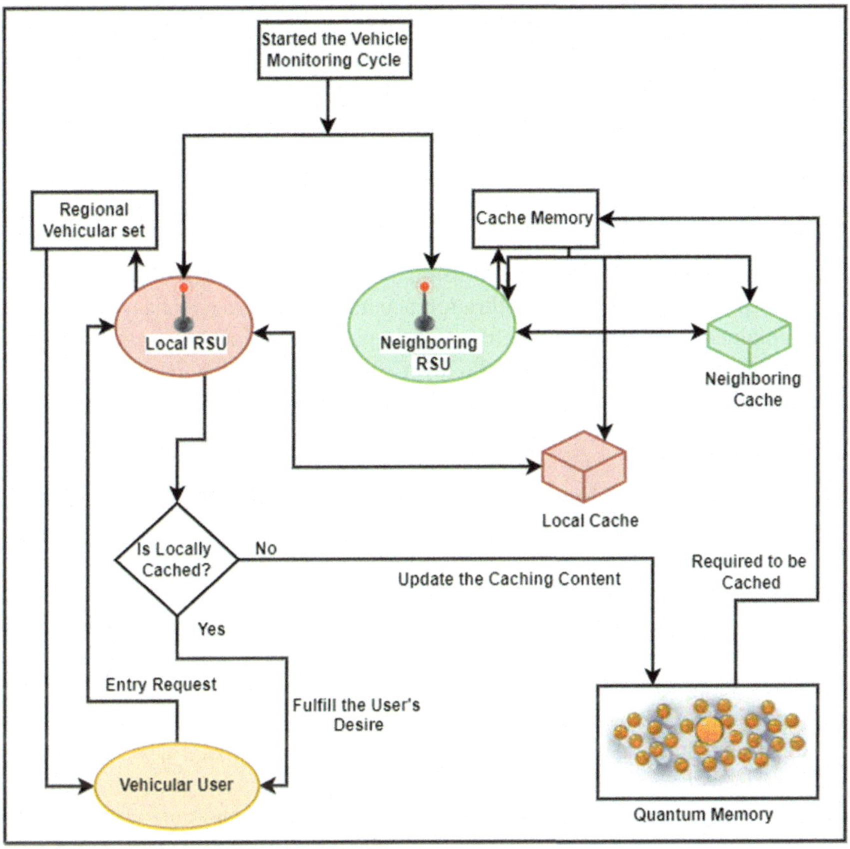

Fig. 5 Architecture of quantum vehicular dew computing

11 Some Proposed Future Application Areas

Dew Computing-based Vehicular Routing in Disaster Management: During calamities, communication is crucial for traveling from one place to another place via vehicle. Based on the dew computing concepts, we communicate through the V2V via messaging protocol that is resistant to the degraded network conditions brought on by disasters. This communications protocol aims to force messengers to adapt to both typical and terrible situations. The client-server design facilitates effective communication under normal circumstances; in the event of an emergency, messengers will make full use of P2P and other available communication channels to ensure that crucial information may be communicated whenever it is possible [55].

Dew Computing-based Intelligent Vehicular Intrusion Detection: In a widely spread IoIs, Edge of Things (EoT) facilitates the seamless transmission of services, storage, and data processing from the cloud layer to edge devices (IoT) ecosystems (e.g., Industrial systems). The EoT paradigm's dispersed privacy and security concerns are brought up by this shift. In order to defend the underlying resources from intruders, intrusion detection systems (IDSs) are built in EoT ecosystems. The present IDSs can't, however, prevent false data transfer through inter-vehicular communication, which greatly reduces the time, cost, and inconvenience of route selection of a vehicle using IDSs. In that scenario, Dew computing as a service (DaaS) for EoT ecosystems enables intelligent intrusion detection. An intelligent routing filtration system in DaaS is designed using a deep learning-based classifier. By utilizing deep belief networks, the filtration accuracy in this mechanism is increased (or maintained). Offloading EoT chores in the past have involved using cloud-based approaches, which adds to the middle layer's workload and lengthens communication delays. This section present the basic computing features that were employed in the creation of the intelligent vehicular routing system.

Dew Computing-based Smart Car Parking: Knowing the status of parking spaces in real time in the smart parking system helps address a lot of issues. According to the current trend, a well-implemented cloud image recognition system can satisfy this need, provided that the training and adjustment of the recognition model have been given to the cloud service provider. As a result, the system is now more scalable and economical, and the business owner is also relieved of the strain of managing sophisticated systems. The cloud service has a restriction on its reliance on Internet connections. So, for the smart parking system, we suggest a novel cloud-dew architecture. The benefits of this architecture include improved dew computing's micro-service characteristic, function independence without Internet, automatic service proxy to promote horizontal and vertical scalability, and commitment to the development of smart parking systems [60]. Many applications that are comparable to the concept of dew computing have been around for a long time before the concept was proposed. The authors proposed a dedicated dew computing architecture for the smart parking system without taking into account the broader definition because the theoretical foundation of dew computing is still being developed [18, 43]. The entire cloud-dew architecture is set up as a hybrid on-premises/hosted cloud environment. Additionally, it can connect the cloud of a smart city to create a communal cloud alliance of smart parking systems. The key elements as per following: Private clouds often fall into two categories: internally hosted (or on-premise hardware) solutions and externally hosted cloud solutions. Scaling of features, overall cost, and required level of support are the primary variations. The external cloud can provide the components required for enormous batch processing or long-term persistence, while the on-premise one can give the components required for performance-critical missions. As a result, the client-side of the smart parking system and dew computing system are located on the on-premise private cloud, while the server side is located on the external cloud. The smart parking system's client-side system handles communication with the server side. It is in charge of network administration and parking lot hardware. However, a cloud image recognition service is used by default when performing the picture recognition function. Normal operation of a local dew service

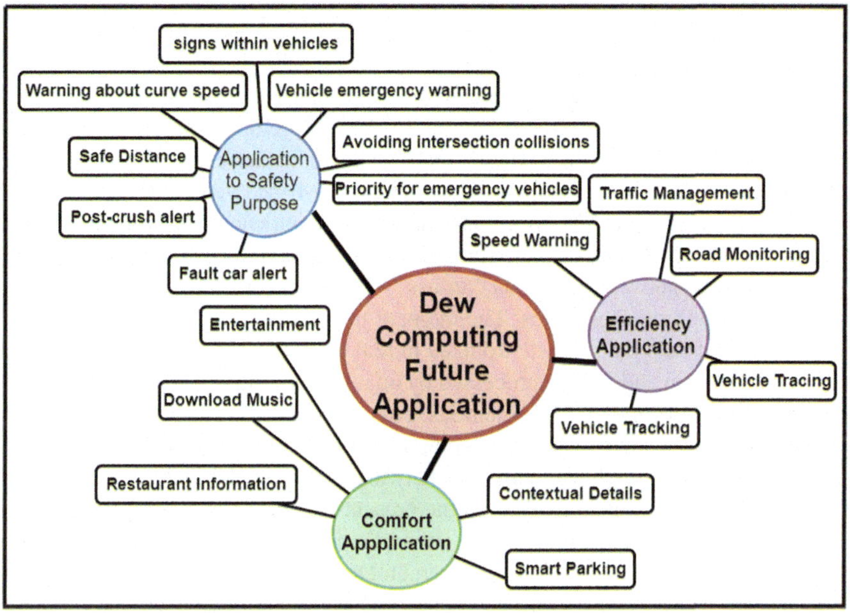

Fig. 6 Some future scope of dew computing in vehicular network

as a cloud image recognition proxy when the Internet is down. Applications include parking spot detection, licence plate identification, and visual surveillance. When the system is deployed into the hosted cloud, certain computing operations must remain local to meet real-time requirements. The smart parking system's server side gathers data on parking lot operation transactions for management analysis and data backup. The private cloud service or API is made available for the client-side system to use. Additionally, it functions as an administration portal for managers. It can provide parking services to consumers via the web interface by merging with the public cloud. Services for object detection in videos, celebrity face recognition, and speech-to-text conversion are all offered by cloud service providers (CSP). Some of these vendors even provide cutting-edge computer platforms (ex: AI Platform as a Service; AI Paas). Some parking systems that are currently in use have started to assemble their image recognition systems utilizing third-party AI cloud services rather than employing a customized machine-bundle software package. A smart city's efficiency may be increased by using the cloud. Stakeholders and governments can use data for storage, analysis, and sharing in the Smart Parking cloud alliance field to take the best possible decisions. The various facets of the ecosystem for the Internet of Vehicles are discussed in this study, along with some future scope that is given in Fig. 6.

12 Conclusion

The intelligent transportation system is big challenges for the industry and government in terms of sustainability. It can be difficult to predict a path when there is a heavy traffic or other unexpected happenings on the road. Currently, routing strategies are carried out using cloud infrastructure. Thus dew-based vehicular networks be a remedy for that system. Also dew-cache architecture may new research paradigm in terms of limited Internet connectivity which is reflected in the present investigation. Vehicles in bad network locations can use V2V to transmit the most recent high-definition map to other vehicles. The moment that vehicle connects to the Internet, high-definition map data is synchronized in real time with the cloud server through dew-cache system and will be processed by quantum computing. The safety of automobiles can be ensured by these methods. There are various research methodologies in vehicular routing. We invoke dew computing under cache in our methodology. In most of the urban area have facing the network is not available everywhere and it is not stable though it is available. So the present investigation gives some research directions over dew-cache protocol for vehicular networks using routing mechanism that can provide service without the Internet. Dew-cache architecture improves the vehicular network system's ability to scale both horizontally and vertically.

References

1. Ahmed, Z.E., Hasan, M.K., Saeed, R.A., Hassan, R., Islam, S., Mokhtar, R.A., Khan, S., Akhtaruzzaman, M.: Optimizing energy consumption for cloud internet of things. Front. Phys. **8**, 358 (2020)
2. Amadeo, M., Campolo, C., Molinaro, A.: Priority-based content delivery in the internet of vehicles through named data networking. J. Sens. Actuator Netw. **5**(4), 17 (2016)
3. Alrawais, A., Alhothaily, A., et al.: Fog computing for the internet of things: security and privacy issues. IEEE Internet Comput. **21**(2), 34–42 (2017)
4. Behbehani, F.S., El-Gorashi, T.E., Elmirghani, J.M.: Optimized processing placement over a vehicular cloud. IEEE Access **10**, 41411–41428 (2022)
5. Bhatia, T., Verma, A.K.: Data security in mobile cloud computing paradigm: a survey, taxonomy and open research issues. J. Supercomput. **73**(6), 2558–2631 (2017)
6. Chai, S., Lau, V.K.N.: Online trajectory and radio resource optimization of cache-enabled UAV wireless networks with content and energy recharging. IEEE Trans. Signal Process. **68**, 1286–1299 (2020)
7. Chen, C., Xiang, H., Qiu, T., Wang, C., Zhou, Y., Chang, V.: A rear-end collision prediction scheme based on deep learning in the internet of vehicles. J. Parall. Distrib. Comput. **117**, 192–204 (2018)
8. Chen, M., Mozaffari, M., Saad, W., Yin, C., Debbah, M., Hong, C.S.: Caching in the sky: proactive deployment of cache-enabled unmanned aerial vehicles for optimized quality-of-experience. IEEE J. Sel. Areas Commun. **35**(5), 1046–1061 (2017)
9. Chen, Z., Kountouris, M.: D2D caching versus small cell caching: where to cache content in a wireless network? In: 2016 IEEE 17th International Workshop on Signal Processing Advances in Wireless Communications (SPAWC), pp. 1–6. IEEE (2016)
10. Dai, Y., Xu, D., Maharjan, S., Qiao, G., Zhang, Y.: Artificial intelligence empowered edge computing and caching for internet of vehicles. IEEE Wirel. Commun. **26**(3), 12–18 (2019)

11. Das, M., Roy, A., Maity, S., Kar, S.: A quantum-inspired ant colony optimization for solving a sustainable four-dimensional traveling salesman problem under type-2 fuzzy variable. Adv. Eng. Inform. **55**, 101816 (2023)
12. Dijkstra, E.W.: A note on two problems in connexion with graphs. In: Edsger Wybe Dijkstra: his Life, Work, and Legacy, pp. 287–290. Springer (2022)
13. Fu, L.: An adaptive routing algorithm for in-vehicle route guidance systems with real-time information. Transp. Res. Part B: Methodol. **35**(8), 749–765 (2001)
14. Goudarzi, F., Asgari, H., Al-Raweshidy, H.S.: Traffic-aware VANET routing for city environments-a protocol based on ant colony optimization. IEEE Syst. J. **13**(1), 571–581 (2018)
15. Guo, Y., Yang, Q., Yu, F.R., Leung, V.C.: Cache-enabled adaptive video streaming over vehicular networks: a dynamic approach. IEEE Trans. Veh. Technol. **67**(6), 5445–5459 (2018)
16. Gusev, M.: A dew computing solution for IoT streaming devices. In: 2017 40th International Convention on Information and Communication Technology, Electronics and Microelectronics (MIPRO), pp. 387–392. IEEE (2017)
17. Gusev, M.: What makes dew computing more than edge computing for internet of things. In: 2021 IEEE 45th Annual Computers, Software, and Applications Conference (COMPSAC), pp. 1795–1800. IEEE (2021)
18. Gusev, M., Wang, Y.: Formal description of dew computing. In: Proceedings of The 3rd International Workshop on Dew Computing, pp. 8–13 (2018)
19. Gushev, M.: Dew computing architecture for cyber-physical systems and IoT. Internet Things **11**, 100186 (2020)
20. Han, K.-H., Kim, J.-H.: Quantum-inspired evolutionary algorithm for a class of combinatorial optimization. IEEE Trans. Evol. Comput. **6**(6), 580–593 (2002)
21. Fida Hasan, K., Kaur, T., Mhedi Hasan, M., Feng, Y.: Cognitive internet of vehicles: motivation, layered architecture and security issues. In: 2019 International Conference on Sustainable Technologies for Industry 4.0 (STI), pp. 1–6. IEEE (2019)
22. Hasan, M.K., Ismail, A.F., Abdalla, A.H., Ramli, H.A., Hashim, W., Islam, S.: Throughput maximization for the cross-tier interference in heterogeneous network. Adv. Sci. Lett. **22**(10), 2785–2789 (2016)
23. Hindustantimes.: FDI rules in food retail (2022)
24. Hou, L., Lei, L., Zheng, K., Wang, X.: A q-learning-based proactive caching strategy for non-safety related services in vehicular networks. IEEE Internet Things J. **6**(3), 4512–4520 (2018)
25. Javed, M.A., Zeadally, S.: Ai-empowered content caching in vehicular edge computing: opportunities and challenges. IEEE Netw. **35**(3), 109–115 (2021)
26. Karp, B., Kung, H.T.: GPSR: greedy perimeter stateless routing for wireless networks. In: Proceedings of the 6th Annual International Conference on Mobile Computing and Networking, pp. 243–254 (2000)
27. Li, H., Zhang, J., Zhao, L.: Vehicular high-definition maps cache based on dew computing. In: 2022 9th International Conference on Dependable Systems and Their Applications (DSA), pp. 1067–1068. IEEE (2022)
28. Lochert, C., Hartenstein, H., Tian, J., Fussler, H., Hermann, D., Mauve, M.: A routing strategy for vehicular ad hoc networks in city environments. In: IEEE IV2003 Intelligent Vehicles Symposium. Proceedings (Cat. No. 03TH8683), pp. 156–161. IEEE (2003)
29. Lv, Z., Chen, D., Wang, Q.: Diversified technologies in internet of vehicles under intelligent edge computing. IEEE Trans. Intell. Transp. Syst. **22**(4), 2048–2059 (2020)
30. Mane, T.S., Agrawal, H.: Cloud-fog-dew architecture for refined driving assistance: the complete service computing ecosystem. In: 2017 IEEE 17th International Conference on Ubiquitous Wireless Broadband (ICUWB), pp. 1–7. IEEE (2017)
31. Marinescu, D.C.: Cloud Computing: theory and Practice. Morgan Kaufmann (2022)
32. Martuscelli, G., Boukerche, A., Foschini, L., Bellavista, P.: V2v protocols for traffic congestion discovery along routes of interest in VANETS: a quantitative study. Wirel. Commun. Mob. Comput. **16**(17), 2907–2923 (2016)
33. Meng, Z., Guan, Z., Wu, Z., Li, A, Chen, Z.: Security enhanced internet of vehicles with cloud-fog-dew computing. ZTE Commun. **15**(S2), 47–51 (2020)

34. Mohammed, C.M., Zeebaree, S.R., et al.: Sufficient comparison among cloud computing services: IaaS, PaaS, and SaaS: a review. Int. J. Sci. Bus. **5**(2), 17–30 (2021)
35. Mulay, P., Kadlag, S., Shirodkar, R.: Smart supply-chain management learning system for homeopathy. Indian J. Publ. Health Res. Dev. **8**(4) (2017)
36. Nguyen, T.N.: The challenges in ml-based security for SDN. In: 2018 2nd Cyber Security in Networking Conference (CSNet), pp. 1–9. IEEE (2018)
37. Ning, Z., Huang, J., Wang, X., Rodrigues, J.J.P.C., Guo, L.: Mobile edge computing-enabled internet of vehicles: toward energy-efficient scheduling. IEEE Netw. **33**(5), 198–205 (2019)
38. Pan, Y., Thulasiraman, p., Wang, Y.: Overview of cloudlet, fog computing, edge computing, and dew computing. In: Proceedings of The 3rd International Workshop on Dew Computing, pp. 20–23 (2018)
39. Parvez, I., Rahmati, A., Guvenc, I., Sarwat, A.I., Dai, H.: A survey on low latency towards 5g: ran, core network and caching solutions. IEEE Commun. Surv. Tutor. **20**(4), 3098–3130 (2018)
40. Patel, H.M., Chaudhari, R.R., Prajapati, K.R., Patel, A.A.: The interdependent part of cloud computing: dew computing. In: Intelligent Communication and Computational Technologies, pp. 345–355. Springer (2018)
41. Pradhan, K., Basu, S., Thakur, K., Maity, S., Maiti, M.: Imprecise modified solid green traveling purchaser problem for substitute items using quantum-inspired genetic algorithm. Comput. Ind. Eng. **147**, 106578 (2020)
42. Ramesh, T.: Traveling purchaser problem. Opsearch **18**(1–3), 78–91 (1981)
43. Ray, P.P.: An introduction to dew computing: definition, concept and implications. IEEE Access **6**, 723–737 (2017)
44. Rindos, A., Wang, Y.: Dew computing: the complementary piece of cloud computing. In: 2016 IEEE International Conferences on Big Data and Cloud Computing (BDCloud), Social Computing and Networking (SocialCom), Sustainable Computing and Communications (SustainCom) (BDCloud-SocialCom-SustainCom), pp. 15–20. IEEE (2016)
45. Skala, K., Davidovic, D., Afgan, E., Sovic, I., Sojat, Z.: Scalable distributed computing hierarchy: cloud, fog and dew computing. Open J. Cloud Comput. (OJCC) **2**(1), 16–24 (2015)
46. Šojat, Z., Skala, K.: Views on the role and importance of dew computing in the service and control technology. In: 2016 39th International Convention on Information and Communication Technology, Electronics and Microelectronics (MIPRO), pp. 164–168. IEEE (2016)
47. Suwansrikham, P., Kun, S., Hayat, S., Jackson, J.: Dew computing and asymmetric security framework for big data file sharing. Information **11**(6), 303 (2020)
48. Talbi, H., Draa, A., Batouche, M.: A new quantum-inspired genetic algorithm for solving the travelling salesman problem. In: 2004 IEEE International Conference on Industrial Technology, 2004. IEEE ICIT'04, vol. 3, pp. 1192–1197. IEEE (2004)
49. Tefera, G., She, K., Shelke, M., Ahmed, A.: Decentralized adaptive resource-aware computation offloading & caching for multi-access edge computing networks. Sustain. Comput.: Inform. Syst. **30**, 100555 (2021)
50. Toth, P., Vigo, D.: Vehicle Routing: problems, Methods, and Applications. SIAM (2014)
51. Wang, H., Liu, T., Kim, B., Lin, C.-W., Shiraishi, S., Xie, J., Han, Z.: Architectural design alternatives based on cloud/edge/fog computing for connected vehicles. IEEE Commun. Surv. Tutor. **22**(4), 2349–2377 (2020)
52. Wang, J., Osagie, E., Thulasiraman, p., Thulasiram, R.K.: Hopnet: a hybrid ant colony optimization routing algorithm for mobile ad hoc network. Ad Hoc Netw. **7**(4), 690–705 (2009)
53. Wang, X., Leng, S., Yang, K.: Social-aware edge caching in fog radio access networks. IEEE Access **5**, 8492–8501 (2017)
54. Wang, Y.: Definition and categorization of dew computing. Open J. Cloud Comput. (OJCC) **3**(1), 1–7 (2016)
55. Wang, Y.: A disaster-resilient messaging protocol based on dew computing. In: 2020 43rd International Convention on Information, Communication and Electronic Technology (MIPRO), pp. 1922–1926. IEEE (2020)

56. Wang, Y., Pan, Y.: Cloud-dew architecture: realizing the potential of distributed database systems in unreliable networks. In: Proceedings of the International Conference on Parallel and Distributed Processing Techniques and Applications (PDPTA), pp. 85. The Steering Committee of The World Congress in Computer Science, Computer (2015)
57. Wang, Z., Liu, Y.Y., Thulasiraman, P., Thulasiram, R.K.: Ant brood clustering on intel XEON multi-core: Challenges and strategies. In: 2018 IEEE Symposium Series on Computational Intelligence (SSCI), pp. 1126–1233. IEEE (2018)
58. Wu, H., Lyu, F., Zhou, C., Chen, J., Wang, L., Shen, X.: Optimal UAV caching and trajectory in aerial-assisted vehicular networks: a learning-based approach. IEEE J. Sel. Areas Commun. **38**(12), 2783–2797 (2020)
59. Yao, L., Chen, A., Deng, J., Wang, J., Guowei, W.: A cooperative caching scheme based on mobility prediction in vehicular content centric networks. IEEE Trans. Veh. Technol. **67**(6), 5435–5444 (2017)
60. Yu, Y.-C.: A dew computing architecture for smart parking system with cloud image recognition service. In: 2021 IEEE 45th Annual Computers, Software, and Applications Conference (COMPSAC), pp. 1805–1809. IEEE (2021)
61. Zeadally, S., Javed, M.A., Hamida, E.B.: Vehicular communications for its: standardization and challenges. IEEE Commun. Stand. Mag. **4**(1), 11–17 (2020)
62. Zhang, K., Leng, S., He, Y., Maharjan, S., Zhang, Y.: Cooperative content caching in 5g networks with mobile edge computing. IEEE Wirel. Commun. **25**(3), 80–87 (2018)
63. Zhang, K., Zhu, Y., Leng, S., He, Y., Maharjan, S., Zhang, Y.: Deep learning empowered task offloading for mobile edge computing in urban informatics. IEEE Internet Things J. **6**(5), 7635–7647 (2019)
64. Zhao, Y., Li, Y., Zhang, X., Geng, G., Zhang, W., Sun, Y.: A survey of networking applications applying the software defined networking concept based on machine learning. IEEE Access **7**, 95397–95417 (2019)

Dew-Computing in Future Telerobotic Applications: An Exploration

Abhijan Bhattacharyya, Ashis Sau, and Madhurima Ganguly

1 Till the Dew Drops: Let's Understand Telerobotics

1.1 What is Telerobotics?

Telerobotics, as suggested by the prefix "*Tele*," is a semi-automatic robotic system which has a human operator in-loop maneuvering the robot remotely over the Internet. *Telerobotics* can be broadly classified into *Telepresence* and *Teleoperation*.

1.1.1 Telepresence and Teleoperation

A telepresence robot is a remote-operated, wheeled device with a video conferencing gadget that can be driven around from remote locations. An operator using his/her computer or personal digital assistant (PDA) can drive this robot at a remote location, see things and speak out to people through the camera and audio devices installed on the robot. Thus, the robot acts like a physical avatar of the remote operator. This type of a mobile video conferencing device enables telecommuters and remote workers to feel more connected to their colleagues by giving them a physical presence in proximity to where they would like to be in person (Fig. 1).

If a Telerobot is equipped with actuators (along with the wheel system to move around) and the operator can actuate remotely in real-time through the robot, then

A. Bhattacharyya (✉) · A. Sau · M. Ganguly
TCS Research, Tata Consultancy Services, Kolkata, India
e-mail: abhijan.bhattacharyya@tcs.com

A. Sau
e-mail: ashis.sau@tcs.com

M. Ganguly
e-mail: ganguly.madhurima@tcs.com

D. De and S. Roy (eds.), *Dew Computing*, Internet of Things,
https://doi.org/10.1007/978-981-99-4590-0_10

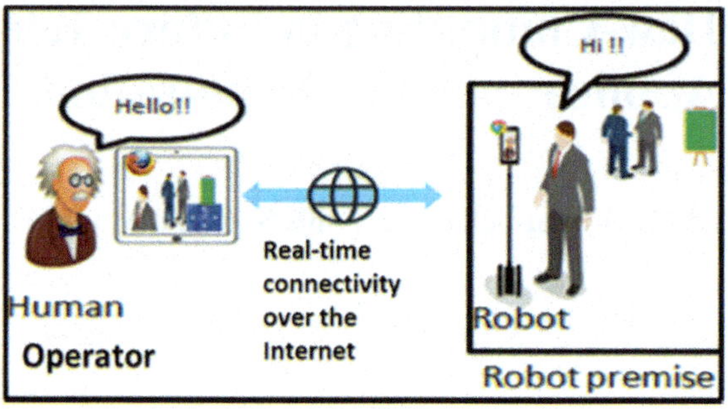

Fig. 1 Illustrating the concept of telepresence

such systems are termed as *Teleoperation* systems (Fig. 2). If the Teleoperation robot too has a display through which the robot-side participants may engage in a multimedia chat during the teleoperation session, then such Teleoperation systems become an extension of Telepresence System with additional actuation capability (Fig. 3).

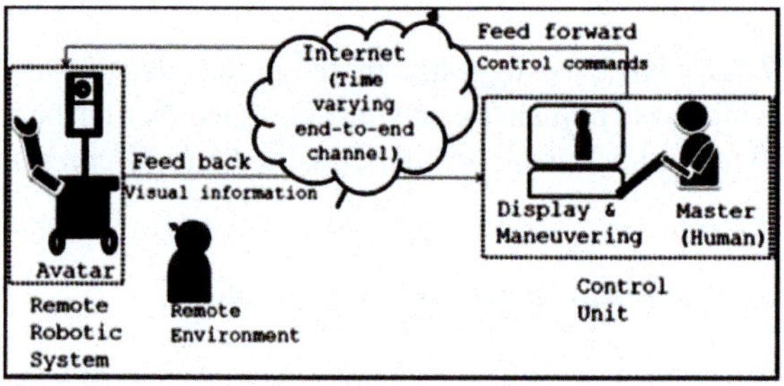

Fig. 2 Illustrating the concept of teleoperation

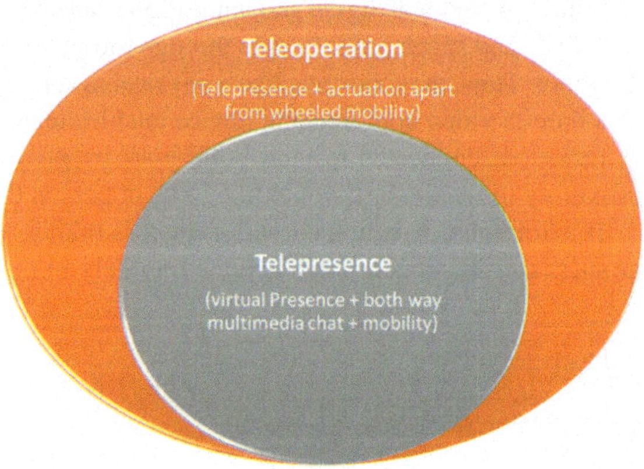

Fig. 3 Teleoperation as an extension of telepresence

1.2 Business Traction Behind Telerobotics-Based Solutions

Demand for such systems is badly felt during situations like pandemic, war, etc. when the civilization struggles to meet the diverging requirements of continuing "social" cooperation in one hand, while maintaining physical "distancing" on the other. The divergence arises due to the impending restrictions on physically traveling or gathering. Besides, even under generic business situations, there are several aspects of such applications that influence the return of investment (RoI) in terms of increased scalability of business operations without jeopardizing desired efficiency, rather than enhancing it (Fig. 4).

It is needless to mention that the above-described paradigm of robotics is appealing to a pervasive spectrum of applications cutting across a wide range of business verticals. Table 1 describes how different futuristic applications across different business domains are enabled through Telerobotics.

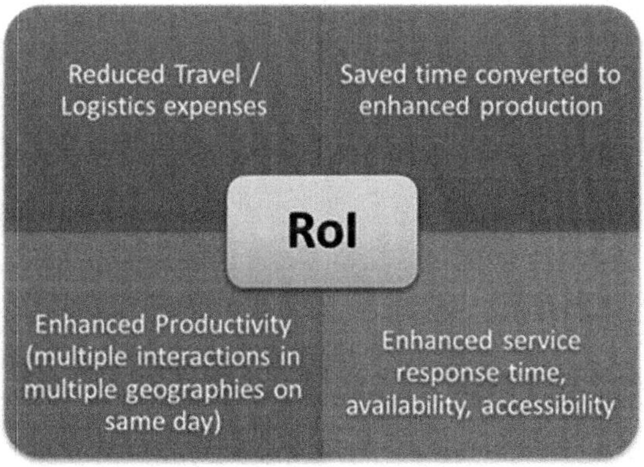

Fig. 4 The general aspects influencing the RoI for a telepresence/teleoperation deployment

Table 1 Fitment of teledrive across business verticals

Business domain	Example application
MICE—Meetings, incentives, conferences and exhibitions	Remotely attend meetings, conferences, exhibitions, etc.
Hospitality	Remote escort/May I help you
Telemedicine	Remote consultation/visit to isolation centers/ remote operation
Education	Remote teaching/lab visits and proctoring
Retail	Remote visit to stores
Insurance	Remote inspection of damage
Surveillance and security	Remote surveillance of premises and property
Manufacturing	Remote live inspection/observation/mentoring of ongoing manufacturing processes

1.3 Some Existing Telerobotic Solutions

There are several off-the-shelve proposals/products for Telerobotics applications. Some of these are for generic applications and some are for dedicated needs. The surveyed offerings are shown in Fig. 5 (Table 2).

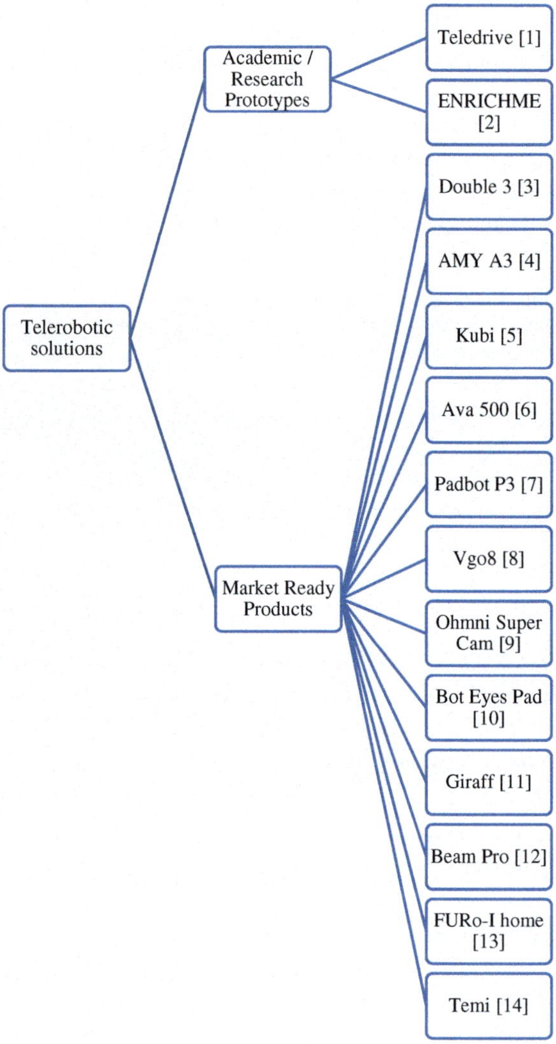

Fig. 5 Different existing telerobotics solutions/ prototypes

Table 2 Feature-wise comparison for different available/proposed solutions

Features	Teledrive (experimental prototype)	Double 3 [1]	ENRICHME [2]	Amy A3 [3]	Kuby [4]	Ava 500 [5]	PadBot P3 [6]	Vgo [7]	Ohmni supercam [8]	BotEyes pad [9]	Giraff [10]	Beam pro [11]	FURo-I home [12]	Temi [13]
Browser GUI	Y	Y	n	n	n	n	n	Y	Y	Y	n	n	n	n
Manual navigation	Y	Y	Y	Y	n	Y	Y	Y	Y	Y	Y	Y	Y	n
AR goal-based navigation	Y	Y	n	Y	n	n	n	n	n	n	n	n	n	Y
Map-based navigation	Y	n	Y	n	n	Y	n	n	n	n	n	n	n	Y
Speech-based navigation	Y	n	n	n	n	n	n	n	n	n	n	n	n	Y
Information mashup	Y	n	n	n	n	n	n	n	n	n	n	n	n	n
Speaker localization	Y	n	n	n	Y	n	n	n	n	n	n	n	n	n
Face identification	n	n	Y	Y	n	n	Y	n	n	n	n	n	Y	Y
Multi-party federated control	Y	n	n	n	n	n	n	n	n	n	n	n	n	n
Dialog-based disambiguation	Y	n	n	n	n	n	n	n	n	n	n	n	n	n

1.4 Understanding the Concept of Multi-party Telerobotics Session

In the above comparative study, only [14] supports the feature for "multiparty federated" robot control. The operator to robot connection works on a "logical" peer-to-peer (P2P) topology. This limits the number of users who may remotely join a session. Thus, most existing Telepresence systems are not conducive to remote collaboration. However, there may be practical situations in which more than one human user may require to connect simultaneously in a telerobotic session to collaboratively perform a combination of tasks remotely through the robot. Along with usual real-time multi-user multimedia conference, each of the human users may have to execute a part of the combination of tasks. So, for a desired duration, each user may assume the role of an operator. During this time span, the concerned user will have exclusive control on the robot. Thus, the robot will act as an Avatar of that concerned user during the given period. During this time, the other human users remain in the session as observers. They can have multimedia chat and observe what is happening at the robot-end, but they cannot have the privilege to maneuver/control the robot. Once done with the required task, the concerned user acting as the operator may release the control on the robot. Another user, who might be interested next to control the robot, may assume the role of an operator and, similarly, gain exclusive control on the robot. Thus *federated-yet-exclusive* robotic operations are performed in a multiparty scenario. Sau et al. [15] elaborate on the communication architecture to enable such system with standardized technologies.

For example, let us consider a telemedicine scenario (Fig. 6). In a typical situation, a specialist doctor and a medical assistant may simultaneously join a telerobotic session from two distant locations to provide consultation to a patient in an isolation ward. Let's assume that the medical assistant is more conversant with the geography of the patient's premise. So, to start with, the medical assistant takes the role of an operator and remotely maneuvers the robot and navigates the robot to the patient. The doctor remains an observer. After this the assistant relinquishes her control, and the doctor becomes the operator by taking control on the robot to continue further interactions with the patient. Now, the assistant continues as an observer in the multimedia conference till the situation needs her to maneuver the robot again and the doctor relinquishes the control on the robot.

Again, in a typical test-witness/quality control and verification stage of a large manufacturing process distributed across places, multiple inspectors from multiple authorities may need to join remotely and control the robot at will. For example, in typical aviation industry scenario, multiple inspectors from aviation authority and certification authority may need to join a session to check the results of the equipment under test as shown in Fig. 7.

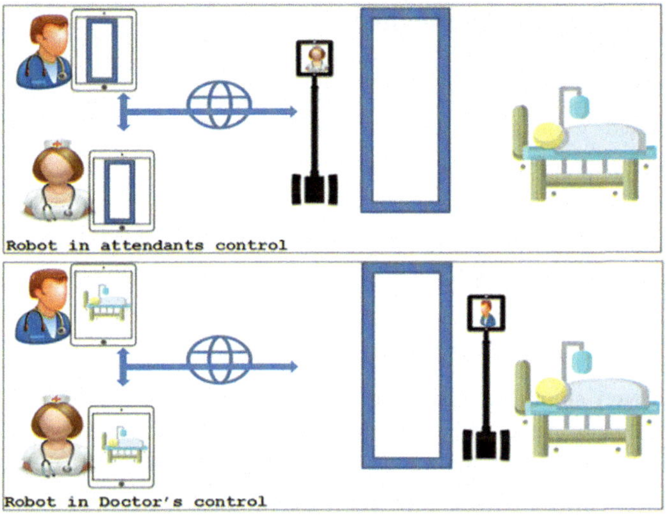

Fig. 6 Depicting the need for multiparty sessions with federated-yet-exclusive robot control [15]

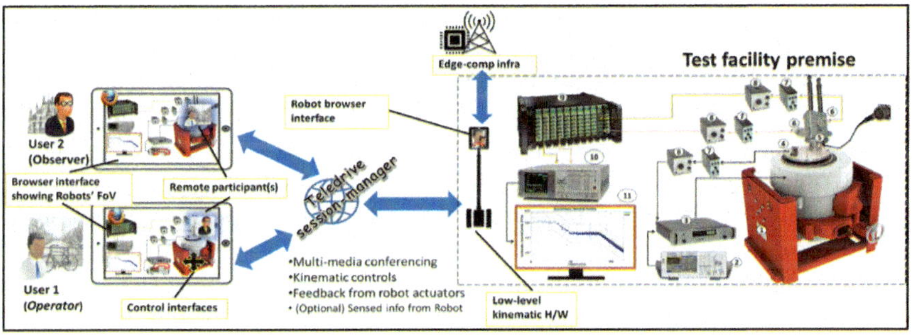

Fig. 7 Functional concept of Teledrive in a multi-user remote inspection scenario with multiple remote inspectors. Note that the users may swap the *operator* and *observer* roles as per situational demands. The illustration exemplifies a typical situation of vibration testing of an equipment under test (EUT). The numerical labels indicate the followings: 1 → Electrodynamic shaker, 2 → Signal generator, 3 → Power amplifier, 4 → Control accelerometer, 5 → Equipment Under Test, 6 → Output accelerometer, 7 → Signal conditioning amplifier, 8 → Tunable bandpass filter, 9 → Data acquisition system, 10 → Dynamic signal analyzer, 11 → The vibration testing output

2 The Prevalent Technologies in Telerobotics

With the above background on what is Telerobotics and the exemplary solutions, let us now discuss the technologies that enable Telerobotics solutions as we see them now. However, before that we need to understand the technical challenges that need to be solved and the design constraints.

2.1 Challenges in Telerobotics

In semi-automatic Telerobotics systems, the key technical domains involved can be categorized as:

1. Communication between the operator and the robot
2. In situ intelligence at the robot
3. The robotic mechatronics and the interfacing between the H/W and S/W
4. User interface design curtailed according to the operator-side and robot-side requirements.

The key constraints/challenges are:

1. **The latency should be extremely low**: Since the commands from the operator depend on the visual feedback from the robot, there must be extremely low delay in the end-to-end communication path between the robot and the operator. In ideal situation the typical delay may range from few hundreds of milliseconds to few tens of milliseconds and can only be achieved in the Tactile Internet era promised in future [16]. Also, given the multitude of network traffic-types with different quality of service (QoS) requirements, a synergy needs to be maintained among the traffics exchanged. For example, usually video feedback is considered to tolerate some losses and delay, while kinematic control commands need 100% reliability and extremely low latency. Thus, the QoS demanded is different for these two types. But, the end-users at both sides must have a synergized feeling of: "What is happening at the remote end as a result of my response." Otherwise the overall Quality of Experience (QoE) is going to suffer.
2. **Robots are constrained by computing and storage capacity**: The robots are mobile devices with limited computing and storage capacity. The form factor of the computing unit impacts the structure and balance of the robot. Hence too large computing unit is not recommended. Also, too heavy computing drains energy from the robot battery. Thus, given present state of miniaturization of computing circuitry, if the robot-end intelligence/service needs too heavy computation and a large amount of storage, that becomes infeasible in practical deployment.
3. **Robots are constrained by available energy**: As discussed above, because of mobility, robots are battery powered with limited capacity. The onboard computer draws energy from this battery. Hence, too heavy computation load may in turn cause too quick drainage of power, and the whole system may be practically unusable.
4. **Democratization of user interface at the operator end**: The remote operating user should ideally be able to maneuver the robot through any standard personal digital assistant (PDA) without requiring to install heavy software installation/hardware augmentation. Otherwise, the telerobotic application cannot be democratized. This is key to ensure that the robot can be connected and controlled from anywhere, at any time, by anybody (of course with authorization) just like the way we access resources in the Web. However, it may not be possible to satisfy the above in typical VR-based telerobotic systems [17].

2.2 Popular Technologies in Use

In the light of the above we now discuss the popular technologies, relevant in the context of this chapter, being popularly used in Telerobotics:

- **Cloud-robotics**: Because of the computing and storage constraints, robots have been designed to be cloud-centric for running the decision engines [18]. The cloud also maintains the session with the robot for global participants [1]. However, too much cloud-centricity not only increases the overall latency in getting the decision outcome, but also adds a certain probability of uncertainty in performance and availability of the cloud-service due to the unreliability in the public Internet backhaul.
- **Edge-cloud hybrid system**: To mitigate the uncertainties and latency due to *cloud-only* system, some infrastructures to use an additional Edge close to the robot/operator premise (in most cases the Edge resides in the same network as the robot to avoid routing delays). It is expected that the latency in communicating to the Edge is significantly lower than the Cloud while the Edge node has enough capacity so that a significant amount of computation load due to cognitive decision-making can be offloaded to the Edge. Also, certain amount of data requiring frequent access may be cached in the Edge as well. The scheme is depicted in Fig. 8. However, such Edge based systems increase the capital expenditure for a system. Also, it adds to the operating expenditure for maintaining and managing the Edge infrastructure.
- **Web Browser-based Real-time communication**: To satisfy the need for democratization of use interface Web browser has been a popular option [1]. Again, the system needs a communication mechanism to ensure low latency. To converge these two requirements Web Real-time Communication (WebRTC) [19] has been a de facto choice for both multimedia and command exchange over the

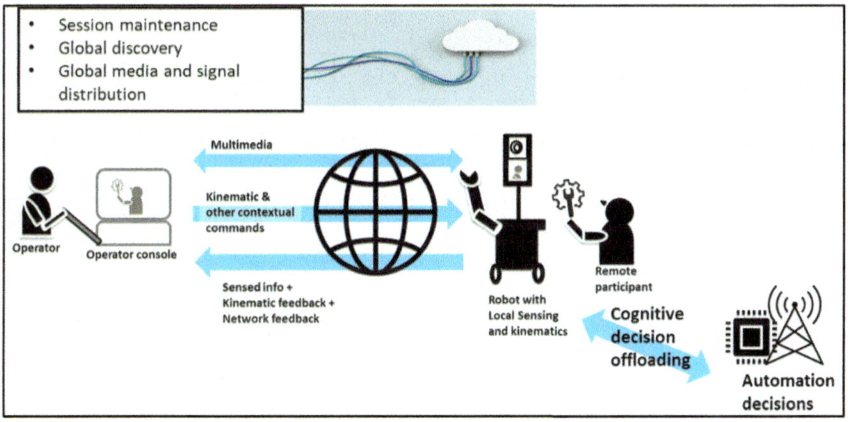

Fig. 8 Edge-cloud hybrid telerobotics

Internet between the operator and the robot for even futuristic robotics-based standardization propositions [20].

2.2.1 A Short Primer on WebRTC

We are now almost set to explore the *dew paradigm* for Telerobotics. However, as a necessary pre-requisite for the forthcoming discussion we describe the WebRTC technology very briefly.

WebRTC is designed as a peer-to-peer (P2P) communication protocol suit for the exchange of multimedia between two endpoints across the Internet. The media channel comprising Secure Real Time Protocol (SRTP) [21] is used to exchange audio/video and any other data (including kinematic control commands and sensory feedbacks in case of robotics) is exchanged over a reliable data channel on Secure Stream Control Transport Protocol (S-SCTP) [22]. WebRTC APIs allow the end-applications to establish end-to-end (E2E) low-latency channel over a P2P association of the robot with the operator (Fig. 9). However, if any one or both robot and operator nodes are behind restrictive NATs then establishing a direct P2P is not possible. In such cases, the P2P connection must be relayed through a special server called TURN (Traversal Using Relays around NAT) server (Fig. 10) [23, 24]. Most telerobotic solutions work in a P2P topology whereby only one Operator connects the Robot. But, for multiparty scenarios all the participating parties, including the robot, need to connect through a central conferencing server residing in the cloud. This effectively leads to multiple P2P connections converging into a star-like topology. This somehow breaks the true sense of P2P that one would like to leverage through WebRTC APIs. The cloud-centric topology is bound to cause delay leading to unpredictable increase in latency while transferring control commands, as well as while exchanging the visual feedback. Both are critical in terms of the Quality of Experience (QoE) for the end-users of telerobotic session.

Fig. 9 The P2P topology for WebRTC

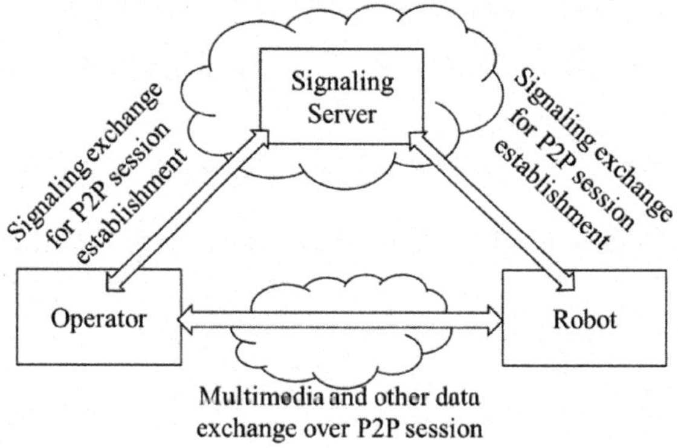

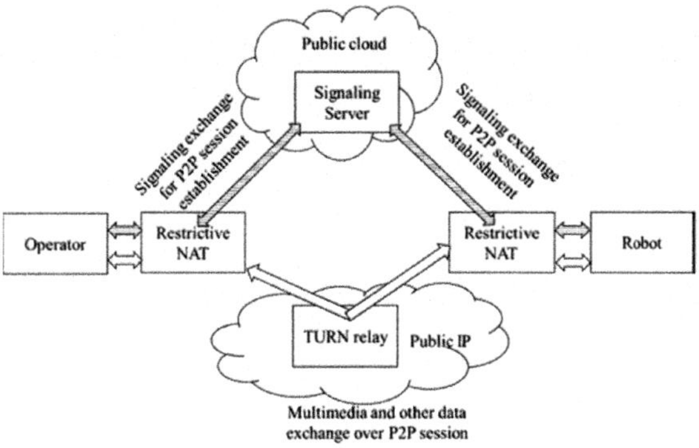

Fig. 10 The P2P topology through TURN relay in case of restrictive NATs

2.3 A Practical Edge-Cloud Based Architecture for Telerobotic Control

Reference [14] provides a practical Edge-Cloud-based hybrid architecture for a multi-user scenario as shown in Fig. 11. They use a *session manager (SM)* and a *media broadcaster (MB)* in the cloud. The signaling server establishes a P2P connection between the MB and each of the peers including the robot. The SM keeps track of state of all the P2P relations over time. Also, [14] uses the protocol described in [15] to create a dedicated on demand P2P channel only for transfer of control commands between an existing operator and the robot. This way only one operator can establish a low-latency path for control command transfer which is much more sensitive to delay than audio/video. Thus a *federated-yet-exclusive multi-user telerobotic system* is achieved in [14]. But it will still have the cloud-centric signaling delay and uncertainty and, also, the delay in multimedia. This is bound to reduce the synergy between the control commands transferred over low-latency P2P path and the visual feedback (showing the effect of the command transferred) transferred via the cloud.

3 A Dew-Based Architecture for Improved Telerobotics Experience

With the above backdrop we would now explore how the *dew* computing concepts can be applied to Telerobotics with an intention to improve the perceived efficiency of the system from an end-user's perspective. We conceive propositions mainly for enterprise scenarios with two broad aspects:

(i) Improving end-application experience
(ii) Improving performance of WebRTC-based infrastructure.

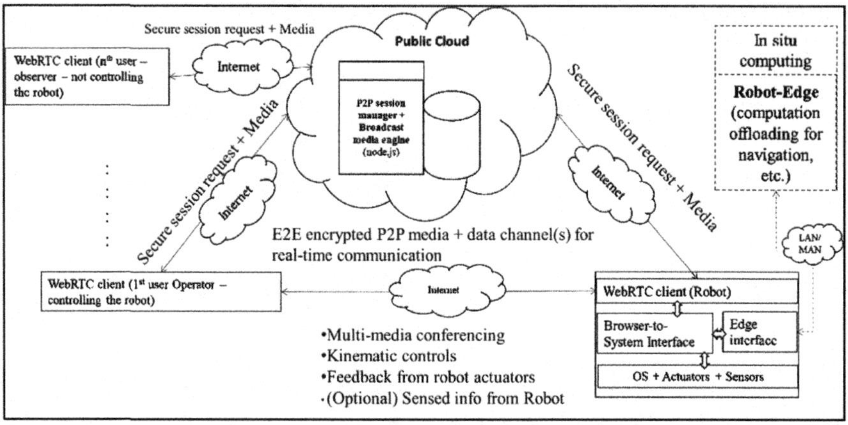

Fig. 11 A practical edge-cloud-based architecture with a unique proposition to create a direct data channel between the robot and the operator user in a multi-user scenario. *Note* The media is still routed through the media broadcaster in the cloud. Edge is used only for local intelligence

3.1 Improving End-Application Experience

As described in the very beginning, Telerobotics can be used for different business verticals. In fact, in some near future, we see an adoption of business models around "*Telerobotics as a Service.*" Under such circumstances an enterprise may employ the same Telerobotics infrastructure in different settings. For example, in a hospital, Telerobotics service may be used in different wards housing different categories of patients. During a session, a remote doctor may be provided with different details in real-time for the patients being examined in a ward. In a cloud-centric system, every-time such queries and related analytics must be routed through the application cloud (this is different from the signaling and media routing through cloud that we discussed in the context of WebRTC) leading to network overhead and delay. So, depending on the context of the session, for which the Telerobot will be used, an instance of application specific services may be deployed on the robot in run time. Thus, an application specific *Dew Service* may be deployed following the *Dew-Computing* concepts described in [25, 26]. The Service will maintain a P2P connection with the application cloud for sync up operations, etc. as shown in Fig. 12. This will also enable close coupling of relevant services residing in the due with the *in-situ* sensing capability of the robot. For example, in the hospital scenario, suppose a telerobot session is conducted in an isolation ward. Then, at the beginning of the session, the robot can offer to use its available resources to institute a *Dew Service* to the application cloud which holds the entire business logic and entire doctor/ patient databases, etc. The application cloud may delegate to the robot some small analytics modules related to the class of disease for which the ward is made of and, along with that, may transfer a snapshot of the cloud database so that only the details of the patients housed in that ward. This way queries from the remote doctor will be locally

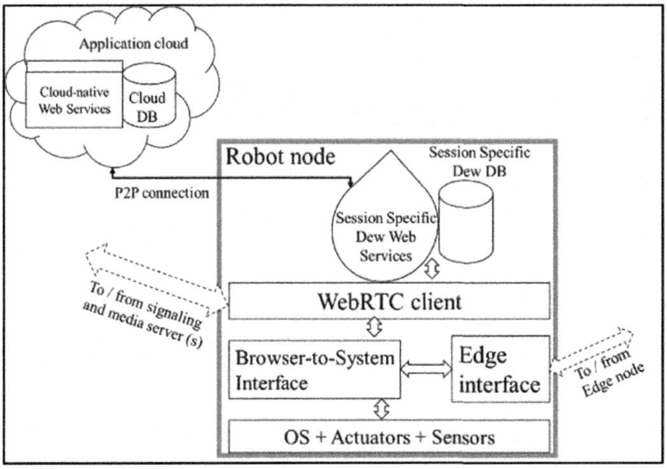

Fig. 12 A representative architectural modification of the robot node to conceive how an application specific Dew service can be deployed for telerobotics as a service business model

resolved at the robot itself without going through the cloud. Also, the local analytics modules may use the robot's sensors (temperature sensors, etc.) and camera feed to deduce valuable insights very fast and feed the doctor. In this case, neither the sensor data needs to travel all the way to the cloud to feed the analytics services and, also, the doctor need not get the analytics outputs from the cloud. The same way the patient may also see the doctor's details from the local dew data base at the robot itself.

We expect that such architectural paradigm would greatly reduce the complexity of distributed S/W design through opportunistic close coupling of the services with the environment in question.

3.2 Improving Performance of WebRTC-Based Infrastructure

As we previously discussed, though WebRTC has become a *de facto* choice for web-based real-time applications including robotics, despite its fundamental P2P topology, it still has latency penalty and uncertainty of experience due to cloud-centricity in the following cases:

- The signaling is cloud-centric
- For a multi-user session, the multimedia is always routed through the cloud.

Previously we have discussed the problems related to these. So, let us now explore how, in an enterprise scenario, we can leverage the *Dew paradigm* for opportunistically improve the perceived experience of the end-users.

We consider that in a typical enterprise scenario the users involved in a session may be classified into two categories: (a) users who are communicating from within the enterprise network and (b) users who are communicating from outside the enterprise network (ex. external stakeholders like clients, vendors, certification authorities, etc.). So the cloud-centricity can be reduced by slicing out a certain responsibilities into the enterprise network as a session specific *Dew Service* as described below.

3.2.1 WebRTC with Cloud-Dew Signaling Service

Assuming that the robot is within an enterprise network, we can consider that the robot may act as a signaling server for the users within the enterprise. Thus, as shown in Fig. 13, the robot may offer to the cloud to act as a dew signaling server through a dedicated URL (say: https://www.mytelerobo.com/dew). This triggers to establish a P2P connection between the cloud signaling server and its dew counterpart on the robot. Over the P2P the cloud server may share the necessary information to instantiate the dew server. Now, let the URL to the actual session is https://www.mytelerobo.com/start (in actual case this URL will be formed by the corresponding meeting ID). Now, let's assume that the enterprise has a proxy that has a policy to redirect all the requests to the session URL to the robot IP. Then all the enterprise-level signaling requests would be directed to the *dew signaling server* on the robot. This way, despite the telerobotic system remaining as a cloud-centric system, the enterprise users are going to experience a very quick session establishment. Of course, number of enterprise users in the session should have a tight upper limit so that robot is not too overloaded.

In case, a user wants to join the session from outside the enterprise network he/she will still hit the same URL but will be served by the public cloud signaling server. The P2P connection between the dew server and the cloud will exchange the necessary

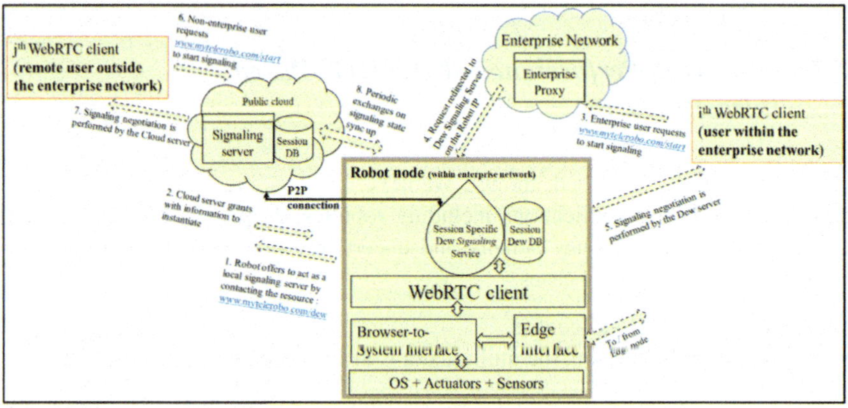

Fig. 13 Robot as a signaling server in a cloud-dew WebRTC signaling

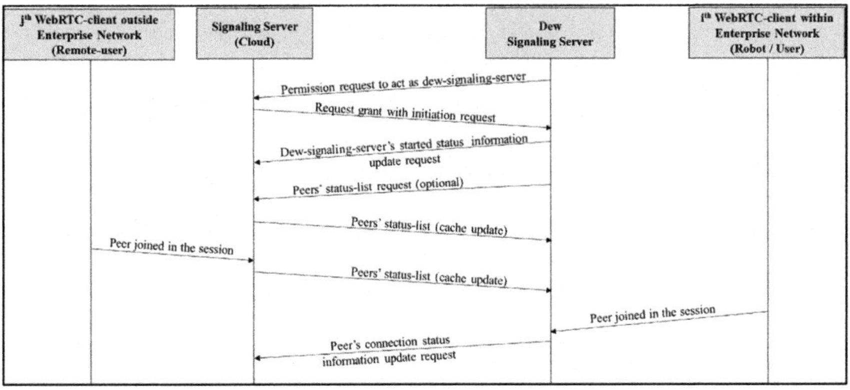

Fig. 14 Possible exchanges between the cloud signaling server in the public cloud and the dew signaling server on robot within the enterprise network

messages for syncing up the present signaling states. A possible handshaking for the dew-cloud exchanges is shown in Fig. 14.

Thus, when the session is limited within the enterprise users only, the signaling bandwidth is not consumed from the public Internet infrastructure. The legitimate signaling server resides in the cloud. Still the users enjoy the advantage of a localized service without undermining the global infrastructure.

3.2.2 WebRTC with Cloud-Dew Media-Exchange Service

In the previous section we discussed on the signaling aspect. In this section we conceive how the live media distribution can be made efficient through a cloud-dew architecture. The proposed scheme is illustrated in Fig. 15. To relieve the robot from the computational load of continuously maintaining multiple P2P connections and performing the video aggregation and distribution, we consider that a resourceful node within the enterprise network offers to act as a dew for media management and distribution. The node may well be the Edge device within the network. Thus, all the nodes connecting from within the enterprise are to establish a multimedia P2P channel with the local media manager dew server. The dew signaling server on the robot establishes these connections (We assume that somehow the dew signaling server knows about the *dew media manager* server). Thus all video feed from the enterprise users are pushed to the *dew media manager (DMM)*. The DMM also receives the external feeds from the public cloud media manager. In the same way the public cloud also receives the composite view from within the enterprise and combines that with what it has from public domain and distributes to non-enterprise users as shown in Fig. 15. This would reduce a lot of cloud-centric overhead. Also, the end-user experience, at least within the enterprise network, would be much more while the system still maintains its cloud-centricity. This will also help quick reaction

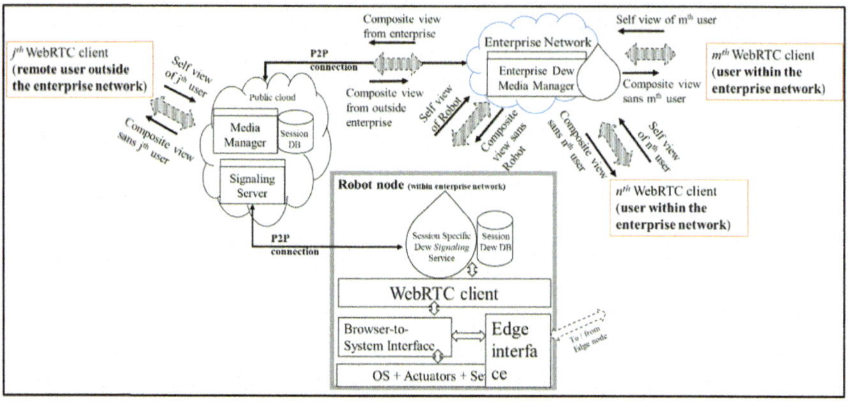

Fig. 15 A possible approach toward distributing the live media during a telerobotic session through cloud-dew paradigm

in case the robot needs to be stopped urgently from within the enterprise, while the remote non-enterprise user may have lost the connection after firing a series of kinematic commands and the robot action may pose a threat.

4 Conclusions

We have described the importance, use cases, and architectures for Telerobotics. We have described how signaling and media manager services can be distributed between the enterprise and the public cloud. Experience from the service may also be enhanced through cloud-dew sharing of application logic and database. We are still in the process of evaluating the conceived architectures. But we believe that the above propositions may open up future cutting-edge research scope. What is of prime importance is to check the tradeoff between the perceived advantage out of use of the dew paradigm against the limits of the robot's computational and energy capacity.

References

1. Double Robotics—Telepresence Robot for the Hybrid Office.: https://www.doublerobotics.com/. Accessed 17th Nov. 2021
2. Serhan, et al.. ENRICHME. perception and interaction of an assistive robot for the elderly at home. Int. J. Soc. Robot. (Springer) (2020)
3. AMY Service Robots: https://www.amyrobotics.com/. Accessed 17th Nov. 2021
4. Kubi Telepresence Robots.: https://www.kubiconnect.com/. Accessed 11th June 2022
5. AVA Robotics.: https://www.avarobotics.com/. Accessed 11th June 2022
6. Padbot Telepresence Robot.: https://www.padbot.com/padbotp3. Accessed 11th June 2022

7. VGO Robotic Telepresence.: https://www.vgocom.com/. Accessed 11th June 2022
8. Ohmni Telepresence Robot.: https://ohmnilabs.com/products/ohmni-telepresence-robot/. Accessed 11th June 2022
9. BotEyes Telepresence Robot.: https://boteyes.com/. Accessed 11th June 2022
10. Casiddu, N., et al.: Robot interface design: the Giraff telepresence robot for social interaction. In: Ambient Assisted Living. Biosystems & Biorobotics, vol. 11. Springer (2015)
11. https://robots.ieee.org/robots/beam/. Accessed 11th June 2022
12. FURo-i Home: Multi-Function Home Telepresence Robot.: https://telepresencerobots.com/robots/futurerobot-furo-i/. Accessed 11th June 2022
13. Temi: The Personal Robot.: https://www.robotemi.com. Accessed 11th June 2022
14. Bhattacharyya, et al.: Teledrive: an intelligent telepresence solution for "collaborative multi-presence" through a telerobot. In: 2022 14th International Conference on COMmunication Systems & NETworkS (COMSNETS), pp. 433–435 (2022). https://doi.org/10.1109/COMSNETS53615.2022.9668466
15. Sau, A., Bhattacharyya, A., Ganguly, M.: Teledrive: a multi-master hybrid mobile telerobotics system with federated avatar control. In: 18th EAI International Conference on Mobile and Ubiquitous Systems: Computing, Networking and Services (MOBIQUITOUS) (2021)
16. Chen, K.-C., Lin, S.-C., Hsiao, J.-H., Liu, C.-H., Molisch, A.F., Fettweis, G.P.: Wireless networked multirobot systems in smart factories. Proc. IEEE **109**(4), 468–494 (2021). https://doi.org/10.1109/JPROC.2020.3033753
17. Omarali, B., Denoun, B., Althoefer, K., Jamone, L., Valle, M., Farkhatdinov, I.: Virtual reality based telerobotics framework with depth cameras. In: 2020 29th IEEE International Conference on Robot and Human Interactive Communication (RO-MAN), pp. 1217–1222 (2020). https://doi.org/10.1109/RO-MAN47096.2020.9223445
18. Beetz, M., Beßler, D., Haidu, A., Pomarlan, M., Bozcuoğlu, A.K. Bartels, G.: Know Rob 2.0—A 2nd generation knowledge processing framework for cognition-enabled robotic agents. In: 2018 IEEE International Conference on Robotics and Automation (ICRA), pp. 512–519 (2018). https://doi.org/10.1109/ICRA.2018.8460964
19. Alvestrand, H.: RFC 8825, overview: real-time protocols for browser-based applications. IETF (2021)
20. Iiyoshi, K., Tauseef, M., Gebremedhin, R., Gokhale, V., Eid, M.: Towards standardization of haptic handshake for tactile internet: a WebRTC-based implementation. In: 2019 IEEE International Symposium on Haptic, Audio and Visual Environments and Games (HAVE), pp. 1–6 (2019). https://doi.org/10.1109/HAVE.2019.8921013
21. Baugher, M., McGrew, D., Naslund, M., Carrara, E., Norrman, K.: RFC 3711: the secure real-time transport protocol (SRTP). IETF (2004)
22. Stewart, R., (ed.) RFC 4960: stream control transport protocol. IETF (2007)
23. Keranen, A., Holmberg, C., Rosenberg, J.: RFC 8445: interactive connectivity establishment (ICE): a protocol for network address translator (NAT) traversal. IETF (2018)
24. Ford, B., Srisuresh, P., Kegel, D.: Peer-to-peer communication across network address translators. In: Proceedings of the Annual Conference on USENIX Annual Technical Conference, ACM, USA (2005)
25. Ray, P.P.: An introduction to dew computing: definition, concept and implications. IEEE Access **6**, 723–737 (2018). https://doi.org/10.1109/ACCESS.2017.2775042
26. Wang, Y.: Cloud-dew architecture. Int. J. Cloud Comput. **4**, 199–210 (2015)

Role of Dew Computing in Smart Healthcare Applications

Kishore Medhi and Md. Iftekhar Hussain

1 Introduction

The advancement of the computer and Internet technologies has begun a new era where an individual and devices can exchange information over the Internet [1]. This information exchange trend has led to the development of a new technology known as the Internet of Things (IoT) [1, 2]. Using IoT technology, any object on the earth can be identified, monitored, and controlled via the Internet [3]. In 1999, Kevin Ashton introduced the term IoT, and he proved that by using IoT, any physical objects could be connected easily with intelligent devices, giving us a remarkable opportunity to create a smart environment [4]. In IoT, the *things* represents the sensors, actuators, tags, camera, smart devices, etc. With the help of these *things*, people can easily interact with their surrounding physical environment.

IoT can be applied in different domains such as industry, healthcare, agriculture, and environmental monitoring. In recent years, the successful use of IoT in healthcare has given us tremendous capacity to improve healthcare facilities and human lifestyles. The market for the Internet of Medical Things (IoMT) and wearable healthcare technology has been growing exponentially and transforming the healthcare industry from treatment to prevention. A range of devices, from smartphones to smart wristbands, are used to capture different healthcare data and monitor the body's condition. All these healthcare devices use various biomedical sensors such as ECG, EEG, EMG, temperature, and pressure. Data from such sensors are collected, stored, analysed, and key decisions are taken from them. The primary purpose of the healthcare device is to analyse the different body parameters of a person and

K. Medhi (✉) · Md. I. Hussain
Department of Information Technology, North-Eastern Hill University, Shillong 793 022, Meghalaya, India
e-mail: kishoremedhirw@gmail.com

Md. I. Hussain
e-mail: ihussain@nehu.ac.in

D. De and S. Roy (eds.), *Dew Computing*, Internet of Things,
https://doi.org/10.1007/978-981-99-4590-0_11

provide correct assistance and guidance. Moreover, using these healthcare devices, we can make some early predictions and perform continuous monitoring of patients, reducing the mortality rate and improving lifestyle.

IoT-based healthcare technologies are being researched extensively. Numerous computing techniques, including cloud, Fog, and edge computing, are used while creating healthcare applications [5]. In general, cloud architectures are typically located at distant Internet locations. Different types of wired and wireless network technologies make it possible to connect to the cloud. Smart healthcare solutions demand quick decisions in order to enhance the user's condition. Because of the overall complexity and variety of components, the network is inappropriate for crucial IoT applications. Transmission of information between the end-user devices and the cloud usually takes a long time. Later, different computing architectures, such as fog and edge computing strategies, are suggested to ease these problems and aid in low-latency decision-making. By processing the frequently requested data near the user's current network's edge, such a computing solution intends to reduce the user's reliance on cloud or fog resources. Consequently, the overall network security is improved. However, Edge computing is unable to offer the user point-of-call service, extreme scalability, or network independence [6].

To address these situations, time-sensitive data are processed near the source compared to the fog or edge layer, and the necessary action is decided and communicated to the consumers. However, because of the user's mobility, the network structure and Internet connection speed extensively affect the overall performance of the healthcare systems [7]. Even in the case of an unstable or disconnected network, a healthcare application needs real-time responses. Massive attempts are being made to overcome these limitations, but they have not been very successful. The current state-of-the-art technologies are insufficient for the IoT-based intelligent e-healthcare system, which requires an extremely intelligent and ultra-low latency decision-making system [8]. The main issues with the current healthcare infrastructure are:

1. devices with limited resources.
2. a need for lightweight analytical modules.
3. a requirement for a smart network and an Internet connection.

The current cloud-based or other computing-based solutions are inadequate since they have a high delay requirement [9]. In regards to this, the industry has recently seen the introduction of the new Dew computing architecture, which offers flexibility in terms of ultra-low latency judgments, mobility, user control, and patient-specific requirements.

The underlying idea of cloud computing is carried through by Dew computing to the end devices (e.g. Mobile phones, gateway nodes, raspberry pi, and personal computers). It addresses the crucial problems related to cloud computing technologies, namely reliance on the internet. Dew computing frameworks can be set up on a tiny, light Raspberry Pi device unlike the Cloud and Fog computing frameworks. Dew computing is substantially different from cloud and fog computing in crucial IoT applications and more feasible due to its qualities like ultra-low latency, zero Internet need, mobility in offline environments, and quick response time.

Table 1 Comparison of some existing e-healthcare technologies

Parameters	Dew computing	Edge computing	Fog computing	Cloud computing
Latency	Negligible	Low	High	Very high
Distance	Very short	Short	Long	Very long
Number of hops	No hops	Single hop	Multi hops	Multi hops
Mobility	High	Moderate	Limited	Very limited
Internet dependency	No	Limited	Moderate	Very high
Service location	Near source	In edge network	Within internet	Within internet
Delay tolerant	Yes	No	No	No
Deployment scenario	PC, laptop, smart phone	Router, gateway	SME	Large enterprises
Hardware	Very limited capabilities	Limited capabilities	Scalable capabilities	Scalable capabilities
Green energy	High	Low	Very low	Very low

Cloud computing on a smaller scale can be called "dew computing", and it is situated close to the data source. Similar to a cloud, it is capable of carrying out all of the aforementioned functions (data preparation, analysis, storage, and transmission), but it is limited in terms of computational power.

A detailed analysis of different computing architectures is given in Table 1. Dew computing can operate even when there is no active Internet connection. The patient bodies are connected to end devices in smart healthcare, which collect numerous health metrics to track various healthcare-related issues. Healthcare solutions based on dew can prepare potential future services and provide necessary user-specific decisions as needed.

Individual fog modules are taken into consideration for an offline solution by extremely localised computing platforms like Dew computing [10]. This platform offers support for location-aware, user-specific, anytime, anywhere access. However, this dew platform needs to be properly designed with smart, lightweight analytical modules, and judgments in order to provide full smart healthcare IoT solutions.

In Dew devices, machine learning-based solutions show remarkable performance in detecting body abnormalities, but they face certain challenges in some instances, like

i. limited amount of training data
ii. complex computational model for lightweight devices
iii. dynamic nature of healthcare data.

The rest of the paper is organised as follows. The background studies of computing architectures, such as the disadvantages of cloud computing and the advantages of dew computing, are explained in Sects. 2 and 3, respectively. Section 3.2 describes the general architectures of the dew device. Literature reviews of various existing dew-based healthcare architectures are described in Sect. 4. The explanation of the conducted survey is explained in Sect. 4.9. At last, the paper is concluded in Sect. 5.

2 Performance Issues in Centralised Cloud Computing

In order to offer distant services through the Internet, cloud computing works in concert with hardware and software. To offer its clients services, cloud computing mainly relies on remote equipment. Due to the increase in sensors and data production, cloud computing is struggling to digest the data swiftly and struggles to respond to users more quickly [11]. Some of the common problems are discussed below.

1. **Large volume of data**: With the increase in the number of IoT devices, the generation of sensing data is also increasing, and it increases the requirement for data processing day by day.
2. **High Latency**: Critical IoT applications like healthcare, smart traffic, etc. require quick responses to provide emergency services to clients. However, cloud computing can't provide quick services because of its long distance between the Cloud and the IoT device [12].
3. **Downtime:** Due to huge volume of data processing requirement the services of cloud become unavailable and increases the network failure [12].
4. **Security**: Cloud server can be accessed globally by different users which increases the risk for the private data shared from different sources.
5. **Limited Flexibility**: Cloud computing servers are totally managed by the third party service provider which leads to limited control to users and reduces the data security.
6. **Bandwidth Cost**: To provide quick services by the cloud computing, heavy communication channels are required which increases the cost of the system.

3 Dew Computing as a Solution

Dew computing was originally developed to reduce the traffic load in the cloud server, which is one of the major limitations of cloud computing [13]. In dew computing, data can be processed near the source of generation, which reduces the response time and can provide required services during offline conditions also. It will work like a mini Cloud, managing tasks and giving responses locally within a region. A diagrammatic representation of Dew computing architecture is shown in Fig. 1.

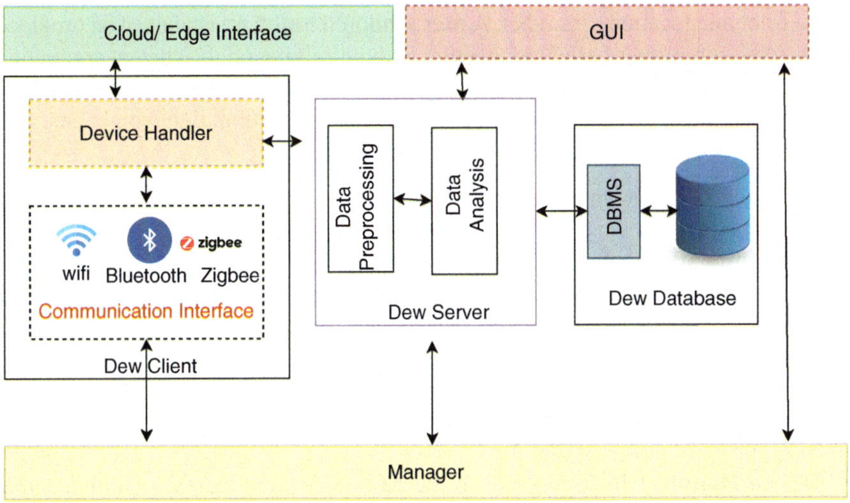

Fig. 1 Architecture of dew device [13]

3.1 *Advantages of Dew Computing*

1. **Low Latency**: As the Dew Computing devices are deployed near the sensors, the data transfer time becomes negligible, which finally reduces the response time.
2. **High Bandwidth**: In dew computing, all the devices are set up locally, which increases the fast connection and availability of devices and reduces the downtime.
3. **High Security**: In dew computing, the data are processed locally, and the dew server can be accessed by local users only; as a result, the security of the data becomes increases.
4. **Better User Experience**: Dew computing provides quick and correct responses to user applications, which increases user satisfaction and reliability.

3.2 *Architecture of Dew device*

To order to set up a fresh dew computing-based system for the healthcare industry, it is essential to perform all execution locally instead of in the cloud, with all the required services. Dew computing's primary goal is to deliver a user-centric, adaptable, personalised, and immediate reaction to a particular task. With the help of dew architecture, we can quickly minimise reliance on the Internet. We can set up a Dew architecture on a resource constraint devices such as raspberry pi and mobile smartphone devices. Portable smartphones have a huge potential to contribute to Dew architecture because they can operate for long period without energy. With a smartphone-based mobile device, it is easier to give the services when the user is travelling from one location to another. A user's tailored health assistant can be provided via the dew computing-based architecture without relying on the current computing model. As shown in Fig. 1, every lightweight dew device uses a specialised module of the dew framework to function, such as cloud synchronisation, lightweight analysis, GUI, a dew-specific database, and dew server.

1. **Manager**: The dew manager starts all the modules of the dew device by calling the functionalities.
2. **Communication Interface**: As shown in Fig. 2, each dew device can communicate with the body sensors (BS) and other computing devices using any of the networking technology like WiFi, Bluetooth, Zigbee, etc. Medhi et. al. [13] use the MQTT message transferring protocol for exchanging the message between BS and dew device, where each BS working as a publisher and sends message to the dew client through the MQTT broker. The architecture of the MQTT message transferring is shown in Fig. 3.
3. **Device Handler**: It discovered all the new nodes and started communication between them. Using the message-passing method, the device handler keeps the connection open between the BS and its server.
4. **Dew Server**: In the dew server, all the receiving data are preprocessed, filtered, and the noises are eliminated. Following the pre-processing, a quick analysis is carried out to look for any irregularities.
5. **Dew Database**: All the results produced by the server are stored in a little database called a Dew database. It mainly helps to store the result during offline conditions. Different dew devices use the MySQL database as the dew database.
6. **GUI**: It manages user communication and presents the findings of all analyses. It also has access to all of the dew software modules' settings.
7. **Cloud Gateway Interface**: It maintains all the links with the edge device or the cloud server to provide and get control instructions for the connected IoT devices.

4 State-of-the-Art Dew Architectures for Healthcare

Different researchers have used IoT in various domains such as healthcare, intelligent traffic, and emergency management. Among all the applications, healthcare needs various time-critical services, and this can be fulfilled by dew devices because of the extreme edge location. A detailed analysis of various dew computing-based healthcare applications is discussed below.

4.1 *Dew-Based Offline Computing Architecture for Healthcare IoT [13]*

A dew health architecture was created by Medhi et al. [13] utilising a mobile smartphone. They used the Pydroid mobile programme to instal the Python environment on an Android smartphone in order to evaluate the data. The mobile device interacted with the MQTT publisher as a MQTT client through the MQTT broker. The ECG dataset obtained from the Physionet repository was used in this experiment.

The ECG dataset remained in the MQTT publisher throughout the experiment and, like the sensor device it continuously provides data to the MQTT client at regular intervals. The MQTT client executed the trained CNN module as soon as a signal was received, reported the outcomes, and saved the information to a nearby MySQL server. Each single device gathers real-time streaming health data from several users, does the necessary analysis, and then provides the user current health status, the architecture is shown in Fig. 1.

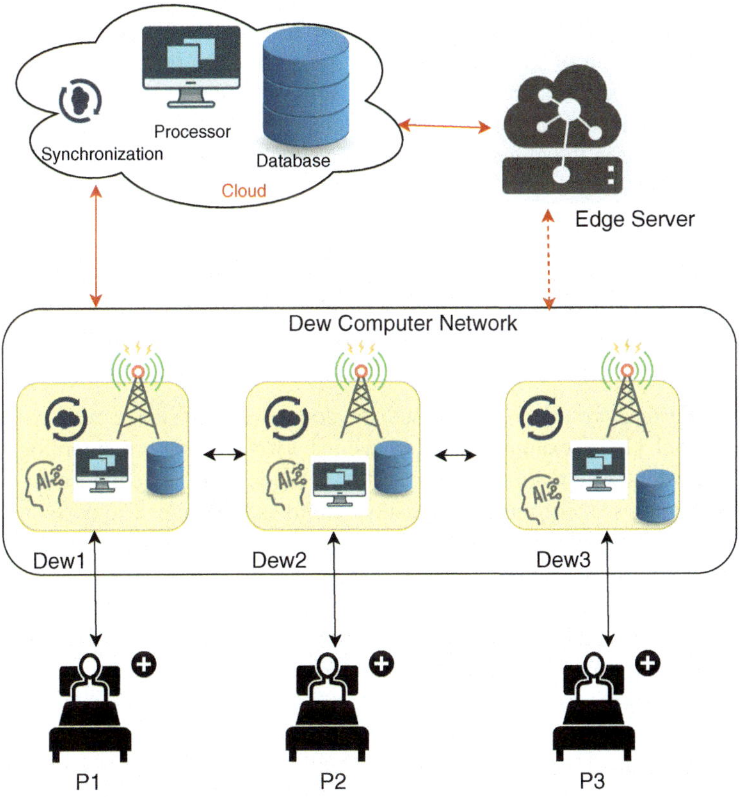

Fig. 2 A dew-based healthcare architecture developed by Medhi et al. [13]

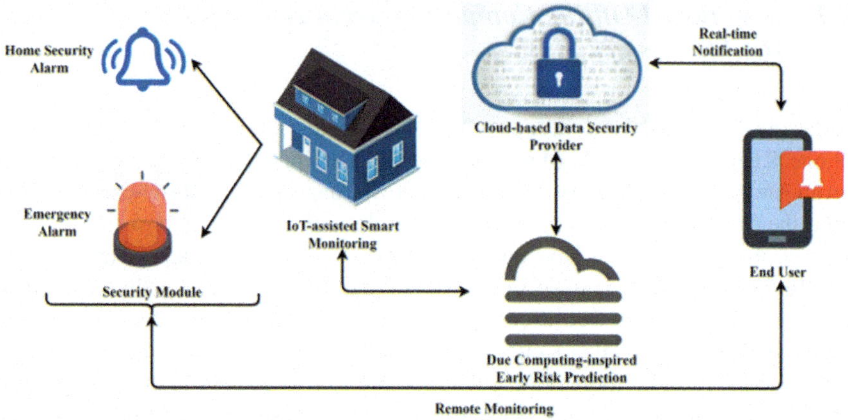

Fig. 3 A dew-based healthcare architecture developed by Manocha et al. [14]

Each dew device gathers and analyses BS data from the linked patients, and it is shown in Fig. 2. With the aid of the dew GUI, the analysis module provides the analysis information of the patient's condition. When the dew server is offline, it connects to the local database management system to update the database with the most recent details on current medical problems. The dew synchronizer updates the data that is saved in the cloud for later use by synchronising the dew device with it at predefined intervals.

4.2 *Dew Computing-Inspired Health-Meteorological Factor Analysis for Early Prediction of Bronchial Asthma [14]*

A cyber-physical system (CPS) based on dew-cloud assistance was created by Manocha et al. [14] to examine the relationship between people's meteorological and health indicators. The main goal of the effort is to identify the health issues brought on by the erratic scale of meteorological elements in real time. The development of a dew-based smart gadget allows for the use of sensors to gather information from the interior environment that is omnipresent and has a significant direct or indirect impact on an individual's health. Utilising the Weighted-Naive Bayes data classification technique, the data is examined to assess the likely irregular health events. Additionally, the Adaptive Neuro-Fuzzy Inference System (ANFIS) is used to calculate a unifying factor over the temporal scale and assess the association between meteorological and health parameters. The suggested model is put into practice at four different schools in Jalandhar to validate the monitoring performance.

4.3 *MedGini: Gini Index-Based Sustainable Health Monitoring System Using Dew Computing [15]*

Intelligent health apps often include biosignal monitoring. A novel route for monitoring biosignals is offered via the Internet of Health Things (IoHT). The design of data synchronisation with the cloud server is inadequate in the existing dew-based healthcare systems. To address this, a model is created for a sustainable health monitoring system that effectively utilises cloud-dew architecture. Various bio sensor nodes are employed to dynamically monitor the biosignals. To enable seamless accessibility and the best scalability of the data, all acquired data is momentarily stored in the dew database and synchronised with the cloud. Gini index and Shannon entropy were utilised to provide the best possible data synchronisation in order to guarantee (Fig. 4).

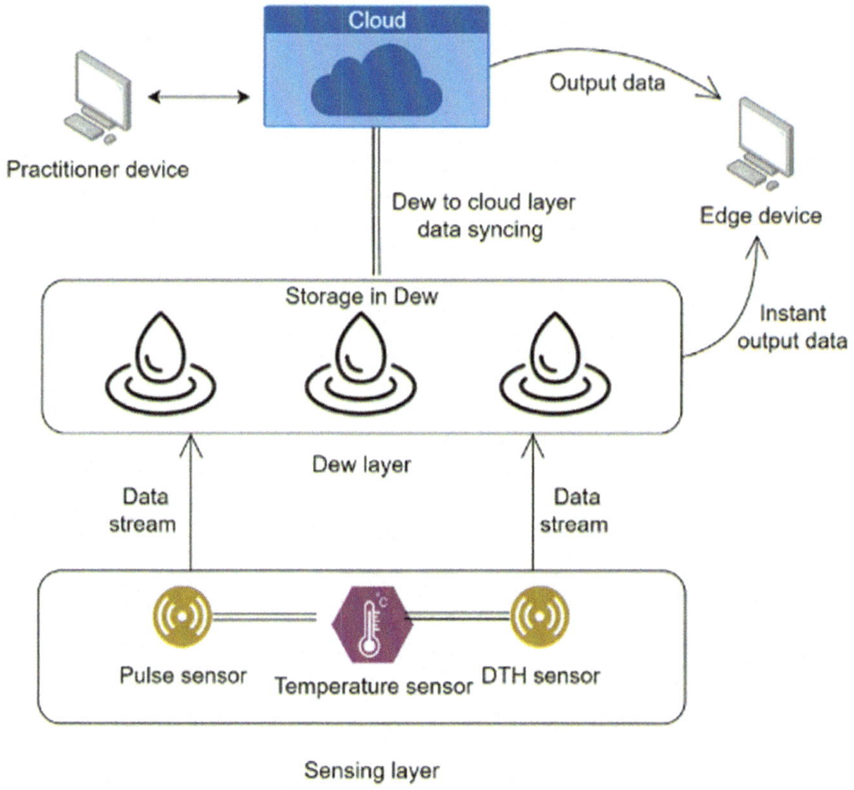

Fig. 4 Dew-based healthcare architecture developed by [15]

4.4 CONFRONT: Cloud-Fog-Dew-Based Monitoring Framework for COVID-19 Management [16]

In recent times, the use of IoT in the healthcare industry has been growing because and the low-cost sensors. Using IoT, we can cover a large area with a limited number of sensors, which makes it an important tool to fight against situations like the COVID-19 pandemic. In this research, a cloud-based monitoring framework for COVID-19 administration was suggested. This cloud, fog, and dew-based healthcare paradigm may help with early diagnosis and monitoring of patients even during quarantine or after undergoing home-based therapies. The reduced bandwidth requirements of the fog architecture guarantee that the model is appropriate for real-time settings. Cloud servers are used because of their scalable computing and storage capacity to analyse big-scale COVID-19 statistics data for obtaining aggregate information on the illness's spread. Faster application uptime is achieved by the dew architecture, which makes sure that the application is still accessible at a small scale, even when cloud connectivity is lost (Fig. 5).

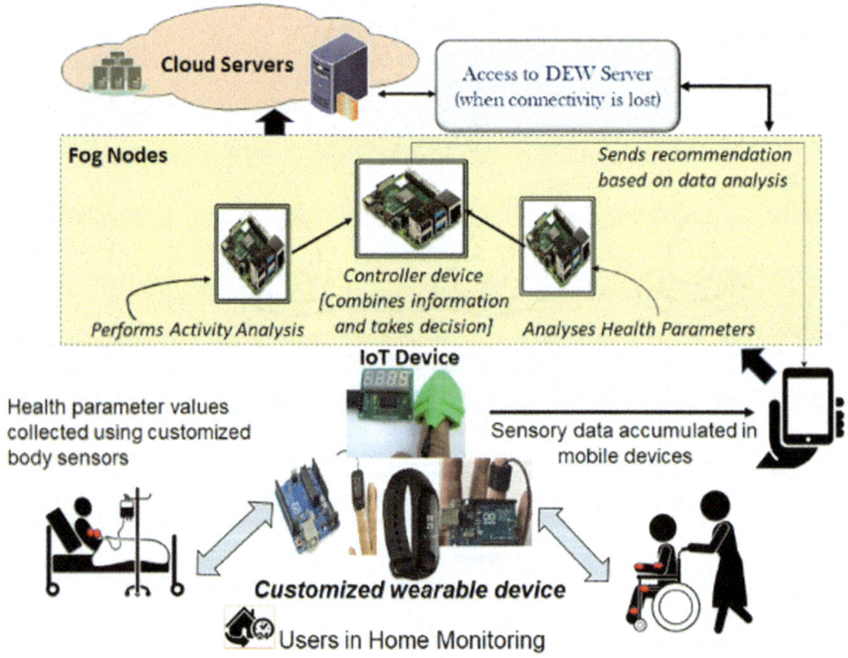

Fig. 5 Dew-based healthcare architecture developed by [16]

4.5 AI Cardiologist at the Edge: A Use Case of a Dew Computing Heart Monitoring Solution [17]

Applications that deal with the IoT, pervasive computing, and ubiquitous computing are what postcloud architectures are known for. The foundation of edge computing solutions is the addition of extra processing layers between end-user devices and cloud data centres to take advantage of quicker processing and smaller data transport requirements. It is categorised as a dew computing solution since these smart devices' edge requirements for autonomous performance even without Internet support. In order to bring computing closer to the user in wearable sensors, edge computing devices should execute trillions of operations per second while consuming incredibly little power must be built. However, this is already being done by different industries with the newest chips used by smartphone makers. Although edge computing has been around for a while, no specific killer application has yet to drive widespread adoption.

4.6 Dew-Cloud-Based Hierarchical Federated Learning for Intrusion Detection in IoMT [18]

Due to the overwhelming impact of the coronavirus epidemic on healthcare facilities, doctors are now forced to treat and diagnose patients from remote places. In addition, COVID-19 has increased human awareness of their health, leading to a significant increase in the purchasing of IoT-enabled medical devices. The rapid expansion of the Internet of medical things (IoMT) sector attracted cybercriminals for different attacks. Nowadays, medical health information data are extremely important and sensitive on the dark web. Despite the fact that it has not been adequately safeguarded, allowing trespassers to misuse the patient's medical information. Due to the resource-constrained network devices' limited storage and processing power, the system administrator is not able to strengthen security measures. Even though many supervised and unsupervised ML algorithms have been created to detect abnormalities, the main goal is to investigate rapidly developing hostile attacks before they compromise the integrity of the healthcare system. Using a Dew-Cloud-based model, hierarchical federated learning is made possible in this research. The suggested Dew-Cloud paradigm offers enhanced IoMT essential application availability along with a higher level of data privacy. At distributed Dew servers, the hierarchical long-term memory (HLSTM) concept is implemented with cloud computing as the backend. In the suggested model, the data pre-processing feature aids in achieving high training accuracy (99.31%) with little training loss (0.034%). The results of the trial show that the suggested HFL-HLSTM model outperforms existing plans in terms of performance indicators including precision, recall, accuracy, and f-score.

4.7 Dew Computing-Assisted Cognitive Intelligence-Inspired Smart Environment for Diarrhea Prediction [19]

Diarrhea is the most prevalent infectious disease influencing people of all ages and poses a severe public health risk worldwide. Diarrhea is primarily brought on by poor food quality, contaminated water, indoor and outdoor weather conditions, and meteorological factors. In order to examine the relationship between a person's health, food condition, and indoor meteorological conditions to anticipate the source of diarrhea with the degree of severity. An adaptive dew computing-assisted surveillance system is developed to gather the individual's diet, indoor weather, and specific health metrics. All the information are gathered in the physical layer using different smart sensors. At the cyber layer, the collected events are categorised using the Probabilistic Weighted-Naive Bayes (PWNB) classification method for estimating anomalous health occurrences. A Multi-scale Gated Recurrent Unit (M-GRU) is also recommended in order to determine the severity of irregular health, food, and environmental occurrences by examining their correlation. Accordingly, the suggested M-GRU model has attained a very high precision accuracy value of (93.26%), while LSTM, RNN, and SVM have attained precision values of (89.13%), (90.43%), and (88.23%), respectively. Additionally, the PW-precision NB's value is (97.15%), which is greater than both KNN (93.25%) and DT (96.91%). In both the dew computing and cloud computing, the suggested methods' results display greater Precision values. Additionally, in terms of event classification, severity assessment, monitoring stability, and prediction effectiveness, a comparative study characterises the proposed solution's prediction efficacy relative to a number of alternative decision-making systems (Fig. 6).

4.8 Real-Time Event-Driven Sensor Data Analytics at the Edge-Internet of Things for Smart Personal Healthcare [20]

For the IoMT-based smart e-Healthcare, real-time service has become an essential requirement to operate effectively. There have been a number of tries to improve this area of technology, but there is a critical need to enhance the incorporation of lightweight IoT-based frameworks. In this research, two separate experiments are presented that deal with real-time visualisation, charting, and analytics while utilising real time and readily available Javascript frameworks, such as Node.js, EON.js, Chart.js, Express Server, and Socket.io. The goal of this research is to examine real-time analytics behaviour based on the IoT in a scenario where e-health sensors are being installed efficiently. The study's findings support the development and use of open-source, IoT-based Javascript frameworks for supplying real-time sensor data analytics in the e-Healthcare industry (Fig. 7).

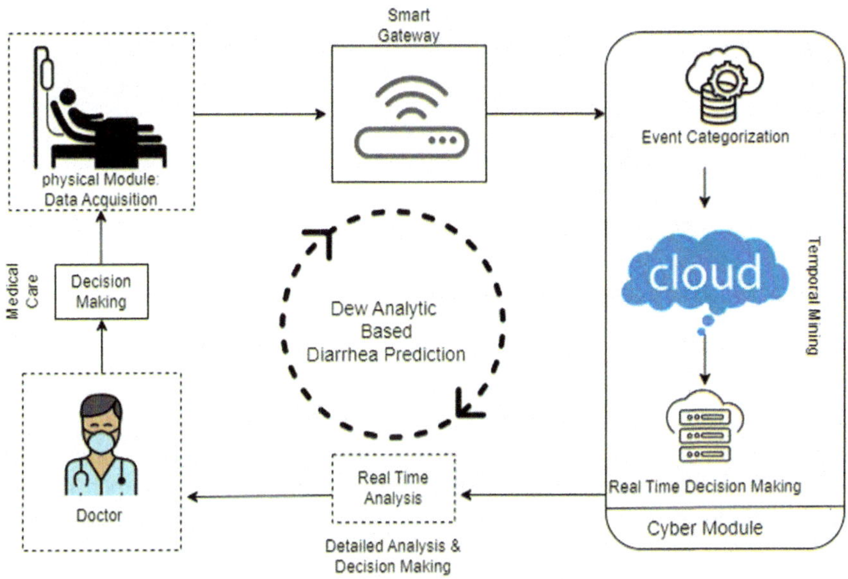

Fig. 6 Dew-based healthcare architecture developed by [19]

4.9 Synthesis

A detailed comparison of all the above-mentioned existing dew-based experiments is mentioned in Table 2. From this literature survey, it is clear that for applications like healthcare, quick response is one of the major concerns in fulfilling the user's

Table 2 A detailed comparison of different dew computing-based healthcare applications

Paper	Application area	Technique used	Device used as dew
Medhi et al. [13]	Heart disease	Light weight CNN	Smartphone
Manocha et al. [14]	Asthma	Weighted-Naive Bayes	Raspberry Pi
Karmakar et al. [15]	Healthcare	MedGini: fuzzy based entropy	Arduino Uno
Poonia et al. [16]	Covid-19	KNN with handcrafted features	Raspberry Pi
Gusev et al. [17]	Healthcare, Arrhythmia	Shallow CNN	Smartphone
Singh et al. [18]	Healthcare	Hierarchical federated learning	Raspbery Pi
Afaq et al. [19]	Healthcare, Diarrhea	Multi-scale gated recurrent unit	
Ray et al. [20]	Healthcare	Lightweight Javascript framework	Arduino Uno

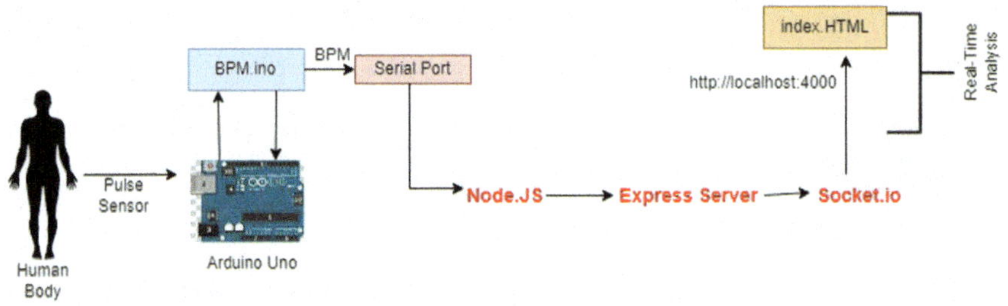

Fig. 7 A dew-based healthcare architecture developed by [20]

requirements. However, due to the remote location, the cloud server fails to fulfil this demand. To solve this, all the above-mentioned experiments used dew computing as an alternative solution to process data near the source and reduced the overall latency. Different researchers have used various devices as dew nodes. For example, Medhi et al. [13], Manocha et al. [14], Karmakar et al. [15] use smartphones, raspberry pi, and Arduino Uno respectively as dew devices. However, due to the lightweight nature of dew devices, the different authors used different lightweight analytical models. Such as lightweight CNN [13], where the number of arithmetic operations is reduced by using the K-mean. From all these experiments, it is clear that using the dew device; it is possible to fulfil all the requirements of healthcare applications by processing data and giving quick responses to the users.

5 Conclusions

This chapter examined various IoT-based healthcare systems critically and covered how dew computing technology can be used to satisfy the needs. The dew solution overcomes major challenges like time-criticality, lightweight services, flexibility, service during offline conditions, etc., which are essential for smart healthcare applications. To complete this literature survey, we have used more than ten different latest dew-based healthcare research articles collected from various reputed journals. All these papers used different devices, such as Raspberry Pi, Arduino Uno, and Smartphone, to instal the dew-based computing architecture. All of the created dew models effectively carry out various duties at the very edge of the network using various lightweight analytical modules. From this critical review, it is clear that using the dew computing architecture, we can easily overcome the issues like delay, response time, power consumption, CPU usage, memory consumption, and accuracy. We plan to establish a real-time testbed for evaluating the performance of dew-based healthcare schemes.

References

1. Atzori, L., Iera, A., Morabito, G.: The internet of things: a survey. Comput. Netw. **54**(15), 2787–2805 (2010)
2. Rezaeipanah, A., Nazari, H., Ahmadi, G.: A hybrid approach for prolonging lifetime of wireless sensor networks using genetic algorithm and online clustering. J. Comput. Sci. Eng. **13**(4), 163–174 (2019)
3. Alaa, M., Zaidan, A.A., Zaidan, B.B., Talal, M., Kiah, M.L.M.: A review of smart home applications based on internet of things. J. Netw. Comput. Appl. **97**, 48–65 (2017)
4. Sobin, C.C.: A survey on architecture, protocols and challenges in IoT. Wirel. Pers. Commun. **112**(3), 1383–1429 (2020)
5. Ray, P.P.: An introduction to dew computing: definition, concept and implications. IEEE Access **6**, 723–737 (2017)
6. Ray, P.P., Dash, D., De, D.: Edge computing for internet of things: a survey, e-healthcare case study and future direction. J. Netw. Comput. Appl. **140**, 1–22 (2019)
7. Ahmed, E., Ahmed, A., Yaqoob, I., Shuja, J., Gani, A., Imran, M., Shoaib, M.: Bringing computation closer toward the user network: is edge computing the solution? IEEE Commun. Mag. **55**(11), 138–144 (2017)
8. Xiong, Y., Sun, Y., Xing, L., Huang, Y.: Extend cloud to edge with KubeEdge. In: ACM/IEEE Symposium on Edge Computing (SEC), pp. 373–377 (2018)
9. Ali, O., Ishak, M.K.: Bringing intelligence to iot edge: machine learning based smart city image classification using microsoft Azure IoT and custom vision. J. Phys.: Conf. Ser. **1529**, 042–076 (2020)
10. Ray, P.P., Dash, D., De, D.: Internet of things-based real-time model study on e-healthcare: device, message service and dew computing. Comput. Netw. **149**, 226–239 (2019)
11. Botta, A., De Donato, W., Persico, V., Pescapé, A.: Integration of cloud computing and internet of things: a survey. Fut. Gener. Comput. Syst. **56**, 684–700 (2016)
12. Saini, K., Raj, P.: Chapter eight-edge platforms, frameworks and applications. Adv. Comput. **127**, 237–258 (2022)
13. Medhi, K., Ahmed, N., Hussain, M.I.: Dew-based offline computing architecture for healthcare IoT. ICT Express **8**(3), 371–378 (2022)
14. Manocha, A., Bhatia, M., Kumar, G.: Dew computing-inspired health-meteorological factor analysis for early prediction of bronchial asthma. J. Netw. Comput. Appl. **179**, 102995 (2021)
15. Karmakar, A., Banerjee, P.S., De, D.,Bandyopadhyay, S., Ghosh, P.: Medgini: Gini index based sustainable health monitoring system using dew computing. Med. Novel Technol. Dev. 100145 (2022)
16. Poonia, A., Ghosh, S., Ghosh, A., Nath, S.B., Ghosh, S.K., Buyya, R.: Confront: Cloud-fog-dew based monitoring framework for covid-19 management. Internet Things **16**, 100459 (2021)
17. Gusev, M.: AI cardiologist at the edge: a use case of a dew computing heart monitoring solution. In: Artificial Intelligence and Machine Learning for EDGE Computing, pp. 469–477 (2022)
18. Singh, P., Gaba, G.S., Kaur, A., Hedabou, M., Gurtov, A.: Dew-cloud-based hierarchical federated learning for intrusion detection in IOMT. IEEE J. Biomed. Health Inform. (2022)
19. Afaq, Y., Manocha, A.: Dew computing-assisted cognitive intelligence-inspired smart environment for diarrhea prediction. Computing **104**(11), 2511–2540 (2022)
20. Ray, P.P., Dash, D., De, D.: Real-time event-driven sensor data analytics at the edge-internet of things for smart personal healthcare. J. Supercomput. **76**(9), 6648–6668 (2020)

Dew as a Service for Intermittently Connected Internet of Drone Things

Amartya Mukherjee, Debashis De, Nilanjan Dey, Rubén González Crespo, and Houbing Herbert Song

1 Introduction

Modern-day computing techniques are significantly depending upon networking and local and cloud-level infrastructures. The introduction of the edge-fog-cloud layer helps the distributed ecosystem to store and process the data in a distributed fashion. The cloud computing philosophy provides us with a convenient way to store, distribute, analyze, and process data efficiently [1]. The usage of cloud infrastructure is also increasing day by day. Now a day billions of devices are getting connected with the cloud services like AWS, Microsoft Azure, and Google Cloud Platform (GCP).

Due to that, the processing load in cloud devices is also increasing exponentially. To overcome this situation, the edge and fog layers have been incorporated. The primary tasks of these two layers are to support load balancing and offloading in the cloud. The edge computing philosophy primarily emphasizes the processing of

A. Mukherjee (✉) · D. De
Department of Computer Science & Engineering, Maulana Abul Kalam Azad University of Technology, Kalyani, India
e-mail: mamartyacse4@gmail.com

A. Mukherjee
Institute of Engineering & Management, Kolkata, India

N. Dey
Department of Computer Science & Engineering, Techno International, Newtown, Kolkata, India

R. G. Crespo
Universidad Internacional de La Rioja, Logroño, Spain
e-mail: ruben.gonzalez@unir.net

H. H. Song
Department of Information Systems, University of Maryland, Baltimore County (UMBC), Baltimore, MD 21250, USA
e-mail: songh@umbc.edu; h.song@ieee.org

D. De and S. Roy (eds.), *Dew Computing*, Internet of Things,
https://doi.org/10.1007/978-981-99-4590-0_12

a small volume of data within the edge of the network [2]. In this case, data that are gathered by the sensor nodes in a sensor network are temporarily buffered in the edge node and some of the basic levels of data preprocessing get done in this level so that some amount of a load of cloud gets reduced. Secondly, the edge nodes must be connected by Internet connectivity so that distributed data processing can also be achieved. The connectivity between edge nodes in this case is either local or global based on the demographic parameters. The fog node, on the other hand, performs medium-level processing and can be considered the extended part of the cloud. In comparison to the edge node, the fog node has more amount of processing and storage capability.

The drone network or the Internet of Drone Things (IoDT) [3] is another emerging platform that we are considering here. IoDT philosophy establishes seamless connectivity between the UAV node to perform real-time and mission-critical applications. This includes aerial surveillance, crowd monitoring, disaster management, and many more. Such IoDT network always requires constant processing of real-time data. Most of the data in this case are images, videos, and other sensor data. Thus, the IoDT infrastructure required strong networking and data storage and buffering capability. In the case of the drone network, there are two major concerns that we need to take care of. The first thing is connectivity and message routing. In the drone network, often drones change their positions very fast. Therefore, nodes experience frequent disconnectivity. To address this problem, we can introduce the store and forward mechanism where the node acts as a Dew node and buffer the messages. Another major challenge, in this case, is the mobility of drones. In the case of mission-critical applications, drone movement highly matters to achieve precise information about the region of interest. The drones, in this case, performs numerous kind of mobility like Gauss-Markov movement, shortest path movement, map-based movement, etc. [4]. The proposed survey primarily emphasizes the application of Dew buffer management utilities as Dew as a Service (DewaaS) for the drone network which is extremely useful in mission-critical applications. The major contribution of this chapter encompasses the establishment of the fundamental idea about Dew computing as a service. Secondly the application perspective of the DewaaS for the intermittently connected IoDT.

The rest of the chapter has been organized as follows. Section 2 provides the related technologies; Sect. 3 describes the background study; Sect. 4 describes Dew as a Service (DewaaS) architecture. Section 5 describes the state-of-the-art classification; Sect. 6 describes the IoDT ecosystem for Dew as a service; Sect. 7 provides the idea about the intelligent learning mechanism. We have concluded the paper in Sect. 8.

2 Related Research

The concept of the Cloud-Fog-Edge-Dew methodology was started lots of years ago with the introduction of mobile computing methodology which is perhaps the harbinger of the modern-day computing philosophy. If we consider the era-wise

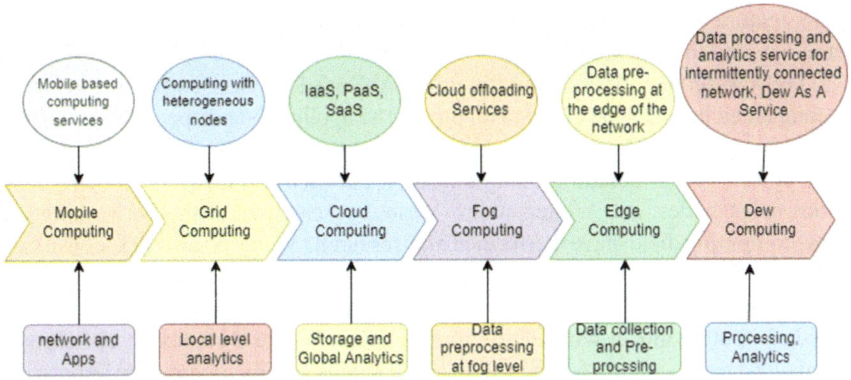

Fig. 1 Evolution of the concept of the dew as a service

progress, then we can easily understand that the primary evolution of distributed computing has been started with mobile computing methodology which starts in the late 90s (Fig. 1). In this case, mobile application-based computing was the predominant research domain [5]. In the early 2000 grid, computing was very popular [6]. This technique primarily emphasizes the distributed and heterogeneous load sharing among computing nodes of various platforms. Between 2005 and 2009, the cloud computing philosophy gained popularity massively [7] accompanied by blockchain. The cloud ecosystem fundamentally ensures the on-demand platform, infrastructure, and software as a service methodology through which the hardware virtualization has to be achieved. In the case of Infrastructure as a Service (IaaS), the major emphasis is on the hardware, processor, and storage space specification. AWS EC2 is one of the major examples of IaaS. The platform as a service (PaaS) on the other hand provides ample support for user interaction and deployment of the software component in a dedicated platform running on a specific OS infrastructure.

The post-cloud era starts in 2010. In this era, the major emphasis is on cloud offloading and load sharing. This part has greatly been achieved from the concept of fog computing and related edge computing platform [8, 9]. The primary objective of edge computing is to grab data from the sensor and perform primarily level data processing at the edge of the network. The edge computing philosophy provides quantitative data processing and always acts on the local-level data hierarchy. The nodes, in this case, majorly ensure the optimized memory and processing power with an energy-aware computing service. Dew computing on the other hand deals with caching, buffering, data processing, and providing analytics support when the whole network is offline [10]. It means that in the absence of a cloud, the Dew node can be able to give support in the form of storage and data preprocessing. This is also useful in an intermittently connected network where a group of nodes may be disconnected due to the unavailability of the network. In such cases, the Dew nodes perform operations independently in a real-time fashion. Energy awareness is another major feature of Dew computing devices. Table 1 summarizes the various computing paradigm and their features.

The Dew service methodology is primarily based on Dew computing which is accompanied by edge and cloud computing. The application perspective of the Dew-enabled Internet of Drone Things primarily amalgamates the dew and edge-level computing aspects including communication between the intermediate dew relay nodes, edge nodes, and broker nodes. Various researchers in the networking and Dew computing field have performed the research. Wang et al. [11], considered to be the pioneer of the Dew computing methodology, described the post-cloud computing model. The major contribution illustrates the classification of the post-cloud computing model with CDEF nomenclature. Where C stands for cloudlets, D stands for Dew, E stands for the Edge, and F stands for Fog computing. According to the author to fully leverage the potential of PCs with cloud computing services, Dew computing is a software organization model for PCs in the cloud computing era. The cloud-Dew architecture is used to organize software in the Dew computing paradigm. In addition to collaborating with cloud services, local computers can offer a wealth of capabilities. The post-cloud computing methodology majorly emphasizes resource management, computational complexity, load balancing, and offloading from one node to another node. Mobile edge computing on the other hand is another predominant research domain that primarily accompanies the cloud, Dew as well as fog. In the case of edge computing, the sensor data is primarily preprocessed at the edge level.

Edge computing with vehicular mobility can provide various sustainable solutions in a smart city and society. Wu et al. [12] proposed a flight trajectory optimization methodology for a UAV-assisted vehicular edge computing framework. A traffic situation awareness system and based on the traffic jam the dynamic path planning has been done in this case. Bao et al. [13] on the other hand proposed an edge-based

Table 1 Summary of the various computation paradigm and its feature

	App-based computing	Connectivity constraints	Offloading service	Resource optimization	Real-time computation	Energy awareness
Mobile computing	✓			✓		
Grid computing				✓		
Cloud computing	✓		✓		✓	
Fog computing			✓	✓		✓
Edge computing		✓	✓	✓	✓	✓
Dew computing		✓	✓	✓	✓	✓

joint client selection methodology for vehicular IoT. In this case, the Edge vehicles act as federated learning (FL) clients. Hochstetler et al. [14] on the other hand provide an application-level implementation of the vehicular deep learning application which illustrates intelligent edge computing. Dong et al. [15] proposed a new energy-efficient task scheduling for vehicular networks based on deep reinforcement learning (DRL). The key application area of the Dew as a service also encompasses the UAV networks. Majorly, this network plays a pivotal role to perform communication between Dew nodes, edge nodes as well as the cloud. Therefore, to develop sustainable Dew as a service framework for the Internet of Drone Things several types of research have been carried out. Khan et al. [16] described the UAV aided in a 5G framework. UAV elevation, UAV horizontal variation, and base station elevation has been discussed in this study. Gushev et al.[17] discusses how the component of Dew-inspired IoT-based cyber-physical systems acts with each other in an interconnected manner. The summary of the recent work on Dew computing with the various aspect of the Dew methodology has been tabulated in Table 2.

Table 2 Summary of the recent work, key features, and the research scopes

Author	Key focus area	Major component	Scope
Wang et al. [11]	Post-cloud era with dew paradigm	Implementation of the dew computing	Application service implementation
Wu et al. [12]	Application of trajectory optimization for edge-enabled UAV network	Edge enables UAV communication	UAV data buffering and caching using dew nodes
Bao et al. [13]	Edge-enabled selection application for vehicular IoT	Client selection methodology for the edge-enabled system	Edge dew coordination in vehicular IoT
Hochstetler et al. [14]	Implementation of intelligent edge	AIoT-based deep learning implementation	Edge dew network for federated learning in AIoT
Dong et al. [15]	Edge-enabled intelligent learning	Edge-enabled deep reinforcement learning	Leverage the dew computing methodology to increase the performance of learning
Khan et al. [16]	UAV-aided 5G framework design	UAV station elevation computing	Edge dew enables 5G-UAV networks
Gushev et al. [17]	Dew inspired IoT-based cyber-physical system (CPS) implementation	Interconnection of the IoT-enabled dew nodes for CPS	Major implementation of the dew as a service for IoT-CPS

3 Background Study

In a computing system, scalability is a crucial concern. Every firm aspires to develop a scalable and flexible computing infrastructure. This problem is connected to the organization's changing needs, which might broaden or narrow as the business develops. Initially, cloud computing was seen as a promising technique to solve scalability and flexibility issues. By choosing service levels based on their capabilities, cloud computing is thought to make it simpler for businesses to manage their computing resources. However, cloud computing technology cannot stand alone at this stage of development. The availability of mobile devices (mobile), followed by the Internet of Things (IoT), which promotes wider technological growth, enables the spread of the advantages of cloud computing technology. Dew computing is one of them. Dew computing is regarded as the implementation's lowest layer for distributed computing systems. The system's proximity to the user is indicated by the lowest layer. Dew computing prioritizes using the internet for data synchronization and empowering user devices as computing resources. Users of Dew computing do not necessarily need to be connected to the internet to enjoy system features.

The DewaaS philosophy is fundamentally based on some of the major and relevant technology which is the major pillar of this methodology. The technologies are cloud computing, mobile edge computing, fog computing, Internet of Drone Things (IoDT) [18], and Flying Ad-hoc Networks (FANET) [19].

3.1 Cloud Computing

Cloud computing is an emerging concept where data can be stored and analyzed remotely in a distributed manner. This is an extension of the distributed and grid computing methodology. Cloud computing is considered to be an effective component in today's distributed and intelligent computing methodology. Generally, the cloud computing methodology comprises various modules like the server, computing infrastructure, data center, analytics engine, and many more. Categorically they can be viewed as (a) Infrastructure as a Service (IaaS) where the cloud platform provides a hardware infrastructure like VCPU, RAM, storage, and GPU for processing the data. We can choose any operating system platform and establish the computing setup in this case. The user has the flexibility to choose the proper infrastructure depending on the data load. (b) Platform as a Service (PaaS) is another way to use cloud resources. In this case, the computing or the hosting platform has to be given to the user as the platform. For example, we can set up a cloud web hosting platform for hosting our website, or suppose we are executing a set of analytics codes in the cloud platform. In that case, we do use a PaaS methodology. (c) The SaaS (Software as a Service) on the other hand emphasizes the user approach by giving support at the application level [20]. In this case, the user cannot interact with the platform and the hardware infrastructure but only interact with specific software tools. Google Docs and spreadsheets are the best examples of SaaS. Often the cloud fog and edge can interact with each other to perform load balancing and load sharing.

3.2 Mobile Edge Computing

This is the major pillar of the DewaaS methodology. We are currently living on the edge of data. The data is mostly generated at the edge of the network. It is a good idea to process the data at the edge of the network as well. There are push–pull mechanisms [21] that are highly useful for edge computing where the cloud pushes some of the major data toward the edge for processing. Similarly, the IoT sensor nodes are also connected with the edge node, thereby pulling the data from the sensor node for processing. In the communication aspect, edge computing prefers simple TCP/UDP communication with sensor nodes and a secure TCP/UDP or lightweight MQTT communication from edge to fog or cloud [22]. As the dynamic network is concerned to achieve near real-time processing, a very low latency network has to be considered. Real-time streaming is also an important phenomenon of edge computing. In this case, the real-time video or audio data broadcasted from the drone or ground vehicle's camera can be processed on board. Another aspect of edge computing that is perhaps the most attractive for the current researchers is the intelligent edge [23]. This philosophy offers a fusion of edge computing with artificial intelligence. This exclusively refers to the edge computing platform which is capable of performing artificial intelligence tasks.

3.3 Fog Computing

The fog computing paradigm is another pervasive computing philosophy that is primarily useful to distribute the computing operation between the various layers to reduce the load of cloud service [24]. This platform primarily emphasizes the low-end and moderately powerful computing infrastructure. The major task of this layer is to perform a low-level computation that requires minimal memory and processing power. The device comprised of the fog computing infrastructure is known as the fog node. The data which are used for rigorous computation are mainly stored in the fog nodes. The major advantage of the fog computing infrastructure is the latency. In the case of fog, the data routing efficiency increases significantly as the message transfer latency decreases. Cloud offloading also is a big challenge that can be highly relevant in fog computing methodology.

3.4 Dew Computing

Dew computing, which is still a component of the development of distributed computing technology, can be seen as a tiny replica of the huge circumstances present in cloud computing systems [25]. Dew computing is a scaled-down form of massive cloud computing architecture. Given the volume of data saved in the cloud computing system, the term "Dew" might be used to analogize the data stored in this system. Several articles on Dew computing define Dew computing as an information technology paradigm that makes use of a user's device's capacity to be integrated with the system in the cloud to receive the most possible information availability (for example, a computer or other mobile device). One of the official definitions of Dew computing is as follows: the paradigm of an on-premise software-hardware organization that operates in a cloud computing environment and that, thanks to its on-premise capabilities, allows computers to operate independently of cloud computing services as well as in collaboration with them. The most effective use of in-house computers and cloud computing services is the core goal of Dew computing.

Some of the major components of Dew Computing Infrastructure are:

A. Dew Virtual Machine (DVM): a computer or personal computer that runs the Dew computing system in a virtualized environment. Several elements, including the Dew Server (DS), Dew Analytic Server (DAS), and Artificial Intelligence of Dew, are placed in the DVM (AID).
B. A Dew Server (DS): It is a representation or miniature copy of a cloud-based server that is connected to a local computer. This device will periodically communicate and sync with data stored on cloud servers.
C. Dew Analytics Engine: Data created when system users use or visit a Dew server is analyzed by the Dew analytic engine (DAE), which is a localized version of the analytics web server. DAE offers the benefit of preprocessing user data before sending it to a cloud server.
D. Artificial Intelligence Components (AIC) are components that use artificial intelligence to perform operational customization on the DS and to direct activities on the Dew server based on data collected from the DAE.

Figure 2 represents the top-to-the-bottom hierarchy of the cloud, fog, Dew, and edge computing methodology. From the picture, it is quite clear that cloud computing is at the top of the hierarchy, whereas fog computing is primarily used for offloading purposes. The Dew computing philosophy ensures caching processing, and computing support even if the network gets disrupted. The Dew and fog coordination are very crucial in case of emergency response where it might so happen that the dew node performs offloading toward the cloud. In this scenario, the load balancing algorithm takes a vital responsibility to offload the dew data to fog. Edge computing node on the other hand also depends on the Dew nodes in terms of load sharing, buffering, and caching.

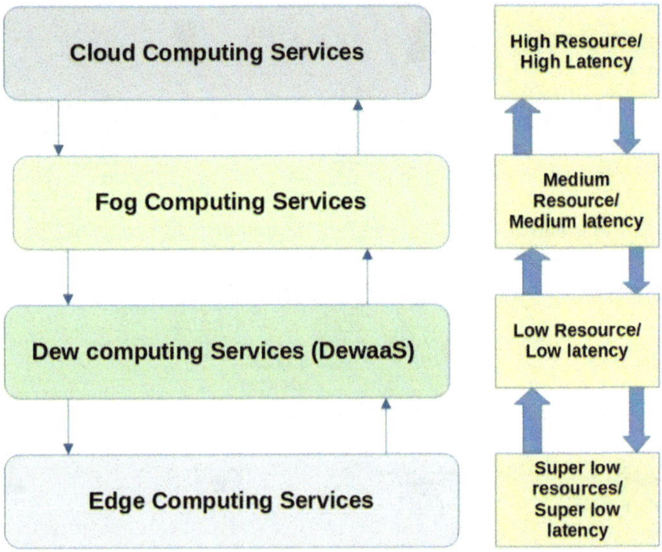

Fig. 2 Cloud dew fog computing hierarchy

4 Layered Architecture of Dew as a Service

A five-layered architecture can be visualized for the Dew as a service mechanism shown in Fig. 3. Layer 1 comprises the sensor nodes that gather the data from the ambiance. These kinds of nodes are having various types like sensor clients, sensors with storage, and sensors with a minimum processing capability. The sensors can be deployed in various items often known as things like cameras, automobiles, traffic signals, electronic mobile devices, etc. The main objective, in this case, is to gather data from various ambient parameters. The gateway UAV nodes are the additional component that helps to route the data in the Dew and edge layer.

Layer two is the Dew layer. The primary task of this layer is buffering and caching. The ambient parameters sensed by the sensor are generally converted by the raw data. This data gets buffered in the Dew nodes. This data majorly comprises crucial parameters. The task of this layer also classifies various functionality such as data filtering, parameter extractor, feature selector, and many more. The 3rd layer is the edge layer which is amalgamated with the network backbone layer. The main job of the edge layer is to perform edge-level processing. This comprises data preprocessing, feature engineering, feature extraction, principal component analysis, etc. The network backhaul which is considered to be the 4th layer majorly acts as a base for the data communication between the edge, cloud, and Dew. The 5th layer comprises the cloud layer where the major cloud services get deployed. In general, there are services like platform as a service (PaaS), software as a service (SaaS), and infrastructure as a service (IaaS). The AI-based big data analysis can be done in this layer.

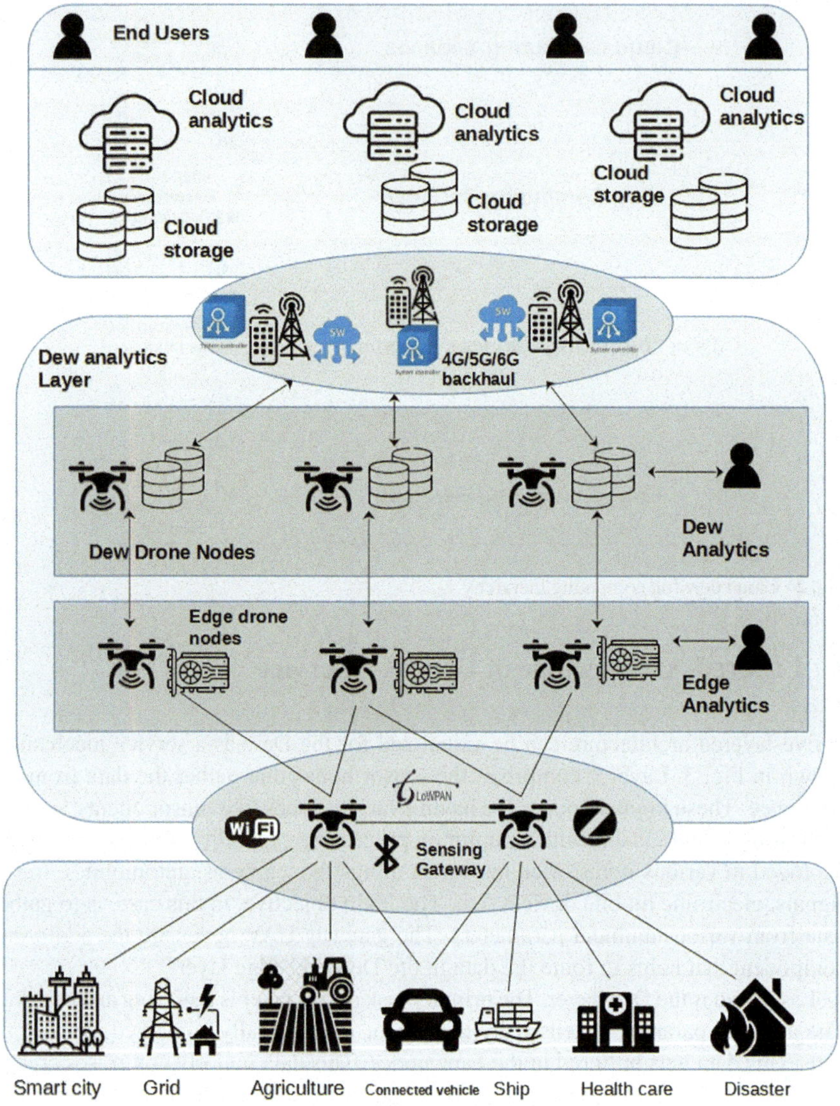

Fig. 3 The fundamental architecture of dew as a service for IoDT

4.1 Dew as a Service Methodology

Dew computing methodology is primarily needed due to unstable internet connectivity. In some situations like a natural disaster, or a vast geographic distance time, the network coverage of the LTE or 5G networks gets compromised. Especially during floods, storms, and earthquakes, the failure of the infrastructure networks is majorly disrupted. As a result, communication is at stake. In this situation, communication with the rescue workers is pretty difficult. Therefore, we need an infrastructure that can able to communicate in an ad hoc manner as well with there must be processing and decision-making capability are there. Therefore, fundamentally, Dew as a service can be achieved in 3 different scenarios:

(i) In this case, the entire network gets connected to the gateway. Hence, the data that has been gathered within the ad hoc network get preprocessed and routed to the cloud infrastructure. In some cases, the raw data is even routed to the cloud as well.
(ii) In the second scenario, the part of the network gets disconnected from the gateway. As a result, part of the network is unable to forward the data to the cloud. Therefore, it required a partially local-level processing infrastructure.
(iii) In the third case, which is perhaps most crucial is the complete disconnection of the network cluster from the gateway. As a result of that, no data can be able to route from the Dew layer to the cloud. This is a crucial issue in terms of data processing. To overcome this issue, the edge Dew-cloud infrastructure combination is highly required.

In order to achieve this an edge-enabled Dew network can be designed that cater to the Dew as a service philosophy. Primarily, Dew computing is an extension of the client–server architecture. The Dew-Cloud architecture on the other hand works as a peer-to-peer (PPP) architecture. The best place to start is with the Dew-Cloud architecture. However, a recapitulation of the typical client–server architecture is worthwhile before that. The foundation of current distributed network application frameworks is a client–server architecture. It is made to divide up allocated or scheduled jobs between service providers and potential service seekers. It was initially intended to offer a client-dispersed service from anywhere in the network.

The classical Dew-Cloud architecture primarily encompasses the major components like the Dew server, Dew database, client-side program in Dew, and client services. Commonly there are 1 to 1 as well as 1-to-many relationships. The primary goals of the proposed Dew Server include (1) to provide users with the services they seek, such as cloud computing, and (2) to synchronize and correlate local and remote data.

Dew Server has the following essential features:

i. It must be compact, meaning that its structure, function, and application should be clear.
ii. Normally, it will only serve one client at a time.
iii. It should typically keep the user information that is utilized frequently in its databases.
iv. The size of the data that is stored must be as small as possible.
v. Because of its irregular interactions with fatal failures such as malware attachment by viruses or system destruction, it may be vulnerable to data loss.
vi. The cloud server should be able to regenerate the missing data.

vii. It will be in charge of all tasks, such as DBMS management, identity mapping, raw data synchronization, rule-based data collection, Dew script analysis, etc. In the end, it will become Personal information.

The most important component of the suggested Dew-Cloud architecture is the Dew Server. Therefore, it needs to include a variety of adaptive technologies. Typically, it includes four parts. necessary co-servers, including a database, a mobile information server, POP/IMAP messaging protocols, and the application. In most cases, when a user accesses or searches for website material online, the "remote cloud/server" all of the responses to such inquiries. According to a theory, the user can browse the same contents using the Server Dew. The existence or absence of the internet won't be a problem. Consider, for instance, the website. This has already been incorporated into the Dew-Cloud architecture, available for surfing. The Dew Server is located on the user's local computing device.

This may be referred to as a "Dew Site," which is similar to the current website. Once such a Dew Site is present on the local computer, the user can browse the internet for information on the Dew Site and perform any necessary adjustments. However, altering a Dew Site is a difficult task. A unique script file created by a web developer language may be used to deal with these activities. It's important to remember that each Dew Site needs to have a unique counterpart in Dew Script. Now, whatever details Dew Script can alter content that already exists in a Dew Site. It did as directed by "Dew Analyzer". Dew Analysis Tool is a device made to handle all potential charges.

4.2 The Deployment Ecosystem for DewaaS for IoDT

In a smart city scenario, the deployment of the Dew nodes can be very crucial. In a city-like scenario, we may visualize numerous applications of the DewaaS such as traffic monitoring, environment condition monitoring, traffic route prediction, and many more (shown in Fig. 4). In a village scenario like India and other South Asian countries, on the other hand, the cultivated land, vegetation, and cattle monitoring can often be done using drones or a swarm of drones [26]. Dew-based IoDT can be achieved in various ways. 1. An ecosystem of a single drone and the base station can be considered. In this case, a single drone performs a search and rescue mission. The data gathered by this drone may directly forward to the base station. 2. A set of 5 to 5 drone units. One of them can be considered a master and the rest of others are the slave. In such cases, the data may be buffered in the slave nodes but it has never been processed by it. 3. In the third case, the system can be considered as the significant number of drone nodes that can perform collaborative processing. Modeling of the communication system for the Dew-enabled IoDT can be achieved in many folds. Based on the service allocation, a classification of the DewaaS can be done as below.

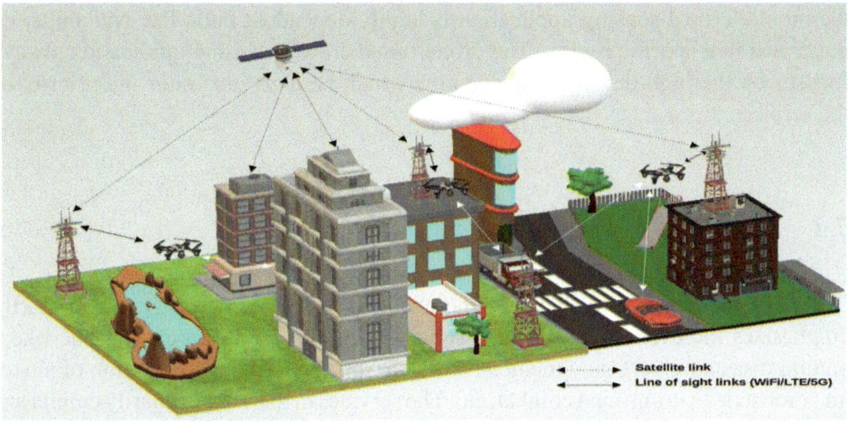

Fig. 4 The ecosystem of the application usage

5 Classification of the DewaaS

Based on the service availability, the DewaaS can be classified into various service categories. These categories are primarily depending upon demand, priority, location, energy awareness, and the coverage of the services based on the network and the resource coverage. In the case of the IoDT application perspective the demand, network, and resource coverage, energy management are very crucial aspect that needs to be addressed based on the drone network deployable in a smart city scenario. The next section has a detail description of the all classes of services.

5.1 On-Demand Service

One of the major Dew computing services that can be realized is the on-demand service. In this case, the service is given based on the emergency demand. Disaster management, real-time application, and mission-critical situations are services that are highly used. The important phenomenon of this service is the availability of the network, computing, and storage resources which are crucial challenges in the on-demand service.

5.2 Priority Aware Service

Priority-aware DewaaS can be useful when we have to select the priority service among all. The services that have top priority in a particular instance can be identified by the utilization of the computing, network, and storage resources. For example, the mobile crowd-sensing application is highly dependent upon the availability of the crowd in a specific region. The drone-based crowed sensing application always focuses on the high-density crowded area which perhaps illustrates the priority of the service.

5.3 Location-Aware Service

Location-aware service is a major phenomenon of the DewaaS. This system primarily emphasizes the event location. This service is highly used in a catastrophe like a natural disaster, emergency medical situation, mission-critical application of air-to-air combat or air-to-ground combat, etc. The services in this case primarily emphasize the region of interest. The location-based Dew services are also categorized into two major parts. In the first case, the outdoor location has been considered. The UAV GPS module takes a pivotal role in this case. The GPS data in this case helps the node to identify its mobility the node. The map-based waypoint mobility model is the best model to choose in this case. Secondly, the indoor location-based service can also be considered. In this case, the drone performs indoor localization.

5.4 Energy-Aware Service

Energy-aware Dew service is a major perspective of the DewaaS. Numerous technologies are involved in implementing energy-aware services. Primarily, in this case, the renewable energy-based system is used to run the low-power Dew computing device. Based on the energy availability, the system has been configured and the computation algorithm has been redefined. The Dew nodes, in this case, perform computation based on the available energy and the power dissipation has also been monitored strictly.

5.5 Coverage-Aware Service

Coverage-aware service signifies the ad hoc distribution of the computation capabilities of the Dew nodes based on the coverage of the nodes. In some cases, when the Dew node is deployed for a Flying Ad-Hoc Network (FANET) ecosystem where the nodes are performing mission-critical tasks, then it might so happen that the Dew nodes are not getting coverage properly to transfer the vital information from one node to the other. In such a case, the Dew network must choose the set of nodes comprised of better coverage so that the major and critical computation can be done on that set of nodes. This service is highly useful for intermittently connected mission-critical applications. In the next section, we are going to describe the concept of the Internet of Drone Things (IoDT) that leverages the UAV-based Flying Ad-Hoc network (FANET).

6 FANET-Based IoDT Ecosystem

The major component of the FANET is the UAV. UAV in this case can be termed as UAV node. We can also consider the heterogeneous nodes within a FANET ecosystem. In this case, the FANET also comprises quasi-stationary balloons or high-altitude platforms (HAP) [27]. This can act as a forwarder or a buffer unit. The UAV node on the other hand performs the movement based on the specified mobility models. Another major factor is routing. In the case of a FANET, the frequent disconnection of the network node is an obvious phenomenon. Due to that, the message transfer may be interrupted. Under these circumstances, the performance of traditional routing protocols like DSDV and AODV gets compromised. Therefore, we have to choose opportunistic routing protocols. The Delay tolerant network (DTN) philosophy, in this case, provides an ample opportunity to leverage the opportunistic message forwarding feature. The use of DTN is crucial in the FANET scenario as it supports store and forward methodology. Each of the DTN-based FANET nodes itself can be considered as a Dew buffer element that can store the message temporarily. We can re-engineer the standard IoT message transfer protocols like MQTT and MQTT-SN with opportunistic forwarding features. The concept of DewDrone in this case is a state-of-the-art implementation of the Dew as a service accompanied by the UAV. There is a layered approach that has been followed by the Dew drone ecosystem in this case. The bottom layer is the perception layer where the data gathering is performed. The Dew and the edge layer perform in a coordinated fashion. The dew nodes majorly perform buffering, whereas the edge nodes perform the processing. The drone node itself has onboard edge processing and dew buffering capability. In the case of implementation strategy in each network, two types of drones must be present. One is the Dew drone which comprises the data storage and the broker instance. And secondly, edge drone nodes comprise of edge processing services and edge brokers. These brokers essentially perform the message routing in this case. Figure 5 depicts the message transfer sequence of the same.

Consider 'm' as a message instance such that '*m*' comprises 3 tuples shown as follows:

$$m = (t_m, C_m, X_{\text{pm}})$$

Here T_m stands for the initial token value, which is only a timer, C_{m} for the number of messages, and X_{pm} for the likelihood that messages will be exchanged. The encounter of the n Dew nodes (Ed) can be expressed as follows:

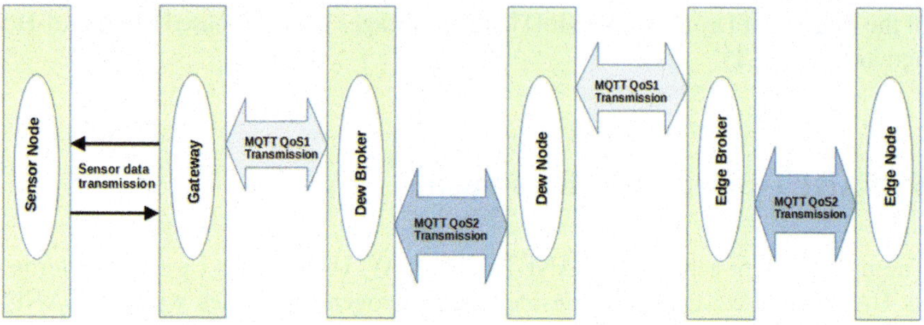

Fig. 5 Message transfer sequence from the sensor to dew nodes and the edge nodes

$$E_d = \sum_{i=1}^{n} (\tau_r + V(m)_i + t) \tag{1}$$

$V(m)_i$ is the ith message vector sent during transmission, τ_r is the latest encounter time, and t is the execution time. The timestamp of the token is compared with the next method when the message is encountered.

$$t_m(n)_i = \sum_{i=1}^{E_d(n)} \tau_r - t_m(i-1) + d(i, i-1) \tag{2}$$

The ratio of the distances between the ith and $(I-1)$th nodes can likewise be used to define $\tau_r - t_m\ (i-1)$. Consequently, we can reword the following expression:

$$t_m(n)_i = \sum_{i=1}^{n} d(i, i-1)\left(\frac{1+v(i)}{v(i)}\right) \tag{3}$$

The message's subscriber selects the message with the lowest timestamp value out of all those encountered, which we can give as

$$t_{\text{sub}} = min(t_m(n)_1, t_m(n)_2 \ldots, t_m(n)_n) \tag{4}$$

Latency is a major issue that needs to be addressed. For the transmission of the sensor data from the sensor to the gateway, from the gateway to the edge node, and then from the edge node to the Dew node, the total latency computation is crucial. When considering the Message Queuing Telemetry Transport (MQTT) of Quality of Service 1 (QoS 1) level for publishing and QoS for subscription, l_d is the delay for conveying the data from the sensor node to the Dew node via the gateway and Dew-broker.

The latency for transmission from the Dew node to the edge node can also be expressed as l_e. In this instance, the overall latency is calculated as follows:

$$l_{\mathrm{db}} = \left(l_{\mathrm{tx}} + l_{\mathrm{cnack}} + l_{\mathrm{puback}} + l_{\mathrm{fwd}}\right) \tag{5}$$

The latency l_{db} between the sensor and the broker is shown here. Furthermore, we can define l_{dn} as the delay from the broker to the Dew node

$$l_{\mathrm{dn}} = \left(l_{\mathrm{cnack}} + l_{\mathrm{pubrel}} + l_{\mathrm{suback}} + l_{\mathrm{fwd}}\right) \tag{6}$$

As a result, the total delay attained from the sensor to the cache node can be stated as

$$l_d = l_{\mathrm{db}} + l_{\mathrm{dn}} \tag{7}$$

By repeating the process, we can calculate the latency for data transfer to edge nodes via the edge broker. Since l_{eb} l_{db} and l_{en} l_{dn}, we can write $l\ e = l_{\mathrm{eb}} + l_{\mathrm{en}}$.

7 Intelligent Processing for Dew

In order to process the data in a mission-critical scenario, the Dew as a service act as a pivotal role in the low-power intelligent processing. Various intelligent methodologies can be adopted in this case. One of the major methodologies is federated learning (FL) [28, 29]. Distributed machine learning is necessary for human-centered mobile traffic. The developed B5G and 6G mobile networks are built on this as their basic framework. Modern machine learning models must have their training data centralized on a single server to be trained classically. Transmission latency is particularly high because a large volume of data needs to be uploaded to the cloud data center. In the cutting-edge application development paradigm, such as unmanned aerial vehicles, AR-VR applications, and driverless vehicles, which employ a centralized machine learning technique. The transmission of the gathered data to the data center is essential due to the restricted communication resources available. Therefore, processing data at edge devices becomes increasingly important (Fig. 6).

A single global statistical model must be learned in the usual federated learning issue using data that is stored on tens to maybe millions of remote devices. We want to learn this model with the restriction that device-generated data are stored and processed locally, with only intermediate updates being regularly exchanged with a central server. The purpose is usually to reduce the following objective function. Several core challenges must be addressed while implementing the federated learning mechanism. Data generated on each device must stay local because communication in federated networks is a major bottleneck, and transferring raw data raises privacy issues. In fact, because of the potentially enormous number of devices in federated networks, such as millions of smartphones, communication in the network may be much slower than local computing by many orders of magnitude. This is because the network may have limited bandwidth, energy, and power. It is crucial to create communication-efficient techniques that iteratively send brief messages or model

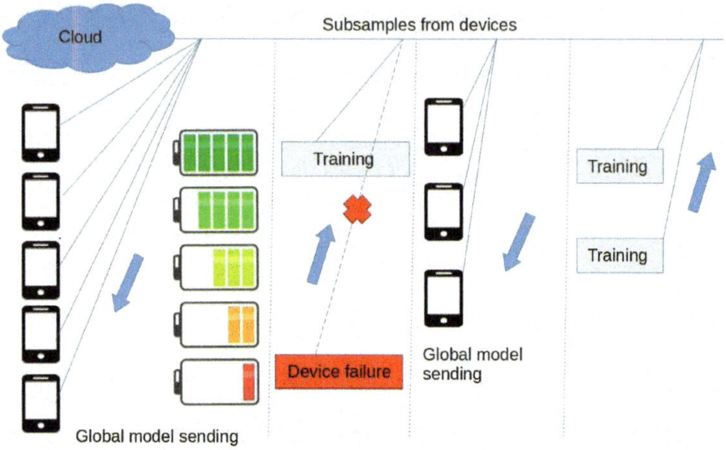

Fig. 6 Federated learning methodology for dew services

updates throughout the training phase rather than sending the complete data set over the network to fit a model to data produced by the devices in the federated network. Reduce the overall number of communication rounds and the size of the transmitted messages at each round to further cut down on communication in this situation. Another challenge in the federated learning environment is device heterogeneity. In the case of a 6G network, the number of mobile nodes is huge. And they may follow different types of platforms and different communication protocols as well. So the integration between the various protocols and the standards is very crucial in this case in order to achieve the proper synchronization of the 6G ecosystem. Federated learning helps to integrate various platforms in this case. In many different areas, including network optimization, clustering, user coverage, system management, and many more, FL can be used to manipulate exceedingly precise optimization issues. Additionally, FL gives users the ability to jointly develop and verify classification and regression models for predicting user behavior, user coverage, authentication, and wireless parameter analysis.

8 Conclusions

The Dew service methodology is a predominant research domain that leverages the implementation of Dew computing in numerous computation domains with the amalgamation of cloud, fog, and edge computing methodology. This chapter primarily studies the various service classifications and the application domain Dew along with the possible deployment strategies. Also, the implementation of the intelligent Dew-enabled decision-making methodology has been focused on conceptualizing federated learning methodology. As a future scope, we can implement this service extensively in real-time, mission-critical applications, serverless applications, and energy-aware computing applications. We can also reflect on the possible solution of communication methodology while deploying the DewaaS for IoDT so that we can leverage the maximum throughput from it. It can be easily said that the Dew as a Service paradigm opens a door to the next level of research for intermittently connected Drones, as well as vehicular networks.

References

1. Dang, L.M., Piran, M.J., Han, D., Min, K., Moon, H.: A survey on internet of things and cloud computing for healthcare. Electronics **8**(7), 768 (2019)
2. Cao, K., Liu, Y., Meng, G., Sun, Q.: An overview on edge computing research. IEEE access **8**, 85714–85728 (2020)
3. Mukherjee, A., De, D., Dey, N.:Dewdrone: dew computing for internet of drone things. IEEE Consum. Electron. Mag. (2021)
4. Mukherjee, A., Dey, N., Kumar, R., Panigrahi, B.K., Hassanien, A.E., Tavares, J.M.R.: Delay tolerant network assisted flying ad-hoc network scenario: modeling and analytical perspective. Wirel. Netw. **25**(5):2675–2695 (2019)
5. Nazir, S., Ali, Y., Ullah, N., García-Magariño, I.: Internet of things for healthcare using effects of mobile computing: a systematic literature review. Wirel. Commun. Mob. Comput. **2019** (2019)
6. Parashar, M.: Autonomic grid computing: concepts, requirements, and infrastructure. In: Autonomic Computing, pp. 73–94. CRC Press (2018)
7. Gai, K., Guo, J., Zhu, L., Shui, Y.: Blockchain meets cloud computing: a survey. IEEE Commun. Surv. Tutor. **22**(3), 2009–2030 (2020)
8. Naha, R.K., Garg, S., Georgakopoulos, D., Jayaraman, P.P., Gao, L., Xiang, Y., Ranjan, R.: Fog computing: survey of trends, architectures, requirements, and research directions. IEEE Access **6**, 47980–48009 (2018)
9. Pham, Q.V., Fang, F., Ha, V.N., Piran, M.J., Le, M., Le, L.B., Hwang, W.J., Ding, Z.: A survey of multi-access edge computing in 5G and beyond: fundamentals, technology integration, and state-of-the-art. IEEE Access **8**, 116974–117017 (2020)
10. Singh, P., Kaur, A., Aujla, G.S., Batth, R.S., Kanhere, S.: Daas: dew computing as a service for intelligent intrusion detection in edge-of-things ecosystem. IEEE Internet Things J. **8**(16), 12569–12577 (2020)
11. Wang, Y.: Definition and categorization of dew computing. Open J. Cloud Comput. (OJCC) **3**(1), 1–7 (2016)
12. Wu, Z., Yang, Z., Yang, C., Lin, J., Liu, Y., Chen, X.: Joint deployment and trajectory optimization in UAV-assisted vehicular edge computing networks. J. Commun. Netw. **24**(1), 47–58 (2021)
13. Bao, W., Wu, C., Guleng, S., Zhang, J., Yau, K.L.A., Ji, Y.: Edge computing-based joint client selection and networking scheme for federated learning in vehicular IoT. China Commun. **18**(6), 39–52 (2021)
14. Hochstetler, J., Padidela, R., Chen, Q., Yang, Q., Fu, S.: Embedded deep learning for vehicular edge computing. In: 2018 IEEE/ACM Symposium on Edge Computing (SEC), pp. 341–343. IEEE (2018)
15. Dong, P., Ning, Z., Ma, R., Wang, X., Xiping, H., Bin, H.: NOMA-based energy-efficient task scheduling in vehicular edge computing networks: a self-imitation learning-based approach. China Commun. **17**(11), 1–11 (2020)
16. Khan, S.K., Naseem, U., Sattar, A., Waheed, N., Mir, A., Qazi, A. and Ismail, M.: UAV-aided 5G network in suburban, urban, dense urban, and high-rise urban environments. In: 2020 IEEE

19th International Symposium on Network Computing and Applications (NCA), pp. 1–4. IEEE (2020)
17. Gushev, M.: Dew computing architecture for cyber-physical systems and IoT. Internet Things **11**, 100186 (2020)
18. Nayyar, A., Nguyen, B.L. and Nguyen, N.G.: The internet of drone things (IoDT): future envision of smart drones. In: First International Conference on Sustainable Technologies for Computational Intelligence, pp. 563–580. Springer, Singapore (2020)
19. Mukherjee, A., Keshary, V., Pandya, K., Dey, N., Satapathy, S.C.: Flying ad hoc networks: a comprehensive survey. Inf. Decis. Sci. 569–580 (2018)
20. Mohammed, C.M. and Zeebaree, S.R.: Sufficient comparison among cloud computing services: IaaS, PaaS, and SaaS: a review. Int. J. Sci. Bus. **5**(2), 17–30 (2021)
21. Fan, Z., Yang, W., Fan, W., Cao, J., Shi, W.: Serving at the edge: an edge computing service architecture based on icn. ACM Trans. Internet Technol. (TOIT) **22**(1), 1–27 (2021)
22. Mukherjee, A., Dey, N., De, D.: EdgeDrone: QoS aware MQTT middleware for mobile edge computing in opportunistic internet of drone things. Comput. Commun. **152**, 93–108 (2020)
23. Zhu, G., Liu, D., Yuqing, D., You, C., Zhang, J., Huang, K.: Toward an intelligent edge: wireless communication meets machine learning. IEEE Commun. Mag. **58**(1), 19–25 (2020)
24. Yousefpour, A., Fung, C., Nguyen, T., Kadiyala, K., Jalali, F., Niakanlahiji, A., Kong, J., Jue, J.P.: All one needs to know about fog computing and related edge computing paradigms: a complete survey. J. Syst. Archit. **98**, 289–330 (2019)
25. Hirsch, M., Mateos, C., Zunino, A., Majchrzak, T.A., Grønli, T.-M., Kaindl, H.: A task execution scheme for dew computing with state-of-the-art smartphones. Electronics **10**(16), 2006 (2021)
26. Hou, X., Ren, Z., Cheng, W., Chen, C., Zhang, H.: Fog based computation offloading for swarm of drones. In: ICC 2019–2019 IEEE International Conference on Communications (ICC), pp. 1–7. IEEE (2019)
27. Mukherjee, A., Chakraborty, S., Azar, A.T., Bhattacharyay, S.K., Chatterjee, B. and Dey, N.: Unmanned aerial system for post disaster identification. In: International Conference on Circuits, Communication, Control and Computing, pp. 247–252. IEEE (2014)
28. Hosseinalipour, S., Azam, S.S., Brinton, C.G., Michelusi, N., Aggarwal, V., Love, D.J., Dai, H.: Multi-stage hybrid federated learning over large-scale wireless fog networks (2020). arxiv:09511
29. Jin, R., He, X., Dai, H.: On the design of communication efficient federated learning over wireless networks (2020). arXiv:2004.07351

Dew Aeroponics: Dew-Enabled Smart Aeroponics System in Agriculture 4.0

Baishali Ghosh, Samarjit Roy, Nurzaman Ahmed, and Debashis De

1 Introduction

Agriculture is the main origin of raw materials for numerous industries all over the world, so it has a great impact on international trade and the country's revenue, apart from this, it is the main source of food for our entire fauna. It creates food security for the entire mankind. So, it is a very significant element for a country's development. Moreover, it helps to heal nature automatically which is a great need in the present age.

India is a major agricultural country. Agriculture is the profession of 58% of the citizens in India. Agriculture contributes a lot to the Indian economy. In India, the share of agriculture in GDP is approximately about 20%. With the increase in food demand and unpredictable natural conditions, there is an increase in labour costs and a decrease in the area of agriculture [1]. There is a need for enhancement in agriculture, which can solve these problems. Here comes the idea of precision agriculture, but the main problem is the yield and energy requirement. So, we need some technology that will give us more yield than our energy consumption. Here the idea of aeroponics is introduced. Aeroponics is the process of cultivation where plants grow without any soil, where the nutrient solution is sprayed on the roots of

B. Ghosh · D. De
Centre of Mobile Cloud Computing, Department of Computer Science and Engineering, Maulana Abul Kalam Azad University of Technology, Haringhata, Nadia, West Bengal 741249, India

S. Roy (✉)
School of Computer Science, Engineering, and Applications (SCSEA), Department of Computer Science and Engineering, D Y Patil International University, Pune, Maharashtra 411044, India
e-mail: samarjit.tech89@gmail.com

School of Computing and Information Technology, Eastern International University, Thu Dau Mot City, Vietnam

N. Ahmed
Donald Danforth Plant Science Center, St. Louis, MO, USA

D. De and S. Roy (eds.), *Dew Computing*, Internet of Things,
https://doi.org/10.1007/978-981-99-4590-0_13

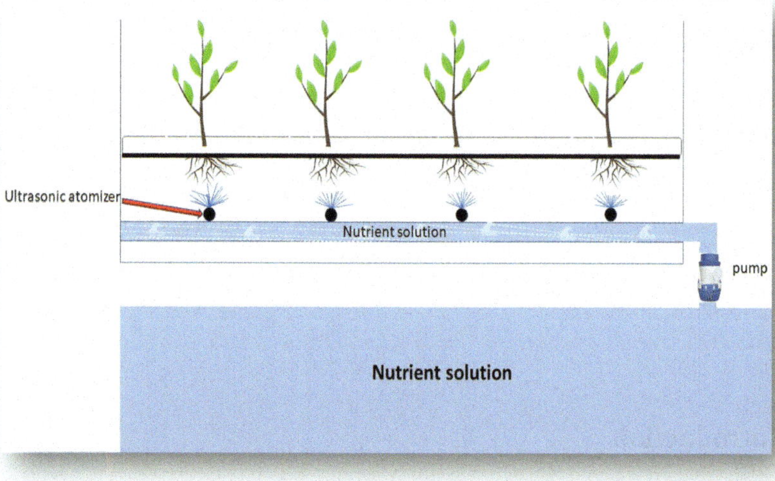

Fig. 1 Basic aeroponic system

the plants, which is represented in Fig. 1. In India, 80% of fresh water is wasted on agriculture, but in this system, plants grow rapidly with very less water and other resources. Generally, in this system, the plants and crops grow two to three times more than in traditional agriculture. In this system, roots are getting alimentation from the air where nutrient solution is misted, but along with the nutrient solution, there is also present oxygen of environment which is accepted by the roots of the plants, for this plant gets more oxygen as compared to roots in soil that is why it is coming out as a more efficient way of cultivation [2].

In 1997, NASA and some companies have experimented with soilless plant growth in microgravity. As a result, they came out with one solution ODC (Organic Disease Control, or Organically Derived Colloidal), which can improve the growth of the plant and stop disease and infection in an aeroponic environment by enhancing plants' immune system. They did another experiment that adzuki bean seeds and seedlings are planted in both piles of earth and aeroponic systems in the microgravity environment. The outcome of this experiment was that the seeds and seedlings which were treated in aeroponics with ODC grew rapidly and got less fungal infection than seeds and seedlings planted on earth [3]. This aeroponic system can be controlled automatically, using some sensors to get the data from the environment of the aeroponic system and one processing unit which processes the data and applied them to the aeroponic system. As a result, they found this automated controlled aeroponic system is more labour-saving and will also increase the economic value of the product [4]

Although, we can use machine learning to make a prediction system, due to the lack of data the machine learning algorithm model's performance is not optimal. So, we can use sensor fusion (Multi-sensor data fusion is the solicitation of respective sensors at once) for evaluating some parameters at a time and predicting the correct decisions of actuators [5].

1.1 Motivations

A sensor can detect whether a plant needs water or not by sensing some parameters from the leaf, which can prevent water wastage in any kind of irrigation system. The temperature difference between leaf and air is measured by the sensor which is related to the plant water stress. These analogue temperature sensors are connected with ADC. One sensor senses the air temperature (T_{air}) and another one measures the canopy temperature of (T_{leaf}). Then it is sent to the receiver where both sensor's voltage value and temperature are calculated as,

$$T_{\text{leaf/air}} = \left\{\left(5.506 - \sqrt{}\left(36.445 - 0.00704\ \text{V}_{\text{leaf/air}}\right)\right) - 0.00352\right\} + 30$$

Also, the temperature difference for a leaf sensor is calculated as $T_{\text{leaf}} - T_{\text{air}}$ [6]. We can get high reliability, lower latency, and less power consumption by using edge intelligence with 5G Narrow Band (NB) IoT in agriculture4.0 with real-time applications. Usage of Multilingual Voice User Interface control-enabled drones can make an automated agriculture system much more advanced which will enable the contactless multilingual voice user interface [7]. But in the fifth-generation (5G) networks, for implementing some end-user applications there is still some existence of interference. That's why some communication protocols are developed, but the problem is still present. To reduce this issue there are some protocols and standards are discussed like non-orthogonal multiple access (NOMA) variants which can solve most of the problems in contention-based grant-free transmissions, it is maintained by the appliance-to-appliance, supportive communication, multiple-input and multiple-output (MIMO), and the heterogeneous networks (HetNets) [8].

In this era there is a lot of wireless technology is present, among them, Li-Fi (Light Fidelity) technology is considered to be cheaper and faster. This system uses LED for sending digital data, we can say it can be the optical version of Wi-Fi (Wireless Fidelity). It is a highly reliable communication system, without complexity like other wireless networks and protocols, and harmonizing technology to the accessible radio frequency (RF) communication technology having high capacity. Its data rate is very high due to the availability of a huge license-free spectrum [9]. A Li-Fi system contains an intensity-contained light-emitting diode (LED) and a receiver device penetrating to precise high-frequency modulations of the luminous intensity. In general silicon photodiode of PIN (P-type intrinsic N-type) or APD (Avalanche photodiode) is used as a receiver, but considering some factors like frequency responses, signal-to-noise ratio, and bit error rate this PIN and APD were found to be inefficient for outdoor Li-Fi communication. In research, it is proved that photodiode works well in the indoor system but when it comes to outdoor systems and sunlight exposure of more than 200 W/m^2 then it is superimposed with a Li-Fi signal. So recent studies suggested photovoltaic (PV) modules, in VOC (open-circuit voltage) condition which can operate a Li-Fi transmission until around 800 W/m^2. Thus, it can easily operate under additional solar illumination, and it can be also considered in indoor conditions when sunlight illuminations are substantial like the system is operating near an open window or door [10, 11]. Edge computing in the Internet of Things activates the unified transmission of services and information processing from the cloud tier to the integrated edges in a all-encompassing distributed Internet of Things. In [12], authors have presented a Dew Computing as a Service for intelligent intrusion detection for the IoT systems. They increased filtration accuracy by using a deep learning-based classifier. They used a simulated environment and got a 60% improvement in latency.

Being motivated by above mentioned automated aeroponic system, we present an aeroponics system with Dew Intelligence, Hybrid Li-Fi and 5G NB-IoT enabled Smart Aeroponics System.

1.2 Chapter Contributions

The substantial contributions of this chapter are stated in the following:

(a) *Internet of Things-induced Intelligent Pumpless Green Aeroponics Systems* A remotely controlled automated aeroponics system without the requirement of a pump is presented.
(b) *Smart Agriculture Using Dew Computing* A new paradigm is introduced by adding the extra layer between cloud and end devices, which will decrease the 24 h Internet dependency.
(c) *Secured IoT-based Aeroponics Using Federated Learning* Secure aeroponics system with a decentralized learning method called federated learning that will provide an extra layer of security to the agricultural data.
(d) *More Reliable Communication System Using Li-Fi* for data transmission purposes Li-Fi is proposed as it has a high data rate and low latency. This will provide a new paradigm to the automated remotely controlled smart aeroponics system.

1.3 Chapter Organization

The remaining chapter has been systematized in the following. The Sect. 2 illustrated the chronological background of this projected schema in the Aeroponics contexts, the Dew computing and IoT approaches. Section 3 represents the benefits of typical aeroponics over hydroponics and traditional cultivation. Hydroponics systems are also elucidated in the prescribed section. In Sect. 4, we state the problems with a manual aeroponics system with a pump. We illustrate the performance of the aeroponics systems with a pump. In Sect. 5 we presented the architecture of the system in terms of advantages. We also illustrated the pumpless aeroponics with voice user interface and data security with federated learning and latency improvement in using dew computing. We have done a comparative analysis and discussions on the speed of different technologies and LASER-based Li-Fi with LED-based Li-Fi. We discussed hybrid Li-Fi implementation in Pumpless Aeroponics with 5G NB-IoT enabled IoDT. Finally, this chapter is accomplished with the conclusion remarks and the scopes for future challenges as presented in Sects. 6 and 7.

2 Related Works

The illustrations of the innovative IoT-based aeroponics system in the domains of the dew computing schema are questionably intermittent. However, numerous substantial technical challenges have been indicated on the viewing platform into three subsections. Proposed section discusses the prevailing assistances by the contributors have been illustrated in diverse domains for assessing the anticipated Dew Aeroponics framework. State-of-the-art illustration of respective approaches to smart aeroponics systems in an IoT-based framework has been depicted in Table 1.

2.1 Aeroponics

Aeroponics is used to generate good characteristic fragrant roots of Hemidesmus Indicus initially [13]. A comparison is made between soilless and soil-dependent archetype. As a result, it had been found that growth in the aeroponic schema is much better for both above and underground plants. For aeroponics, in vitro maintained saplings with health conditions were chosen, and they were allowed to grow for 8 weeks. In this scheme, a Hoagland solution is used. To measure the growth of plants Plant shoot diameter, length, and biomass (fresh and dry weights) were measured manually in 2, 4, and 8 weeks. Roots are analyzed by a rhizovision. The analyses of the image of the root are carried out from manually collected data. This part can be further enhanced by adding Drones for monitoring the system. The growth of leaves is calculated by totaling the number of mellowed leaves, leaf surface region, and chlorophyll contented of leaves. This study stated that an aeroponic system can produce healthy roots by controlling the root environment like temperature, and nutrient echelons and those plants don't have any kind of abnormal plant growth and development. But it would be better if the total aeroponic system was connected to an IoT system where all data can be stored and can be processed for future enhancement. It would be better if some user interface is present in this system. As per their suggestion, the aeroponic system can boost agriculture practices over traditional harvesting.

Some aeroponics systems are implemented in low-light places or indoors [14]. This system has been built up with some additional light for cultivation. In this

Table 1 Comparison table based on features

	Proposed model	[13]	[14]	[18]	[4]	[15]	[19]	[16]
Federated learning	✓	✗	✗	✗	✗	✗	✗	✗
User interface	Voice user interface	✗	GUI	GUI	✗	GUI	GUI	Graphical user-interface
Detect the level of nutrient solution in the preserver	HC sr04	✓	✗	Ultrasonic level sensor	✗	✗	✗	✗
Wi-Fi/GSM/Li-Fi	Li-Fi	✗	Wi-Fi	GSM	✗	Xbee	Wi-Fi	LAN
Humidity	✓	✗	DHT11	✓	DHT11	✓	✓	✓
Temperature	✓	✓	✓	✓	LM35	✓	✓	✓
pH factor checking	✓	✗	✗	✓	✗	✓	✗	✓
Automatic	✓	✗	✗	✓	✓	✓	✓	✓
Manual	✓	✓	✓	✓	✗	✗	✗	✗
Cloud	✓	✗	–	ThingSpeak	✗	ThingSpeak	ThingSpeak	Network-attached storage
Light intensity sensor	✓	✗	BH1750	✗	✓	✓	BH1750	✗
Pumpless water supply	✓	✗	✗	✗	✗	✗	✗	✗

situation grow lights specifically for agriculture) can be used, and for controlling the grow lights, we can use machine learning's prediction system. But a prediction system needs a lot of data, for this purpose sensor fusion concept is used. In this system grow lights are expended as actuators to operate light intensity that can increase the value of light intensity. This will act as a photon source for the plan. In Machine learning, a random forest classifier is used with the data from sensor fusion. The sensor's detected data and the environment's condition which is controlled by the actuators are shown on the user interface. If the data coming from the sensor and the predetermined data does not match, then the user will act for adjusting the actuator. Random forest is a machine learning method which uses the decision tree strategy. After classification method is performed by random forest then accuracy needs to be calculated, using the formula in Eq. (1).

$$\text{Accuracy} = (x + y)\frac{(x + y)}{(x + y + z + t)} \tag{1}$$

where x is truly positive, y is a true negative, z is a false positive, and t is a false negative. Thus, we will get to know how thriving the model from the random forest classification methodology has been carried out. As a result, it is found that multi-sensor information fusion has a good impact on the performance of random forest classification, and it is found that sensor fusion usage is more accurate than the usage of only one sensor, the consequences are 90.62% equated to 85.51% correspondingly. But in all this system data has been collected manually which is a drawback of this system. Instead of this if they had used a dynamic sensor then this system will be more enhanced.

This study is an efficient and simple agricultural data-gathering system [15]. That has been done using IoT and data was processed by big data. This system can be low-cost. They have used GUI as fog computing. But due to this reason, it will need more power as the number of fogs is directly proportional to the power consumption. This problem can be reduced if 5G NB-IoT is used in this system as it requires less power in comparison to another system. In this system, whenever a wireless sensor network receives any environmental parameters or agricultural data then it is transmitted to the cloud using a base station or gateway and gathered data can be downloaded by a cooperative platform designed by the graphical user interface. At first perception layer collects data using Zigbee and Wi-Fi communication, then transmit into the Internet tier that extricates data from the services. Then extracted features are combined with IoT applications for managing IoT devices, there is a reliable management manifesto for the greenhouse background through a wireless sensor network within customary resources. The IoT is used for integrating sensing data from sensors, using big data analysis tools, like support vector machines, machine learning, and stochastic. In this paper Arduino controller is used to implementing three layers, those are GUI, ThingSpeak, controller, and plastic structure cultivation [add citation here]. It will need more power as power consumption is directly proportional to the number of fog nodes. It requires low data-gathering loading, in comparison with other IoT systems.

This system is established for non-agricultural lands like residential and commercial areas in this system temperature, light, relative humidity, and nutrient concentration in the water are controlled for creating an ideal condition for plants. The user, cloud, and all components are connected by a network [16]. This system controls nutrient mixing, temperature setting, maintaining proper humidity, and lighting system. It also retrieves information from sensing devices and transmit information to the cloud where all data are interpreted and then presented to the user through GUI as charts and graphs. Users can give any instruction through mobile applications or computer programs, and according to those instructions, they will adapt and execute. In this system, they have used a water level sensor, EC-pH level sensor, temperature sensor, and relative humidity sensor, and the total system is built using a master–slave structure. Here NAS device is used as a private cloud and data transfer can be faster than in other systems. All components are connected in a star topology. This system will get more enhanced if the gathered data is used for prediction and self-controlling the system.

Different soilless cropping is compared in terms of promising techniques and technologies that reduce nutrient solution loss where aeroponics is mentioned as one of the ways of soilless cropping in southern Europe [17]. Soilless cropping can adapt to the latest technologies. Open-loop soilless can be built with zero drainage where leachate is used as one of the nutrients. In this system, the main thing is the proper application of nutrient solutions following crops requirement. That can be possible by using optimum irrigation control for example moisture sensors and additional sensing knowledges. But for providing representative information of actual growing conditions, a system needs more capacity for making sufficiently reliable decision making, that is the main issue in this system. But the main boundary of the Mediterranean region is limiting nutrient solution recirculation in soilless culture is associated to the brininess of irrigation water. For implementing a soilless system that can minimize water and nutrient losses, need to focus on improving people's comprehension and attentiveness of probable elucidations to diminish emissions by organizing some training programmes and the cost of the system is costly in terms of water and nutrient-saving.

2.2 Dew

Developing IoT with dew computing is more beneficial in terms of latency, energy consumption, data storage and data transfer [20, 21]. In [22] authors offered a dispersed hotspot network design with five tiers which can deliver extensive hotspot coverage. The authors stated that it could provide better network security, optimal coverage, and a reasonable data conduction rate while transmitting information from the IoT peripherals to the remote users and servers. Authors have used dew computing to advance availability, network coverage, and synchronized update capability. They have given an example scenario of a farmer remotely visualizing and analyzing his farm using dew computing. There are many IoT sensors attached to many devices.

IoT gadgets are connected with Wi-Fi and/or LoraWAN according to the data transmission rate. The IoT devices are located nearby hotspots for easy transmission of data. If the farmer doesn't have his hotspot, he can rent a hotspot service near the farm. This is how the authors presented the scenario where a farmer can implement his own dew computing-based IoT system or use other hotspots near his farm [23]. Described a framework with dew cloud computing, music and crowdsourcing by merging dew computing with IoT. Crowdsourcing is applied with sound sensing devices to invoke the acoustic information from the environment, and this data are passed on to fog computing devices through a dew server. The cloud data centres are responsible for aggregating musical information and providing service to the end users.

2.3 IoT

IoT is used for remotely monitoring the sensors and actuators, and here Arduino is used for connecting the system to the Internet. The parameters controlled by this system are temperature, humidity, irrigation time, nutrient level, and pH of the nutrient. Here GPRS communication is used to send the data to the cloud [18]. For getting the level percentage in the reservoir tank an ultrasonic level sensor JSN-SR04T is used. Due to the bad geographical condition of the greenhouse Internet service was not available but remote communication is a very important component for knowing the crop status. so, in this system Shield, SIM900 is used for connecting to the Internet. The thingSpeak platform is used for visualization, analysis, and motorization of the variables. The SIM900 Shield establishes a GPRS connection by connecting to Arduino for transferring the information with ThingSpeak. The disadvantage of the system is Metal tips of the ultrasonic level sensor can be trusted. Machine learning can be applied to the data, which are received by sensors.

A lab-scale aeroponic system is designed and implemented, for online and automated monitoring capability using IoT. For temperature monitoring, they have used DHT-11 which uses a Wemos-D1-mini-integrated microprocessor and Wi-Fi module for connecting to the Internet. Components are used for controlling the temperature and to provide nutrients to the integrated roots, those are Peltier cells, fans, and mist makers. The light intensity and chamber temperature are continually monitored online and to control the temperature for realizing ideal conditions for plant growth, some actuators are also used [19]. The system contains 3 basic subsystems, 1. a temperature control subsystem 2. a humidity & nutrient distribution subsystem 3. a light intensity sensor. On the backside of the container, a Peltier cell is attached for cooling of temperatures. Here for decreasing the temperature of the chamber, water federated learnings through the cool side of the Peltier cell. In this system the temperature and humidity are measured and controlled in a good manner thus this system can grow monsoon crops in desert-like places. But this system does not have any sensor for nutrition solutions if the nutrient solution became acidic or Alkaline then that will harm the plant directly. If the nutrient solution is finished then there is no system is built for notifying that, so it would better if they had added a nutrient level sensor also. Otherwise, Sensor fusion can be created and the data coming from that sensor can be used for further studies.

3 Advantage of Aeroponics Over Hydroponics and Traditional Cultivation

In traditional agriculture, we need lots of resources like soil, Chemical fertilizer, Pesticides, human labour, and lots of freshwaters. It also directly depends on od natural phenomena like a federated learning good, Storm, and many more things. It requires more pesticides and chemical fertilizers to cultivate in the soil.

So, if we want to decrease all of the costs of all of these things then the best way out is soilless cultivation. Apart from this, the world's 70% of fresh water is used in traditional agriculture, whereas food or crop production should be increased by 69% in 2035 [24]. So if we continue this type of cultivation, then it can cause a serious freshwater shortage in the near future. In many cases, soil can contain bad microbes and nematodes which cause any kind of disease in the plant. One of the biggest problems in traditional agriculture is the huge amount of land requirement which can't be found in any metropolitan city. In some areas, we find soils having fewer nutrients due to the geographical location of the area, their traditional agriculture is next to impossible. If the land has all good condition but still can be tough to cultivate due to the inadequate drainage system in erosion probable places, then the quality of crops can be degraded because of land issues. Another biggest problem in this century is providing enough labour for this purpose. As mentioned above, all problems faced by a traditional agriculture system can be easily removed by automated hydroponics farming (Fig. 2).

Hydroponics is one of the most popular soilless cultivation processes where plants don't need any soil for nutrition because their nutrient solution is provided to the roots of the plant. The nutrient solution is passed through the roots of the plant. This is why it is possible to cultivate more plants in a small area which can solve the land requirement problem of traditional agriculture [25]. The basic idea of hydroponics is to provide an exactly suitable environment for the plant, which can be possible by an automated controlled hydroponics system. But in this system, if any kind of disease is found in any one plant's root, then other roots will be affected very easily as the same nutrient solution is passed. Though crop production is increased in this system, this system does not prove to be good to save fresh water.

Hence, we need the aeroponics system where plants are suspended from a stage and roots are kept in a close dark place. Water is sprayed into the roots of the plant and water contains sufficient nutrition for the plant. This system requires less space in comparison to other systems. Roots get more oxygen so maximum nutrition is absorbed and growth is faster than in other systems. In this system, roots don't use any shared medium like water in the hydroponics system so there is a very low chance

Fig. 2 Hydroponics system

to spread disease from one plant to another and the suffering from transplantation shock is also very low [26].

4 Manually Growth Observation of Aeroponic Systems Having the Pump

An aeroponics system is controlled environment cultivation. In this type of system, all-time observation is required. In this system, if any fault or mistake happens, then damages can be caused very quickly, therefore, uninterrupted and errorless observation are very important things for the projected system. If the observation is done manually by a human, then it is next to impossible to observe it for 24 h continuously. If a measurement is done by any human, then there is a very high chance to have an error, and this error can cause a lot of problems for the system. A wrong observation can cause a wrong interpretation of the system. This may harm the plant's health or sometimes it can cause the death of the plants also. It can be laborious work to measure the growth of the tree one by one so it will increase labour costs. One of the important elements of the aeroponics system is the pump. It will be federated learning-inspired nutrient solution from solution tank to atomizers, and then the nutrient solution is sprayed to the roots of the plant. It takes a huge amount of electricity to make a pump work, in this era of shortage in power supply, this is a serious problem. Apart from this, if the pump gets any damage and stopped working, then the total system will fall instantly as it is a very important part of the system. Even if it is working, is not properly then also the system can fall and plants can get damaged.

5 Cloud-Dew Architecture in Aeroponics

The monitoring and control system and the solar panel system make up the entire system. With user-defined settings, this system will provide automation for the relevant actuators. A $50 \times 40 \times 20\ cm^3$ box was employed as the growing chamber for this study. The coverage of sensor nodes for agricultural applications, according to [27], requires dense (1 sensor/m^2), requiring only one sensor per parameter in this chamber.

Data from sensors (pH, temperature, light, etc.) used for online monitoring will be sent to a server used without any interruption using the dew computing and shown in our system model. The system, which consists of sensors, a microprocessor, actuators, and a communication module, monitors and regulates the temperature, water pH, and light intensity in the growing chamber. The temperature sensor, pH sensor, DHT-11, etc. were the sensors employed in this study (light intensity sensor).

The Cloud-Dew architecture is expanded upon by the client–server architecture. Figure 3 presents an illustration of this architecture. When we use the system, it can be seen that in the Cloud-Dew design, the term "server" from the client–server architecture has been swapped out for "cloud server." The Cloud-Dew design was put up to address the issue of data accessibility in cloud computing services and we want to be clear that these servers are offering cloud services. A cloud server can also be referred to as a server when there is no chance of confusion when we use dew computing. Another distinction between the Cloud-Dew architecture is the use of a new type of server called a dew server. A dew server is a web server that is installed in our system model. The dew server's databases serve two purposes: first, they synchronize dew server databases with cloud server databases while also offering clients the same services that the cloud server does.

6 Advantages of Intelligent Hybrid Li-Fi and 5G NB-IoT Enabled Smart Aeroponics System in Agriculture 4.0

In this section, we discuss the advantages of Li-Fi and 5G NB-IoT-enabled smart Aeroponics systems for Agriculture 4.0 environment.

6.1 Pumpless Green Aeroponic System with Voice User Interface

The idea of green Aeroponics concept originated from making an eco-friendly aeroponic system. We can easily make an eco-friendly aeroponic system by reducing energy consumption. In an aeroponic system, maximum energy is needed for federated learning. Flowing the water from the nutrition tank to the atomizer. So, if we can

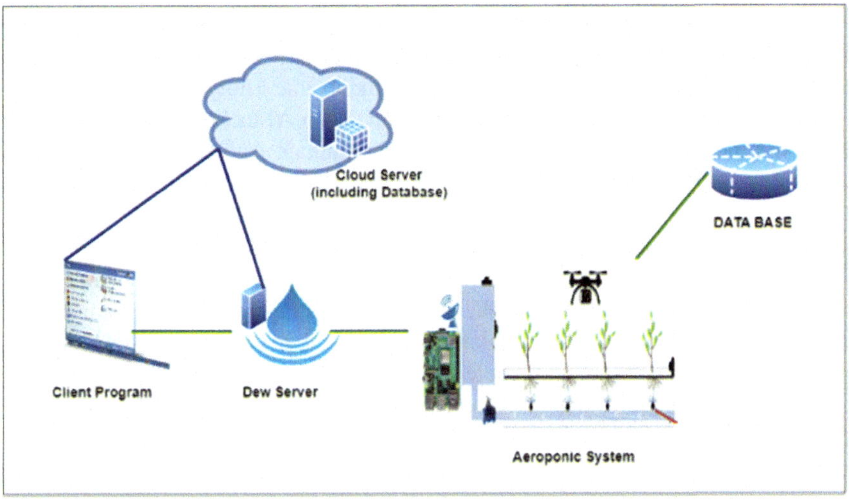

Fig. 3 Demonstration of Dew Aeroponics system architecture

send water to the atomizer without using the pump, then we can remove the pump, and reduce the energy consumption of the total system. We can use an anti-siphon system to send the water to the ultrasonic atomizer.

In Fig. 4, we can see how an anti-siphon system can remove the requirement of the pump. Here additional solenoid valve is used. This solenoid valve is controlled by a raspberry pi. This raspberry pi will be connected to the server. The on-time and off-time will be decided by a raspberry pi. It is connected to the solenoid valve through a relay. This system will function as a pump in an aeroponic system.

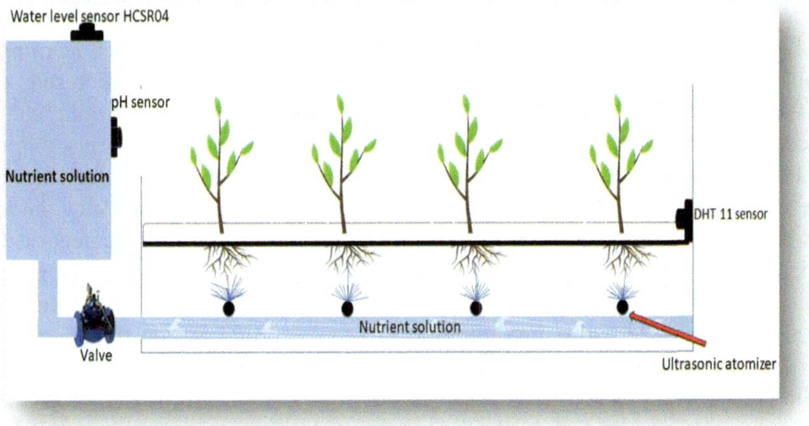

Fig. 4 Pumpless aeroponics system

The voice user interface is one of the revolutionary technologies of this era. It interacts between the user and the system without any hand or eye function. It is implemented by using some AI technologies like Name Entity Recognition Speech Synthesis, and Automatic Speech Recognition. Then AI technology tries to understand the meaning of the voice command of a human. And returns the command to the device. In general speaking speed of a common human is much higher than typing, so it makes this system much faster than others. It allows users multitasking ability when users need it, they just need to say some words while doing some other work. For the implementation of the voice user interface, we need to understand the details of human communication. We can implement this system or we can just use some services that are already popular in the market, like Google Assistant, Amazon Alexa, Microsoft Cortana, and Apple Siri. Many of these services are multilingual, so it would be easy for the user to control the total system in their naïve language.

6.2 Security with Federated Learning in Dew Aeroponics

Federated learning is a branch of machine learning that works on decentralized data. Where a pattern is created on the data of every node or edge device as the local pattern. Then these local patterns are sent to the cloud, all local patterns are combined and a new global pattern is created. But in the total procedure, the identity of a single node is not disclosed. The main benefit of this system is that a user can get the advantage of a huge trained machine learning model without sending their private data. This is how the privacy of each node is preserved, and their federated learning is applied in the fields where security and privacy are major factors. Federated learning-based mobile crowdsensing is used for privacy-preserving in virtual tourism. For sharing geospatial information in a privacy-preserving manner, federated learning is used. Federated learning-based geostatistics gives access to the tourist of many unknown tourist places that will be beneficial for the local people of that tourist place [28]. In [29], a learning-enabled framework for privacy-preserving federated learning is implemented for smart agriculture; as a result, better accuracy, detection rate, etc. is found over some non-federated learning techniques. Federated learning can be used for resolving the privacy protection of data [30]. Security and privacy issues of federated learning are discussed and the challenge or obstacles that are faced in preserving privacy and security at the time of designing federated learning are also discussed [31]. They have categorized the protection method into 3 categories: 1. Client-side privacy protection. 2. Server-side privacy protection. 3. Security protection for federated learning. 1. Client-side privacy protection: they have used some protocols like perturbation and dummy for privacy protection on the client side. Perturbation is noise addition by the clients to the uploaded parameters. Moreover, differential privacy can be used to hide sensitive data unless the third party can distinguish the individual for information restoration and user privacy protection. All these processes are balancing between privacy and accuracy which should be adjusted. In the Dummy method, the model parameters are sent with dummy model parameters, which is how the client's contribution can be hidden. 2. Server-side privacy protection: after averaging all uploaded parameters of the clients, the operating server sends the accumulated constraints to clients to synchronize the model. In this process, while the server is broadcasting, information can be leaked due to the existence of eavesdroppers. To protect from this kind of information leakage, privacy protection needs to be implemented on the server side. Aggregation is a procedure for server-side privacy protection, that is collecting information from clients, where a client sends information in such a way that the server will fail to inspect the client's information from that given data. Securing computation of multiparty is another protocol that can be described as encrypting each device's update to make it uninspectable. After gathering all the information, the server generates an independent response using the set device message, and transmits to each device in this protocol. Then the last stage of this protocol is the devices that uploaded cryptographically masked data, they reveal the cryptographic secrets to the servers, for unmasking the information. 3. Security protection for federated learning framework: here, model stealing attack can happen, any node can be introduced with back-door functionality into the Global joint model. For this, an attacker's level has been chosen by the image classifier. To design a federated learning, other security measures should be maintained. Homeomorphic encryption can provide some security. In this method, nodes encrypt the information and then upload it to the server. For decrypting, public–private keys are sent to the server. For this reason, it can cause extra cost. another method used for federated learning frameworks security is back to defender. Apart from this, they have discussed some challenges of private and Secure federated learning and they have clarified four main problems of a secure and private federated learning system and proposed some solutions to these issues. The issues are (1) Convergence, (2) Data poisoning, (3) Scaling up, and (4) Model aggregation.

6.2.1 Convergence is an Issue that is Caused by Privacy Protection

Previous works stated that approximate convergence can be provided to some extent, but those are done by assuming unrealistic scenarios. For example, either the information is distributed with all devices or shared in an autonomous and indistinguishably disseminated way. Each of the devices takes part in communication at each round. In a real scenario, learning parameters will be in a non-independent-and-identically distributed manner if client-side Perturbation is applied. For satisfaction, convergence learning performance should be characterized properly by proposing appropriate measures. If convergence is guaranteed by adding artificial noise into a deep learning network, then MNIST classification-solving accuracy decreases by about 40%. So some steps should be taken. 1. Theoretical results about convergence should be provided. 2. While considering privacy protection, the learning performance should also be investigated. 3. Algorithms that are used for privacy protection should be devised practically and theoretically. Moreover, the classification accuracy and learning performance in communication rounds should be analyzed. They have proposed a solution to the convergence problem, as in a federated learning system that consists of noise, there should be a perfect balance between privacy level and learning performance. Higher privacy levels can cause more noise and will lead to wickeder learning implementation. So learning performance analysis is required to scrutinize the theoretical relations among the number of clients, noise scale, restricted training repetitions, and the amount of communication sequences.

6.2.2 Data Poisoning

In federated learning, clients notice the Intermediate Model states and give irrational updates. This is only possible for those clients who previously worked as the passive data providers. It gives the scope for malicious clients for manipulating the training process in the wrong way. In other words, we can say that adversaries who pretend to be honest clients can give wrong updates that will influence the model performance maliciously, this is called model poisoning. Researchers show that adversarial participants can give the membership and properties that are related to the training data subset. Some of the malicious clients can update parameters also which can be harmful to the system performance. Also, there can be the presence of eavesdroppers while the server is broadcasting. So, it is required to measure the loss performance when any client is making poisonous data and a secure procedure for preventing and recognizing it. It also needs a secure communication system that can prevent eavesdroppers. They have proposed 3 kinds of ways for solving the data poisoning problem; the first one is identifying the malicious client at system setting time using machine learning techniques like supervised learning. The second one is the quality of the uploaded parameter is checked and the aggregation weight is adjusted with the quality of the parameter. The third one is updating weights with the social network, which is used in each communication round. It is stated that this is how the data poisoning problem can be overcome.

6.2.3 Scaling up

Owing to the obtainability of elevated-performance and low-cost system gadgets, it is easy to extend the current federated learning system to a larger federated learning system. For this, some error may occur, that is, getting proper devices that relate to the distribution of local data in complex ways, another one is unsecured connectivity between devices and interruption during execution, the third one is the coordination of lockstep execution across devices with different availability, the last one is the limitations of computer resources and device storage. But if the number of clients is increased then, less communication round will be performed for more computation. Increasing the number of clients can reduce the effect of the poisonous client in the federated learning model, (e.g., If a wrong parameter is given to a system that consists of 10 clients, and another system that consists of 100 clients, then obviously the wrong parameter will leave a less wrong impact on the system that has 100 clients). It will also increase privacy protection as it is tough to find a single client's data in a large data set than in a small data set. But it is still unknown whether increasing the number of clients can reduce learning time and accuracy or not, so it has been experimented with in this paper. It is necessary to optimize the resource allocation for multiple communication modes as most work does not consider wireless transmission. To solve the scaling-up problem, they have proposed some way out. The first one is about focusing on the long waiting time, which can be overcome by setting up and uploading a deadline for every client. For each learning epoch, the server needs to collect clients' constraints beforehand going to the subsequent rounded of federated learning. If it crosses the deadline, then the learning epoch will be cancelled. For handling enormous number of clients, the user clustering methodologies can be applied from the Game Theory. Where the clients are clustered by some uploading methods, common interests, and similar physical locations, and each cluster will compete with another cluster to complete the learning goal.

6.2.4 Aggregation

It is the process of collecting parameters from the client and then updating to the Global model. Which can be defined as it is absorbing the benefits of the client and calculate the termination of the learning paradigms. That aggregation would not be a simple averaging process if the protection methods such as perturbation are applied to the client, because for some reason that is, the server should know which client is privacy-sensitive and which is not, to design the proper aggregation method. In privacy protection methods such as Perturbation, the noise is increased as the number of clients increases. An intelligent aggregation can be created if some characteristics are maintained, that is, the weights of the participating client's parameters should be updated during the communication round, it should be able to resolve the noise problem that is added by the privacy protection of clients and it should recognize the difference between clients and put different aggregation strategies as a solution. They have proposed some details that are, a smart aggregation method which consists of two parts, one is adding a test progression to the server, which will apprise the aggregation depending on the testing implementation. The second algorithm is to intensify the local epochs for each client system, which could overcome the performance degradation caused by the malicious client. As increasing the number of malicious clients, more local epochs are needed. In this work, they mainly pointed out privacy protection and security issues with the solution. The overview of a federated learning system and its types are discussed with associated regularity requirements and performance evaluation is the agenda of this paper [32]. They presented the federated learning standard which is proposed by (C/AISC/FML) the federated machine learning working group. They have provided the application guidelines and details from the depth of federated machine learning. Through this, they give a better understanding of the usage and standard of federated learning across the organization for privacy and security purpose. In [33, 34], the authors provided a summarisation on the federated learning and its future possibilities. The types of federated learning with its architecture are also discussed in (Fig. 5).

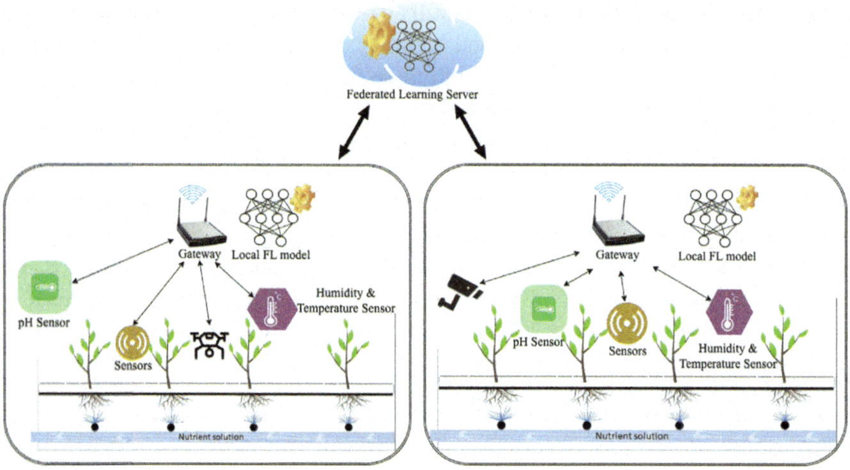

Fig. 5 Federated learning with aeroponics system

6.3 Comparison of Download Speed of Wi-Fi, 4G, 5G, 6G, and Li-Fi Network

Network speed is one of the most important elements of a system. The speed of the total system depends on it. If a very fast processing system is attached to a low-speed network then it would have no worth. So, at the time of implementation, we will need to compare the speeds of different networks. The average download speed of cellular technology. In this table, download speed of Wi-Fi, 4G, 5G, 6G, and Li-Fi is compared. Where the download speed of Wi-Fi and 4G, is observed in South Korea [7]. Based on the frequency range, the 5G can be characterized into 2 types. Type 1 uses the sub-6 GHz frequency band which means it uses only Radio frequency bands below 6 GHz. Whereas type 2 uses a range frequency assortment of 24.25–52.6 GHz. In this Table 2, we have used the average download speed of 5G from Saudi Arabia. As of now 6G, is not implemented in the market, at this moment it is going through experiments so, we put the speed of the laboratory environment. The same 6G Li-Fi is also going through experiments so, here we have considered the value of download speed in a laboratory environment. In Table 2, we can see the download speed of cellular technologies and Wi-Fi and Li-Fi technology. In Fig. 6, we can see graphically the average download speeds of different technologies. Where we can see how much Li-Fi speed is greater than all other technologies. So, if we can use Li-Fi with 5G NB in a hybrid manner in our Aeroponics system then it can make the system more efficient, by providing a high-speed communication system.

Table 2 Average download speed of different networks [7, 35, 36]

Transmission media	Area of testing	Speed (Mbps)
Wi-Fi	South Korea	74.5
4G	South Korea	53.7
5G	Saudi Arabia	291.2
6G	Laboratory	11,000
Li-Fi	Laboratory	24,000

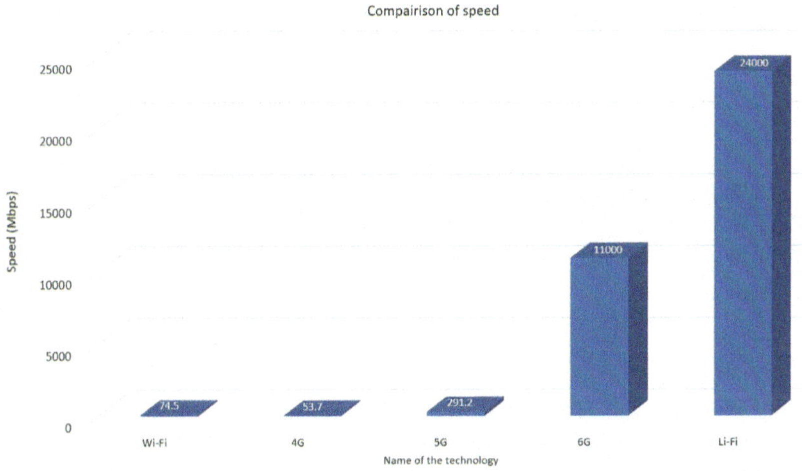

Fig. 6 Graphical representation of average download speed of different technologies

6.4 *Comparison of Laser-Based and LED-Based Li-Fi Network*

Considering the speed of a network A Li-Fi network can be more efficient than Wi-Fi and other communication technologies. A Li-Fi system means communication through light waves. So, it consists of a light source and a photodetector. For enhancing the Li-Fi system, we can improve the photodetector and source of light. In general, Li-Fi system is made with LED but it has several disadvantages. It has a limited carrier lifetime. And it is highly dependent on RC which is the cause of the limiting bandwidth in a few MHz [37]. By using a laser-based Li-Fi system, data rate can be above 20Gbit/s and channel capacity is increased by using QAM-OFDM (Quadrature amplitude modulation—orthogonal frequency division multiplexing) modulation. In [38], it is stated that laser-based Li-Fi can provide a 5X higher data rate than LED-based Li-Fi. It is found that laser-based Li-Fi is safer for the eyes. It has a higher bandwidth also which makes it more energy efficient. A shorter packet duration is also responsible for the energy efficiency of the system.

Table 3 Comparison of LED and laser diode [39, 40]

Sl. no	Characteristics	Led Li-Fi	Laser Li-Fi
1	Current	Drive current (50–100 mA) (peak)	Threshold current (5–40 mA)
2	Working principle	It emanates light by spontaneous emission	It emits light by stimulated emission
3	Coherent/non-coherent	Non-coherent	Coherent
4	Output power	Linearly proportional to drive current	Proportional to current above the threshold
5	On–off switching speed	Comparatively slow	Faster than LED
6	Spectral spreading	Higher	Less than LED approximately 1 to 2 nm
7	Bit rate	Low	High
8	Coupled power	Moderate	High
9	Modulation bandwidth	Moderate, tens of KHz to tens of MHz	High, tens of MHz to tens of GHz
10	Available wavelength	0.66–1.65 mm	0.78–1.65 mm
11	Monochromatic/ non-monochromatic	Non-monochromatic	Monochromatic

From the comparison Table 3, it can be seen that a Laser-based Li-Fi system using a Laser diode is much more efficient than a Li-Fi system. In [41], good bit error rate (BER) performance and a clear constellation map were achieved. SNR (Signal-to-noise-ratio) is also improved. A 20 Gbps of speed is achieved by external light injection technique using 16-QAM-OFDM modulating signals over a 6 m free-space link. More advancement can be possible in the Li-Fi system with this Laser-based Li-Fi system.

6.5 Hybrid Li-Fi Implementation in Pumpless Aeroponics with 5G NB-IoT Enabled IoDT

The Internet of Drone Things (IoDT) is one of the most emerging technologies in this era. In IoDT drones are used as dynamic sensors in IoT systems. In the agricultural system, we can remove static sensors with drones for implementing IoDT in the existing agricultural IoT system. We can say a sensor is, a static sensor if its location is fixed and uses a fixed battery supply until the battery has charged, which is one of the biggest problems and it makes maintaining this type of system more hectic. Whereas dynamic sensors don't have any fixed location or specific place so it is easier to handle. For implementing a dynamic sensor, we can attach a camera (like a night vision camera, IR camera, or thermal imaging camera) or sensor with it to capture images and image processing will be applied to that image, and analysis is done with its GPS location so that system can find the problem and its location. It can also be used in a forest or any other field so that it can save the plants from any unwanted situations. For example, if someone is cultivating some expensive plants that need a lot of care, in a huge area, then it is next to impossible to guard or protect the whole farm's every plant. This task will become very easy with 5G NB-IoT enabled IoDT. With the help of this system, the farmer will be notified if some unusual or unwanted situations take place. Drones at first capture the data as an image, then location and image are sent, which we can analyze by Federated Learning with dew computing. Federated learning is a compassionate of distributed representation learning approach that protects the unprocessed or preprocessed data. It will be very beneficial to all farmers [42].

Hybrid Li-Fi is produced by combining Li-Fi and Wi-Fi technology. It doesn't create any noise in another system. Here Li-Fi produces high data transmission speed with more security. Here LED is used for communication. A Li-Fi system is also called visible light communication (VLC) system. It uses a specific optical band range for the transmission and reception of data. As the Li-Fi system doesn't pass through the wall, it can give more security to a system. It consumes very low power which will increase the battery life of an IoT device sensor node. When current is applied to the LED then, photons are released from it, and at the receiver, the end photodetector is present to receive the data which converts it to the electrical signal. By using Li-Fi, we can reduce the latency of a system. If the LED is on it transmits digital 1, and if it is off then it transmits binary 0. To build a Li-Fi system, we need an LED and a microcontroller that will control the Federated Learning flicker of the LED following the data stream. The data processing rate is depending on the number of LEDs in the lamp. Alike in an IR remote if we remove IR LED with a LED array then it can send thousands of streams very fast. In general, for indoor regions, the Li-Fi is mixed with radio frequency networks. It can be used in radio frequency-sensitive environments. The data transmission in this is energy efficient. In [43], a hybrid communication system is introduced and experimented with. As a result, it is found that it works better in crowded environments with better handover overhead and a rise in average throughput. So it can be derived that this can enhance the performance. In terms of performance, hybrid Li-Fi is better than only Li-Fi or Wi-Fi [44]. A duplex communication link of visible light and infrared has been designed and implemented [45], where visible light is used for downlink and infrared for uplink. This is compatible with a universal asynchronous transmitter/receiver. In this system, 2 nodes are presented as an access point and a user device. Where infrared light-emitting diode and infrared CMOS camera are used for the access point and visible LED is used for the access point. For increasing the light intensity of the LED, an array of LEDs is used. They have modulated the LED by changing the input current of the LED. For providing the data to the KED array, a driver circuit is attached which provides the switching (data) current to the led. For receiving the broadcasted downstream data, the receiver should have photodetectors ranging from 700 to 400 nm. Here PIN photodetector is used. For focusing the light on the small area of the photodetector, a concentrator or lens is required. The change of the light intensity is translated to current by the photodetector. For getting the access point data at the user, this current is amplified. Infrared light (invisible to the human eye) is used for sending the data used to the access point. The circuits used for transmitting and receiving the infrared light data are similar to the visible light transmitter and receiver additionally just an infrared light-emitting diode is used instead of a visible light-emitting diode. One single infrared LED is used for uplink, as it does not provide illumination. The visible light transmitter contains the driver circuit and visible light-emitting diode. For visible light, 100 ma and 2.7 V are given. The data signal sends by the universal asynchronous receiver/transmitter (UART) to the transmitter circuit, the range is 0–3.3 V. For minimizing the loading effects, a buffer circuit is connected between (UART) and the LED. A bipolar junction transistor (BJT) is employed in an LED driver circuit because only one LED can be driven by a buffer and the access point contains the array of LEDs. For making the transmitter suitable for on–off keying, they have used BJT as a switch. The outputs of the buffer are connected to the multiple BJT. Then BJT drives the LEDs. They have used an operational amplifier LM7171 as a buffer, amplifier, and inverter. For modification, the metal–oxide–semiconductor field-effect transistor (MOSFET) can be used by eliminating the BJT, and the drain of MOSFET can be connected with LED in series. The receiver part is made with 4 components, a trans-impedance amplifier (TIA), a high pass filter, an amplifier, Photo Detector. The photodetector generates a current signal by following the change in Light intensity. This current signal is converted to a voltage signal by a trans-impedance amplifier (TIA). Some undesired light is also received but Photodetector, to remove the noise signal, a high pass filter is used with a low cut of frequency. Then 2 step of amplification is done to maintain the receiver bandwidth. Instead of a high pass filter, that is, a comparator that can be used for filtering out the noisy signals. Rise time is one of the important parameters for optical components. The bandwidth of the component can be determined by it. It is the time that is taken for shifting output from 10 to 90% of its peak value. The relation between rising time and Bandwidth is as Eq. (2):

$$\Delta f = 0.35 \div \text{rise time} \tag{2}$$

For getting higher bandwidth, we must choose low-rise time LED or PD. The IR transceiver and VLC transceiver are the same, for IR, the current and voltage levels are needed to be adjusted for driving the IR model. As they have used only one IR LED, it needs only a buffer circuit. To interface the analogue front end and desktop computer, FT4232H is used, which is a USB to UART converter. UART has been used for sending and receiving the analogue signal because it works on the analogue signalling technique. The physical layer of UART is used over VLC, to maintain the synchronization. A custom MAC layer is designed for a VLC-compatible communication link. The Python language is used for implementing the MAC layer. Successful streaming of real-time audio and video between an access point and a user is shown. A duplex VLC system is developed with a MAC layer. For handling, multiple users (Time Division Multiplexing) are used. A channel data rate of 4 Mbps at a link length of 1.2 m has been archived. By using optical emitters or detectors, which have lower rise or fall time, this system can be improved. This kind of duplex life communication

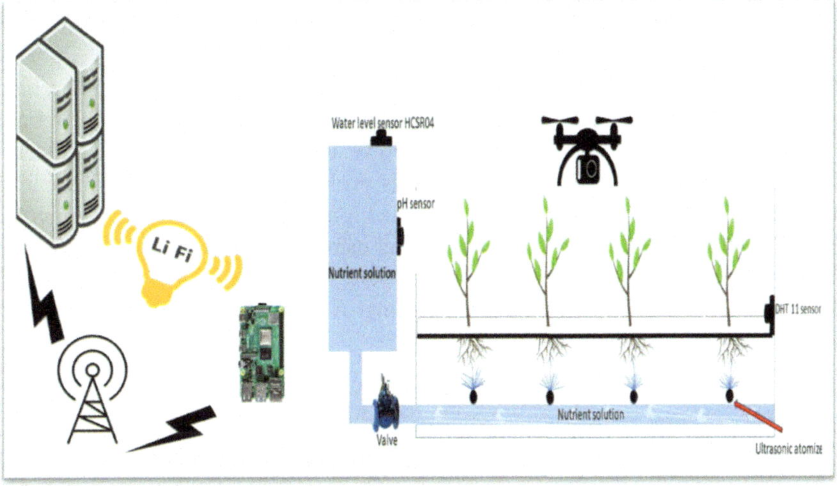

Fig. 7 Hybrid Li-Fi and 5G NB-IoT enabled aeroponic system

can be used in our system for communicating under daylight. Here noise cancellation is done by a high pass filter, which will cancel the undesired frequency range from the light signal (Fig. 7).

7 Future Research Challenges

7.1 Challenge 1: Trustworthy and Decentralized AI for Security

In Federated Learning, the training of the model can be misguided by introducing malicious clients to the network. So, we need some steps to prevent this problem to ensure the security of the system. The blockchain is a decentralized and secured technology, in which tampering with data is near too impossible. In this system, the data is stored in blocks, and this block contains single or multiple transactions. All transactions are validated or agreed upon by a consensus mechanism for checking the transactions are true. So, this blockchain can be merged with federated learning for better and more secure data. This can be truly beneficial for the federated learning network, where the model is misguided by malicious data providers. It will become more trustworthy. So, this Dew Intelligent Hybrid Li-Fi and 5G NB-IoT enabled Aeroponic System can be improved by adding the blockchain system to it.

7.2 Challenge 2: Intelligent Li-Fi

The maximum advantage of any Li-Fi system can be fade if any obstacle comes in between the transmitter and receiver, it will stop receiving data. So for reliable Li-Fi communication, this point also should be under focusing area. If a receiver can move towards the transmitter, then this problem can be solved. This can be designed, as an intelligent Li-Fi receiver. By using artificial intelligence, it is possible. A mobile receiver follows the light transmitter using artificial intelligence. The intelligence receiver is integrated with an intelligent camera that moves the receiver in the direction of the LED transmitter. This improvement will enhance the output of the system.

7.3 Challenge 3: Reconfigurable Network and Services

Concerning the remote connectivity and placement of most of the components of a farming network, it is challenging to adjust the service and network-related policies and configurations automatically. Moreover, the hybridizations between the Li-Fi and 5G NB-IoT will create more complexities in the system. These future-generation network systems may use Software-Defined Network (SDN) and Network Function Virtualization (NFV) approaches to orchestrate different configurations and simplify. Moreover, future works are required to decease which function to select concerning the resource-constraint nature of Dew devices.

8 Conclusions

From ancient times to the present time, people are so much dependent on agriculture. This is not the food provider for a human. It also provides food for animals, which is useful for businesses like dairy and poultry. The raw materials of many industries are produced by agriculture. Thus, it has a great impact on the economy, we can say it is the backbone of the economy. This is why it is necessary to apply the latest technology in this field. So, we have proposed some technological modifications to the agricultural system. In this paper, we have proposed an IoT-based pumpless aeroponic system enabling hybrid Li-Fi and 5G narrow bands with drones. In this system, the advantage of aeroponics like low water consumption and high yield extra is combined with it. Here we have proposed an anti-siphon system instead of a pump, it can supply adequate water to the aeroponic system without any energy requirement for the pump. Multiple sensors (e.g., humidity sensor, pH sensor, temperature sensor, water level sensor) are used to collect real-life data for analyzing the system. Drones are used as dynamic sensors for measuring growth and observing other visual parameters. Federated Learning is used for analyzing the purpose. The communication between server nodes is established using hybrid Li-Fi and Narrowband 5G which provides a faster and more reliable communication system. This system can improve the agricultural system by producing more yield and automatic control can decrease the human labour cost 5G narrowband and Li-Fi communication will enhance this system with high-speed communication with the server and edge devices.

References

1. Vanipriya, C.H., Maruyi, S., Malladi, Gupta, G.: Artificial intelligence enabled plant emotion Xpresser in the development hydroponics system. In: Mater. Today: Proc.**45**, 5034–5040 (2021). https://doi.org/10.1016/j.matpr.2021.01.512
2. Halgamuge, M.N., Bojovschi, A., Fisher, P.M.J., Le, T.C., Adeloju, S., Murphy, S.: Internet of things and autonomous control for vertical cultivation walls towards smart food growing: a review. Urban For Urban Green **61**, 127094 (2021). https://doi.org/10.1016/J.UFUG.2021.127094
3. Publications and Graphics Department NASA Center for AeroSpace Information (CASI).: Spinoff 2006 (2006)
4. Rahmad, I.F., Tanti, L., Puspasari, R., Ekadiansyah, E., Fragastia, V.A.: Automatic monitoring and control system in aeroponic plant agriculture. In: 2020 8th International Conference on Cyber and IT Service Management, CITSM 2020 (2020). https://doi.org/10.1109/CITSM50537.2020.9268808
5. Chung, S., Lim, J., Noh, K.J., Kim, G., Jeong, H.: Sensor data acquisition and multimodal sensor fusion for human activity recognition using deep learning. Sensors **19**(7), 1716 (2019). https://doi.org/10.3390/S19071716
6. Daskalakis, S.N., Goussetis, G., Assimonis, S.D., Tentzeris, M.M., Georgiadis, A.: A uW backscatter-morse-leaf sensor for low-power agricultural wireless sensor networks. IEEE Sens. J. **18**(19), 7889–7898 (2018). https://doi.org/10.1109/JSEN.2018.2861431
7. Bhattacharya, A., De, D.: AgriEdge: edge intelligent 5G narrow band internet of drone things for agriculture 4.0. Lect. Notes Data Eng. Commun. Technol. **67**, 49–79 (2021). https://doi.org/10.1007/978-3-030-71172-6_3/COVER
8. Budhiraja, I., et al.: A systematic review on NOMA variants for 5G and beyond. IEEE Access **9**, 85573–85644 (Institute of Electrical and Electronics Engineers Inc.) (2021). https://doi.org/10.1109/ACCESS.2021.3081601
9. Anbalagan, R. Hussain, M.Z., Jayabalakrishnan, D., Naga Muruga, D.B., Prabhahar, M.: Vehicle to vehicle data transfer and communication using LI-FI technology. Mater. Today Proc. **45**, 5925–5933 (2021). https://doi.org/10.1016/J.MATPR.2020.08.786
10. Lorriere, N., et al.: Photovoltaic solar cells for outdoor li-fi communications. J. Lightwave Technol. **38**(15), 3822–3831 (2020). https://doi.org/10.1109/JLT.2020.2981554
11. Reynaud, C.A., Clerc, R., Lechêne, P.B., Hébert, M., Cazier, A., Arias, A.C.: Evaluation of indoor photovoltaic power production under directional and diffuse lighting conditions. Sol. Energy Mater. Sol. Cells **200**, 110010 (2019). https://doi.org/10.1016/J.SOLMAT.2019.110010
12. Singh, P., Kaur, A., Aujla, G.S., Batth, R.S., Kanhere, S.: DaaS: dew computing as a service for intelligent intrusion detection in edge-of-things ecosystem. IEEE Internet Things J. **8**(16), 12569–12577 (2021). https://doi.org/10.1109/JIOT.2020.3029248
13. Gangopadhyay, M., et al.: Evaluation of growth response for mass production and accumulation of 2-hydroxy-4-methoxybenzaldehyde in endangered Hemidesmus indicus by an aeroponic system. Ind. Crops Prod. **172**, 114072 (2021). https://doi.org/10.1016/J.INDCROP.2021.114072
14. Karuniawati, S., Gautama Putrada, A., Rakhmatsyah, A.: Optimization of grow lights control in IoT-based aeroponic systems with sensor fusion and random forest classification. In:

Proceeding—2021 International Symposium on Electronics and Smart Devices: Intelligent Systems for Present and Future Challenges, ISESD 2021 (2021). https://doi.org/10.1109/ISESD53023.2021.9501863
15. Chang, H.Y., Wang, J.J., Lin, C.Y., Chen, C.H.: An agricultural data gathering platform based on internet of things and big data. In: Proceedings—2018 International Symposium on Computer, Consumer and Control, IS3C 2018, pp. 302–305 (2019). https://doi.org/10.1109/IS3C.2018.00083
16. Belista, F.C.L., Go, M.P.C., Lucenara, L.L., Policarpio, C.J.G., Tan, X.J.M., Baldovino, R.G.: A smart aeroponic tailored for IoT vertical agriculture using network connected modular environmental chambers. In: 2018 IEEE 10th International Conference on Humanoid, Nanotechnology, Information Technology, Communication and Control, Environment and Management, HNICEM 2018 (2019). https://doi.org/10.1109/HNICEM.2018.8666382
17. Massa, D., Magán, J.J., Montesano, F.F., Tzortzakis, N.: Minimizing water and nutrient losses from soilless cropping in southern Europe. Agric. Water Manag. **241**, 106395 (2020). https://doi.org/10.1016/J.AGWAT.2020.106395
18. Lucero, L., Lucero, D., Ormeno-Mejia, E., Collaguazo, G.: Automated aeroponics vegetable growing system. Case study lettuce. In: 2020 IEEE ANDESCON, ANDESCON 2020 (2020). https://doi.org/10.1109/ANDESCON50619.2020.9272180
19. Jamhari, C.A., Wibowo, W.K., Annisa, A.R., Roffi, T.M.: Design and implementation of IoT system for aeroponic chamber temperature monitoring. In: Proceeding—2020 3rd International Conference on Vocational Education and Electrical Engineering: strengthening the framework of Society 5.0 through Innovations in Education, Electrical, Engineering and Informatics Engineering, ICVEE 2020 (2020). https://doi.org/10.1109/ICVEE50212.2020.9243213
20. Gusev, M.: What makes dew computing more than edge computing for Internet of things. In: Proceedings—2021 IEEE 45th Annual Computers, Software, and Applications Conference, COMPSAC 2021, pp. 1795–1800 (2021). https://doi.org/10.1109/COMPSAC51774.2021.00269/VIDEO
21. Gusev, M.: Edge and dew computing for streaming IoT. dewcomputing.org (2018). http://www.dewcomputing.org/publications/IoT.pdf
22. Ray, P.P., Skala, K.: Internet of things aware secure dew computing architecture for distributed hotspot network: a conceptual study. Appl. Sci. **12**(18), 8963 (2022). https://doi.org/10.3390/APP12188963
23. Roy, S., Sarkar, D., De, D.: DewMusic: crowdsourcing-based Internet of music things in dew computing paradigm. J. Ambient Intell. Humaniz. Comput. **12**(2), 2103–2119 (2021). https://doi.org/10.1007/s12652-020-02309-z
24. Is the world running out of fresh water?—BBC Future. https://www.bbc.com/future/article/20170412-is-the-world-running-out-of-fresh-water. Accessed 22 Nov. 2022
25. Khan, S., Purohit, A., Vadsaria, N.: Hydroponics: current and future state of the art in farming. J. Plant Nutr. **44**(10), 1515–1538 (2020). https://doi.org/10.1080/01904167.2020.1860217
26. AlShrouf, A.: Hydroponics, aeroponic and aquaponic as compared with conventional farming. Am. Acad. Sci. Res. J. Eng. Technol. Sci. **27**(1), 247–255 (2017). https://www.asrjetsjournal.org/index.php/American_Scientific_Journal/article/view/2543. Accessed 22 Nov. 2022
27. Jagadesh, M., Karthik, M., Manikandan, A., Nivetha, S., Prasanth Kumar, R.: IoT based aeroponics agriculture monitoring system using raspberry Pi. **6**(1), 601 (2018). www.ijcrt.org. Accessed 05 Dec. 2022
28. De, D.: FedLens: federated learning-based privacy-preserving mobile crowdsensing for virtual tourism. Innov. Syst. Softw. Eng. **2022**, 1–14 (2022). https://doi.org/10.1007/S11334-021-00430-6
29. Kumar, P., Gupta, G.P., Tripathi, R.: PEFL: deep privacy encoding-based federated learning framework for smart agriculture. IEEE Micro **42**(1), 33–40 (2022). https://doi.org/10.1109/MM.2021.3112476
30. Li, Z., Sharma, V., Mohanty, S.P.: Preserving data privacy via federated learning: challenges and solutions. IEEE Consum. Electron. Mag. **9**(3), 8–16 (2020). https://doi.org/10.1109/MCE.2019.2959108

31. Ma, C., et al.: On safeguarding privacy and security in the framework of federated learning. IEEE Netw. **34**(4), 242–248 (2020). https://doi.org/10.1109/MNET.001.1900506
32. Zhang, T., Mao, S.: An introduction to the federated learning standard. GetMobile: Mob. Comput. Commun. **25**(3), 18–22 (2022). https://doi.org/10.1145/3511285.3511291
33. Pandey, M., Pandey, S., Kumar, A.: Introduction to federated learning. In: EAI/Springer Innovations in Communication and Computing, pp. 1–17 (2022). https://doi.org/10.1007/978-3-030-85559-8_1/COVER
34. Singh, P., Singh, M.K., Singh, R., Singh, N.: Federated learning: challenges, methods, and future directions. In: EAI/Springer Innovations in Communication and Computing, pp. 199–214 (2022). https://doi.org/10.1007/978-3-030-85559-8_13/COVER
35. Kalbande, D., Khan, Z., Haji, S., Haji, R.: 6G-next gen mobile wireless communication approach. In: Proceedings of the 3rd International Conference on Electronics and Communication and Aerospace Technology, ICECA 2019 (2019). https://doi.org/10.1109/ICECA.2019.8821934
36. Yu, S., et al.: Si-substrate LEDs with multiple superlattice interlayers for beyond 24 Gbps visible light communication. Photon. Res. **9**(8), 1581–1591 (2021). https://doi.org/10.1364/PRJ.424934
37. Lee, C., et al.: Advanced Li-Fi technology: laser light, vol. 11302, pp. 116–123 (2020)
38. Jungnickel, V., et al.: Laser-based LiFi for 6G: potential and applications Marie Skłodowska-curie multi initial training network (ITN) on visible light based interoperability and networking view project OWICELLS view project laser-based LiFi for 6G: potential and applications. https://www.eliot-h2020.eu/. Accessed 22 Nov. 2022
39. LED versus Laser diode | Difference between LED and Laser diode. https://www.rfwireless-world.com/Terminology/LED-vs-Laser.html. Accessed 22 Nov. 2022
40. Peng, S.-L., Hoang, L., Suseendran, S.G., Balaganesh, D.: Inadequacy of Li-Fi disentangles by laser, polarizing beam, solar, and formation. Springer. https://link.springer.com/chapter/10.1007/978-981-15-3284-9_79. Accessed 22 Nov. 2022
41. Ying, C., et al.: 20 Gbps optical Li-Fi transport system. opg.optica.org. https://opg.optica.org/abstract.cfm?uri=ol-40-14-3276. Accessed 22 Nov. 2022
42. Durrant, A., Markovic, M., Matthews, D., May, D., Enright, J., Leontidis, G.: The role of cross-silo federated learning in facilitating data sharing in the agri-food sector. Elsevier. https://www.sciencedirect.com/science/article/pii/S0168169921006657. Accessed 22 Nov. 2022
43. Selvi, S., Rajesh, R., et al.: An efficient communication scheme for Wi-Li-Fi network framework. IEEE Accessed (2019). https://ieeexplore.ieee.org/abstract/document/9032650/. Accessed 22 Nov. 2022
44. Mugunthan, D.S.: Concept of Li-Fi on smart communication between vehicles and traffic signals. irojournals.com (2020). https://doi.org/10.36548/jucct.2020.2.001
45. Gupta, A., Singh, V., Gautam, M., Dixit, A.: Design and implementation for a duplex visible light communication link. In: 14th International Conference on COMmunication Systems & NETworkS (COMSNETS) (2022). https://doi.org/10.1109/COMSNETS53615.2022.9668350

Internet of Things and Dew Computing-Based System for Smart Agriculture

Somnath Bera, Tanushree Dey, Shreya Ghosh, and Anwesha Mukherjee

1 Introduction

The "Internet of Things (IoT)" facilitates data collection, processing, and exchange by connecting devices and communication paths. The goal of the IoT is to create a smart environment by using intelligent things, items, and devices with communication and sensing capabilities, which can generate data on their own and communicate the data over the Internet [1]. IoT is defined as a network of numerous "things" or "objects" around us, such as sensors, mobile phones, and Radio Frequency Identification (RFID) tags, which enable them to communicate with one another and successfully carry out their responsibilities. These choices are made in order to deal with challenges relevant to human living, such as energy conservation, climate change, mobility, healthcare, business logistics, building automation, etc. IoT is a technology that naturally supports the establishment of autonomous federated services and applications. It is distinguished by a high level of autonomy in data collection, event transmission, network connectivity, and interoperability. Things concentrate on combining generic objects into a framework that is easy to use, whereas the Internet works on creating an integrated network. IoT uses the Internet as a communication and information-exchange medium with the goal of fusing the real world with the virtual one. IoT is described as a network of interconnected computing devices, objects, animals, mechanical and digital machines, or people, with unique identifiers [1], and can transmit data over a network. The tremendous influence the IoT idea will have on a number of facets of daily life and potential consumers' behaviour is

S. Bera · T. Dey · A. Mukherjee (✉)
Department of Computer Science, Mahishadal Raj College, Mahishadal, West Bengal, India
e-mail: anweshamukherjee2011@gmail.com

S. Ghosh
College of Information Sciences and Technology, The Pennsylvania State University, University Park, USA
e-mail: shreya@psu.edu

D. De and S. Roy (eds.), *Dew Computing*, Internet of Things,
https://doi.org/10.1007/978-981-99-4590-0_14

without a doubt its greatest strength. IoT is a technological innovation, and according to the idea of innovation diffusion, consumer acceptance of new technology is highly influenced by how it is introduced and how beneficial the user perceives it to be. Users will more readily embrace technologies in their daily lives if they believe that they are simple to use and allow them to contribute more. The most evident effects of the implementation of IoT will be visible from the perspective of individual users in the working place as well as at home. It is also widely utilised to access multimedia information and services as well as other diverse tasks via social network applications, with statistics showing that about two billion people use the Internet daily [2, 3]. In addition to being successfully used in various fields, IoT is successfully implemented in the agriculture field also. IoT-based solutions are being created to autonomously monitor and maintain agricultural crops with the least amount of human participation. IoT has significantly changed the environment for agriculture by examining many difficulties the farmers face. In the field of agriculture, technological applications are employed to lower costs and increase agricultural yields or quality. The use of WSN in satellite agriculture provides farmers with statistical support, enabling them to make more informed decisions. With the development of technology today, it has been predicted that agriculturalists and technologists will use IoT to solve difficulties the farmers face, such as water shortages, concerns with cost control, and productivity problems. By continuously monitoring the field, IoT-based smart farming enhances the overall agricultural system [4]. Modern IoT technologies have identified all these problems and fixed them to raise productivity while decreasing costs. The IoT in agriculture has reduced the lavish use of resources like water and electricity and saved farmers' time by utilising sensors and connections [1, 5].

The IoT is producing an unprecedented volume of data, which in turn strains the Internet's infrastructure severely. Large corporations are therefore trying to figure out how to lower that pressure and address the data problem. In particular, by enabling all of the connected devices to collaborate, cloud computing will play a significant role in that. Aggregating data and gaining insights from it require the use of the cloud. Data comparison over larger areas is substantially more challenging without the cloud. Additionally, high scalability is another benefit of using the cloud [6, 7]. Sharing resources is a major component of cloud computing, which is essential for IoT platforms. In addition to sharing resources, cloud computing also makes the most of them. Additionally, cloud services are accessed by users from any area and on any device with an Internet connection. When discussing the IoT platform, it should be accessible 24/7 from any location [8]. The mission of passing down human knowledge to future generations can be accomplished very effectively with cloud computing. Farmers all over the world use IT tools to manage and distribute their crop-related data quickly and easily. The farming sector is also deploying additional hardware and software resources for measuring luminance as well as temperature, humidity, and soil moisture. Agriculture's modernisation lessens its reliance on the climate, improves the use of already-available resources, and quickly disseminates information about new methods and resources. Additionally, it offers contemporary agricultural machinery, planting and breeding techniques, weather observation and forecasts, and production management and organisation techniques [9].

Building a network of interconnected smart devices is made possible by cloud computing and IoT. The computing problems cannot yet be solved using these two approaches. Inextricably linked to IoT, the new paradigm of fog computing expands the cloud by extending some services to the network edge near the end users in an effort to lessen the computational load on the network. Additionally, fog computing is mainly introduced for applications that require low latency real-time processing. Large-scale, geographically dispersed fog nodes placed at the edge of networks make up the fog layer [10]. Fog is dependent on the availability of computing, storage, and networking resources, just like the cloud. Servers or networking hardware with more processing power may be used as fog nodes. Fog computing seeks to bring processing power closer to the customers while preventing the overuse of cloud resources and further reducing computational burdens. The most crucial pieces of information are transmitted while any extraneous data is filtered out and deleted according to the fog computing paradigm. Utilising end-device resources to give the network a better intelligence distribution and improve performance is one of the key objectives of fog computing. Servers or networking hardware with additional processing power can act as fog nodes. Even wireless access points could include them. The majority of the time, fog nodes will be placed near end users at the network's edge [11]. Additionally, by providing storage, processing, networking, and data management on network nodes close to IoT devices, fog computing serves as a bridge between the cloud and the edge. For farmers and other agricultural stakeholders, agriculture is made simple by these distinctive qualities. For instance, it may be necessary to swiftly and locally process sensitive data collected by all devices [12].

There is another emerging area called edge computing. Edge computing has been proposed to increase efficiency and address cloud-related problems by providing local data processing and storage at the end devices. In order to meet the high computation and low-latency needs of deep learning on edge devices, edge computing is a viable option. It also offers additional advantages in terms of privacy, bandwidth efficiency, and scalability. Edge computing allows data to be examined near the source, sometimes by a local trustworthy edge server, avoiding the public Internet and minimising vulnerability to privacy and security assaults. This helps to address privacy challenges [13, 14]. In the agricultural sector, edge computing is mainly used for pest identification, product safety tracking, autonomous agricultural equipment, agricultural technology promotion, intelligent management, etc. Furthermore, by delivering cloud capabilities close to end users, edge computing supports the transition to fifth generation (5G). To maximise the benefits in agriculture, edge must be combined with other computing technologies like cloud and fog [15, 16].

For edge, fog, and cloud computing, Internet connectivity is required. However, if the network connectivity is interrupted, then the quality of service is reduced. Here, dew computing is one of the feasible options. Dew computing enhances the offering of human-aware customised services at the network's edge. Dew computing is an innovative form of computer organisation and architecture that builds on the practical cloud and conventional client–server architecture. In order to complete a task, cloud computing offers service models, fog computing permits "on-the-spot" on-network processing, and edge computing makes use of computation of tasks that were started at network edges. All the mentioned generic model is expanded upon by dew computing to a sub-platform paradigm [17, 18].

In this chapter, we specifically focus on the utility of dew computing in agriculture. In Sects. 2 and 3, we discuss the use of machine learning and deep learning in agriculture, respectively. In Sect. 4, we demonstrate the use of IoT in agriculture. Section 5 briefly discusses cloud, fog, edge, and dew computing. Section 6 illustrates the use of dew computing in agriculture. Finally, Sect. 7 concludes the chapter.

2 Machine Learning in Agriculture

In recent years, machine learning (ML) techniques are largely used in the agriculture domain for improving the productivity and quality of the crops [19, 20]. ML algorithms are used for a particular crop to determine the condition in which the best yield would be produced. ML models have the ability of decision-making and can take actions in a real-world framework with minimal human interaction. ML, a subset of artificial intelligence (AI) that focuses on learning, can estimate yields more accurately utilising a variety of features. ML may extract information from datasets by finding patterns and correlations. The models must be trained using datasets depicting the outcomes based on prior knowledge. Multiple features are used to build the predictive model, and as a result, the model parameters are selected using the historical data during the training phase. A part of the previous dataset is used in the testing phase for evaluating the performance. Depending on the research issue, the ML model can be categorised as descriptive or predictive. For gaining and explaining knowledge from collected real-time data, we use a descriptive model, whereas for predicting the data for the future, we use a predictive model. Hence, there is a challenge for ML studies to choose the appropriate algorithm for solving the problem at hand. Single sample spaces can be predicted using traditional statistical methods. Additionally, using ML techniques, many predictions can be made. Unlike ML approaches, where we must take the structure of data models into consideration, traditional methods do not require this [19, 20].

In precision agriculture, crop production prediction is a challenging issue, and various models have been proposed. Since agricultural production varies on a wide range of variables, including soil, weather, climate, seed variety, and fertilizer use, this problem makes the use of several datasets essential. This shows that predicting agricultural yields involves a number of challenging processes and is not a simple operation [21].

The three most important parameters to consider when estimating the amount of water needed in each agricultural crop are moisture, temperature, and humidity. The sensors for measuring temperature, humidity, and moisture are placed in an agricultural field, and the collected data is analysed inside the cloud. ML techniques such as decision tree are applied to the collected data [22].

Machine learning (ML) facilitates extracting useful insights from huge volume of data by analyzing, correlating with contexts, and finding interesting but previously unknown patterns. According to the learning type (supervised or unsupervised), models are used to accomplish the chosen goal. ML tasks are often divided into a variety of broad categories discussed as follows:

- **Categorized tasks of ML**: ML techniques are usually categorised as supervised and unsupervised learning. In supervised learning, the software is trained using training examples before being used to draw an accurate conclusion from the incoming data. Supervised learning techniques include artificial neural networks, decision trees, Bayesian networks, support vector machines, Interactive Dichotomizer3 (ID3), and k-nearest neighbour algorithms, among others. Unsupervised ML involves feeding software with huge data, and the programmes will uncover patterns and connections between them. Therefore, to find hidden patterns in the data, unsupervised learning can be used. K-means clustering, self-organising maps, partial-based clustering, and k-nearest neighbour are a few examples of unsupervised learning algorithms. Supervised learning algorithms employ labelled training data for inference (classification, regression). Unsupervised learning algorithms use unlabelled data to find hidden patterns in the existing data (clustering) [23].
- **Interpretation of learning**: In order to reserve information as much as possible from the original data, dimensionality reduction (DR), an approach that is used in both supervised and unsupervised learning, aims to provide a more condensed, lower-dimensional representation of a dataset. In order to mitigate the effects of dimensionality, it is typically carried out before using a classification or regression model. Principal component analysis [24] and linear discriminant analysis [25] are the popular DR algorithms.
- **ML models**: ML techniques can be utilised for agricultural purposes to forecast crop production, soil monitoring, crop growth monitoring, etc., because the data amount is vast and growing daily. The process of classifying is the transformation of an input set of instances into a specific set of attributes, also referred to as target attributes or labels. Numerous applications employ categorization methods such as bayesian classifiers, decision tree classifiers, artificial neural networks, nearest neighbour classifiers, random forests, and support vector machines [26]. Various ML models are described as follows:

    (i) **Decision Tree**: Decision tree is a very popular and simple classifier for handling classification issues. Decision tree models incorporate nodes, branches, terminal values, strategy, payout distribution, specific equivalents, and the rollback procedure. In a decision tree, objects are arranged in a graph according to the values of their features in order to be classified. This algorithm is built in two steps: first, a massive decision tree grows, then its size is reduced and it is pruned in the second phase if it is overfitting the data. The depicted classification tree is the trimmed decision tree that is employed for categorization. The prediction is influenced by a number of things. To construct the yield mapping and anticipate the production, agronomic data, nitrogen treatment, and weed management utilise ML algorithms like artificial neural network (ANN) and decision tree [27]. They come to the conclusion that ANNs can produce high prediction accuracies. Decision tree was used for detailed modelling of soybean productivity [28]. They took into account environmental elements like evaporation, the highest possible temperature, the highest possible relative humidity, rainfall, and crop yield for soybean. They used the ID3 algorithm, an information-based technique that is predicated on two presumptions. The soybean crop production is significantly influenced by the relative humidity. There are certain criteria developed that aid in the low and high soybean production prediction since relative humidity has a significant impact on soybean yield.

(ii) **Bayesian classifier**: Sometimes prediction of class labels for a given set of input attributes can be challenging. Even when matching some of the attributes from the training data set with values from the input attribute set, class variables are nondeterministic. This is plausible given the existence of certain noisy data and perplexing elements, which are ignored during processing. In such applications, it is necessary to describe the probabilistic correlations between the attribute set and the class label; the bayesian classifier focuses on explaining such tasks. This kind of model falls under the field of supervised learning and can be used to address classification or regression issues. Some of the most well-known algorithms in the literature are Naive Bayes [29], Gaussian Naive Bayes, and a mixture of Gaussians [30].

(iii) **ANN**: ANN, a supervised learning method is based on the biological process of our brain. Once trained, a neural network may predict patterns in future data that are comparable, for example, producing meaningful solutions to issues even when the input data is inaccurate or incomplete. ANN is a connectionist system; given a weight, each connection is in charge of sending a signal from one node to another. Before sending a signal to another node, a node examines the signal it has just received. In ordinary ANN implementations, the signal at each artificial neuron's connection is essentially a real number, and each neuron's output is obtained by a nonlinear function of the sum of all of its inputs. ANNs can absorb complexity without being aware of the underlying concepts. Any process can use ANN to determine the relationship between input and output [31]. ANN was used to forecast potato production in Iran based on input energy and to design output energy and greenhouse gas emissions (GHG) [32]. In terms of energy usage, power, chemical fertilizer, and seed were the three main factors. To confirm the symptoms of the tomato crop, artificial intelligence and ML algorithms (particularly ID3) were used to create a web-based expert system with Java at the front end and SQL at the back end [33]. The professional arrangement had two key components: a tomato knowledge system and an expert system, where the applicant may acquire all the reliable information on various topics such as varieties, symptoms of pests and illnesses, cultural practises, and a mosaic of tomato fruits and plants via message. ANN was used to present the intelligent control system for efficient irrigation scheduling [34]. Models were created using input variables including air temperature, soil moisture, radiation levels, and humidity. The amount of water required for irrigation was then assessed using the proper method, ecological conditions, evapotranspiration, and crop type, after which associated effects were simulated. The suggested system was contrasted to an ON/OFF controller, and it was demonstrated that due to these constraints, the ON/OFF controller-based system failed. However, ANN-based approaches have made it possible to adopt stronger and more effective control. Convolutional neural networks and generative adversarial networks (GAN) were combined in the suggested ANN-based approach [35]. Due to its capacity for knowledge accumulation and forgetting reduction, the suggested method was able to recognise all categories from both new and old tasks with good performance. In [36], the artificial neural network was used to estimate the crop utilising soil characteristics including soil type, nitrogen, pH, phosphate, potassium, calcium, organic carbon, magnesium, manganese, sulphur, copper, and iron as well as climate parameters like temperature, rainfall, and humidity. Crops like sugarcane, cotton, bajara, jawar, soybeans, corn, rice, wheat, and groundnut were used in the experiment.

(iv) **Regression analysis**: Regression is a type of supervised learning model that seeks to produce an output variable based on known input variables. The most popular algorithms are stepwise regression, logistic regression, and linear regression. Additionally, more sophisticated regression methods have been created, including multiple linear regression and multivariate adaptive regression splines. For agriculture-related organisations, consultants, producers, etc., crop output is crucial. Crop forecasting can be done using a variety of data types, including soil, remote sensing, agro-metrological, and agricultural statistics. Marketing, storage, and transportation decisions depend on accurate and timely forecasting. The construction of a crop prediction model was examined in [37], and it was shown that planting methods, particularly the application of the proper quantity of fertilizer, had a significant impact on corn production rather than climate-related variables as the key predictors of yield. For the purpose of predicting the yield, a model was built by employing the fortnightly weather variables, such as average daily maximum and lowest temperatures, morning and evening relative humidity, total fortnightly rainfall, and the yield data of the sugarcane [38]. Their forecast model was able to account for 87% of the fluctuation in the sugarcane output. They also concluded that two months prior to harvest, the sugarcane production might be successfully forecasted using the regression technique. Although there are several statistical methods for agricultural production, regression analysis is one of the mostly used methods [39].

(v) **Clustering**: The practise of finding things that are similar to one another but distinct from individuals in other groups is known as cluster analysis or clustering. As the similarities between things in one group and the differences between objects in various groups increased, the clustering would get better. The fundamental technique of data mining is clustering, which has numerous applications in areas including agriculture, image analysing, pattern recognition, data compression, machine learning, etc. In addition to classification, segmentation, and partitioning, clustering can also be used to categorise items. Cluster analysis and classification are comparable; thus we can define clustering as unsupervised learning. Classification and cluster analysis are distinct from one another since clustering does not preserve class information whereas classification does. Additionally, cluster analysis suggests categories based on data patterns as opposed to classification, which classifies fresh samples into known classifications [40]. There are many clustering algorithms, including k-means, the expectation maximisation technique, k-medoid, hierarchical clustering, and others, but k-means is the most popular and significant one [26]. Numerous factors contribute to the popularity of hierarchical clustering, including the following:

- Unlike k-means clustering, which demands a particular value, it does not.
- The tree that was constructed has useful taxonomy.
- Only the distance matrix is needed to calculate the hierarchical clustering.

  To evaluate the prediction of crop properly, k-means clustering algorithm was demonstrated in [41]. The modified k-means clustering technique is given the determined number of clusters and starting cluster centres. The same determined value of the number of clusters is provided and initial cluster centres are uniformly picked because the number of clusters (k value) is required at the beginning for standard k-means and k-means++. By contrasting k-means and k-means++ methods, the modified k-means clustering algorithm is evaluated and found to produce the greatest number of high-quality clusters, accurate crop predictions, and the highest accuracy count.

(vi) **Support Vector Machine (SVM)**: The general discriminant classifier or SVM is frequently employed in the field of pattern recognition. The SVM model uses the concept of a surface called a hyperplane to represent a number of classes, with the border between the data examples being drawn and shown in multidimensional space. By employing the "kernel trick", it is possible to significantly improve the classification skills of conventional SVMs by converting the original feature space to a feature space with a larger dimension [42]. Non-separable problems are converted into separable problems [43]. SVMs are based on global optimization and handle overfitting issues raised in high-dimensional spaces, which makes them appealing in a variety of applications [44]. The support vector regression algorithm [45], least squares support vector machine [46], and successive projection algorithm [47] are three popular SVM algorithms.

(vii) **k-Nearest Neighbour (k-NN)**: The k-nearest neighbour (k-NN) method is a supervised learning classifier, which applies proximity for generating classifications or predictions regarding the grouping of a single data point. Though, for classification or regression it can be used, it is generally used for classification as it is based on the concept that comparable points close to one another are discoverable. It is computed how far away every data point in the training collection is from the specified test example. The k points which are closest to the data point are known as the k-Nearest Neighbour. The classification of the data point then depends on the class labels of its neighbours [48]. If a data point has more than one neighbour with a class label, the class label that contains the most class labels is applied to the data point. The exact value of k's closest neighbours must be established. The noise that is present in the training data may cause misclassification if the value of the k is too low. However, if the value of k is too high, there is a risk of misclassification since the collection of nearest neighbours can include data points, which were already located distant from the area surrounding the test characteristic [49]. In [50], real-time environmental measurements were taken in Mangalore, Kodagu, Kasaragod, and other districts of the state of Karnataka. These measurements included soil type, rainfall, humidity, etc. The values of the closest known neighbours can be used to predict the crop yield, which is an unknown value. Calculating the Euclidian distance between those places makes this achievable. Thus, for the specified input parameters, crop yield may be predicted. Different distance functions, including the most often used one, the Euclidean distance function, could be used to calculate the distance between points in a feature space. Let's assume that p and q are shown as feature vectors. The Euclidean metric is typically used to calculate the distance between p and q. If $a = (a_1, a_2)$ and $b = (b_1, b_2)$, the distance is given by Eq. (1).

$$\mathrm{d(a, b)} = \sqrt{(b_1 - a_1)^2 + (b_2 - a_2)^2} \tag{1}$$

(viii) **Random Forest**: The supervised machine learning technique known as "random forest" consists of many decision trees created using random vectors, each of which makes a decision. Regression techniques and classification problems can be addressed using this strategy. It is based on ensemble learning (EL), a method of integrating various classifiers to deal with critical issues. EL models create a linear combination of simpler base learners in order to enhance the prediction performance of a specific statistical learning or model-fitting method. The probability of receiving a more accurate result grows as the number of trees in the forest increases since the random forest's output is inversely proportional to the number of trees it integrates into the forest [51]. It is crucial to understand that building a forest differs from building decision trees. In case of random forest (RF) classification, the process of locating the root node and dividing the feature nodes will be random [52]. RF classification is common because of its advantages; enough trees are available so that the overfitting issue will be avoided. Again, in this classifier, missing values are also handled suitably. RF classifier is used in various sectors such as the stock market, banking, e-commerce, medicine, agriculture, etc. [53]. In banking, it is used to distinguish between legitimate and dishonest consumers. In the stock market, an RF classifier is used to watch a stock's activity and then to determine its gain and loss. It can also be applied in e-commerce to predict product recommendations [54].

3 Deep Learning in Agriculture

The term "smart agriculture" refers to the extensive use of AI in agriculture, which includes big data, IoT, deep learning, and many other smart systems. For analysing a huge amount of data, deep learning [55–57] comes into the picture. It has immense potential, has shown promising outcomes, and has been successfully applied in a variety of industries, including agriculture [55]. The foundation of deep learning is a set of machine learning algorithms, which model high-level abstractions in data through a variety of nonlinear transformations. Deep learning (DL) has a number of benefits, including feature learning, which refers to the autonomous extraction of features from the raw data. There are different common deep learning networks used in agriculture, such as recurrent neural networks and convolutional neural networks, briefly described as follows:

- **Recurrent Neural Network (RNN)**: In contrast to standard neural networks, RNNs are neural sequence models, which make use of the sequential information in the network. RNN is an adaptation of ANN, which indicates that the network's current input and output are connected. The unique expression is that the network will remember the previously learned information and utilise it to calculate the output of the current network; in other words, RNN may be thought of as a Backpropagation (BP) neural network whose output is used as the input to the next network. Even though RNN theoretically solves time series problems, it is challenging to do so in practise since the amount of information fluctuates, which can lead to gradients disappearing or exploding. Input layer X, hidden layer S, and output layer Y of an RNN can be thought of as a short-term memory unit [56]. A development of the recurrent neural network, the long short-term memory (LSTM) network is primarily intended to address time series problems with long intervals and large delays. To selectively alter the present state of the RNN, LSTM relies on the structure of a few "doors" [58, 59]
- **Convolutional Neural Network (CNN)**: A deep learning method called CNN, which contains numerous convolutional layers, pooling layers, and fully connected layers, has made significant advancements in speech recognition, face identification, natural language processing, etc. [60]. As long as there are sufficient

large data sets available for defining the problem, classifications will have a higher possibility of being accurate in CNN. Convolutional layers and pooling layers make up the structure for feature extraction, while fully linked layers serve as a classifier [61]. While convolutional networks first transform signals into features and subsequently map the features to a specified target value, Backpropagation neural networks primarily map features via the network to specific values [61]. The fully linked layers often take advantage of the high-level features learned at the final layer to classify the input images into predetermined groups. CNN models perform classification and predictions particularly effectively due to their highly hierarchical structure and great learning capacity. They are also flexible and adaptive in a broad variety of situations [62, 63].

- **Fully Convolutional Network (FCN)**: FCN is a CNN-based architecture that generates a semantic mask as output by down-sampling (convolution), followed by up-sampling (deconvolution). A predicted label for the input image is often produced by downscaling the input image and passing it through numerous convolution layers. FCN networks do not downscale the image, therefore, the output is not a single label, allowing for an upsampling of the output and the prediction of the class's pixel-by-pixel characteristics [64].

CNN indicates that classification accuracy is 1–8% higher than SVM [65, 66], 41% advancement compared with ANN [67], and 3–11% higher than unsupervised learning [68]. Additionally, CNN performed better than SVM regression [69], Large Margin Classifier (LMC), and Naïve Bayes Classifier [70, 71]. In case of RNN, LSTM achieved comparatively 1% better performance than SVM and RF [72], and 44% advancement than SVM [73].

4 IoT in Agriculture

The traditional approach to agriculture is to advance modernised farming while researching relevant IoT areas in the agricultural sector. Systematic evaluation provides current and future trends in the agriculture sector. IoT has several uses in the field of digital agriculture, including crop growth monitoring, fertiliser selection, irrigation decision support systems, etc. With a wide variety of sensors being utilised for diverse smart agricultural goals, the IoT has also recently made a major influence on the agriculture sector. Utilising the Internet to connect numerous networked devices, such as numerous sensors, drivers, and smart objects, to mobile devices, provided efficient production to the smart agriculture industry.

The IoT sensors interact with actuators and need wireless connectivity. Microprocessor, memory, input/output interfaces, and communication components make up the embedded system. There are various types of sensors including airflow sensors, mechanical sensors, optical sensors, location sensors, and mechanical sensors used to track and measure various parameters (for instance, soil nutrients and weather information) and temperatures of the atmosphere, different depths of the soil, precipitation, relative humidity, atmospheric pressure that affect production [74].

The development of mobile technology, wireless communication networks, and the ubiquity of services has made it possible for a vast amount of people to be connected to the Internet. The core network layer, or Internet, provides pathways for the transmission and exchange of data and network information between various subnetworks. Data is accessible anywhere and at any time due to IoT devices' connection to the network. Therefore, adequate security, real-time data support, and accessibility are required for data transfer over the Internet [75]. Managing user interfaces, services, network node organisation and coordination, computing, and data processing are all part of cloud computing, which collects massive amounts of data for archival and analysis. IoT middleware and networking protocols are being created to enable the Internet-based connectivity of heterogeneous systems and devices. IoT middleware, like actor-based and cloud-based, is basically used for supporting the IoT [76].

A crucial component of the effective implementation of IoT systems is communication technology. Standards, spectrum, and application scenarios can be used to categorise the current communication technology.

- Standard: Long-range and short-range communication standards are two categories of the communication standard. Short-range standards like Bluetooth, ZigBee, and Z-Wave can span a range of 100 m or less, whereas long-range communication protocols like LoRa, and NB-IoT can transmit data over a distance of up to 10 km. Long-range communication standards are within the low power wide area category (LPWA) that takes low power but gives a large area coverage [77].
- Spectrum: One can divide the communication spectrum into licenced and unlicensed spectrum. The industrial, scientific, and medical (ISM) band, a radio frequency spectrum, is used in the unlicensed spectrum. On the other side, the licenced spectrum that has been assigned to the cellular network provides consumers with improved traffic management, less interference, higher reliability, increased quality of service (QoS), a high level of security, more coverage, and lower infrastructure costs. The cost of membership for data transmission as well as the transmit power consumption on IoT devices are disadvantages of using licenced spectrum. On the other side, the downsides of using unlicensed spectrum include interference, infrastructure costs, and security concerns.
- Application Scenario: Application requirements for the IoT device influence the choice of communication technology. A backhaul network or an IoT device that serves as a node can employ communication technology. Low-data-transmission nodes use less power and travel a very little distance. The backhaul network can be used over very long distances and supports high data speeds. Additionally, the kind of topology that will be deployed will also influence the communication technology that is selected for the IoT device. An IoT device performs various roles and tasks in each of these topologies. The function ((full function device

(FFD) or a reduced function device (RFD)) can either be an end device or a personal area coordinator (PAN).

Every aspect of conventional farming processes can be drastically altered by integrating the most recent IoT technologies. The IoT may now be seamlessly integrated into smart agriculture. IoT can help to deal with many traditional farming problems, such as drought response, yield optimization, land appropriateness, irrigation, pest management, etc.

4.1 Soil Sampling and Mapping

The major goal of soil analysis is to ascertain a field's nutritional status so that appropriate action can be taken when nutrient deficits are discovered. The cropping history, soil type, application of fertilisers, irrigation level, topography, etc., are important variables to consider when analysing the levels of soil nutrients. These variables provide information on the soil's chemical, physical, and biological conditions, allowing for the identification of the constraints on the crops and the subsequent management of those constraints. In order to better match soil qualities, such as seed compatibility, sowing time, and even planting depth, as certain crop types are deep-rooted while others are not, soil mapping makes it possible to sow various crop kinds in a given area. Manufacturers are offering a variety of sensors and toolkits to help farmers to monitor the quality of the soil. These devices make it possible to keep track of soil characteristics including texture and water-holding capacity. Moderate resolution imaging spectroradiometer (MODIS) sensor was used in [78] for mapping different soil characteristics, and calculating the probability of land degradation in sub-Saharan Africa. Technology based on sensors and vision is useful in determining the distance and depth for effective seeding. In [79], the authors described the development of the autonomous Agribot seeding robot, which is sensor and vision-based.

4.2 Irrigation

Since, fresh water must be retained in lakes, rivers, and similar reservoirs to sustain it, humanity depends on 0.5% water to meet all of its needs and to maintain the ecosystem [80, 81]. It is important to note that the agricultural sector alone uses over 70% of the available freshwater [82]. Due to the shortage of water worldwide and the rising demand for water in various sectors of the economy, water should be delivered only when required and in the required amount. To address the water wastage issues, several controlled irrigation techniques, such as drip irrigation and spray irrigation, are being encouraged. When there is a water deficit, both the quality and quantity of the crops suffer because irregular irrigation, even excessive irrigation, reduces the nutrients in the soil. Building a precise soil and air moisture control system by employing wireless sensors not only makes the best use of water, but also improves crop health because it is not always simple to take crop yield metrics into account. By using developing IoT technologies, it is anticipated that the current state of irrigation methods would change [83]. With the adoption of IoT-based solutions, such as irrigation management based on the crop water stress index (CWSI), a sharp rise in agricultural efficiency is anticipated. A wireless sensor network is utilised to collect the aforementioned measures, followed by transmission to a processing hub, in which the related intelligent programmes are used to analyse the farm data. Water consumption efficiency is finally increased through CropMetrics' Variable Rate Irrigation (VRI) optimization, which takes topography or soil variations into account [84].

4.3 Fertiliser

In order to reduce the adverse effects of nutrients on the environment, fertilisation is a key component of intelligent agriculture. Based on a variety of variables, including soil type, crop type, soil absorption capacity, fertility type and utilisation rate, product yield, meteorological condition, etc., fertilisation requires site-specific measurements of soil nutrient levels. The issue is that measuring soil nutrients is time- and money-consuming because it normally necessitates the analysis of soil samples from each location. With greater accuracy and with less labour input, new IoT-based fertilization techniques enable the assessment of spatial patterns of nutrient requirements [85]. The NDVI is a tool for estimating crop health, vegetation vigour, and density, and it also helps to determine the level of soil nutrients. The IoT-based smart fertilisation is being greatly aided by a number of current enabling technologies, including Global Positioning System (GPS) accuracy [86], Variable Rate Technology (VRT) [87], and autonomous robots [88].

4.4 Crop Disease and Pest Management

By precisely identifying agricultural pests, modern IoT-based intelligent equipment like wireless sensors and drones are enabling growers to significantly reduce pesticide usage. Pest management based on IoT is more successful since it offers real-time monitoring, modelling, and disease predictions. Modern IoT-based pest management is more successful since it offers real-time monitoring, modelling, and disease predictions [89]. Sensing, evaluating, and treating are often the three factors to determine how effectively crop diseases and pests are managed. Advanced disease and pest recognition methods are built on image processing, whereby field sensors, unmanned aerial vehicles, or satellites are used to collect raw images from all over the crop area. Remote sensing imagery typically covers broad areas and provides better efficiency at a reduced cost. On the other side, field sensors have a greater capacity to support data collection tasks including environmental sampling, plant health, and pest situations across the entire crop cycle. Automated traps powered by the IoT can collect, count, and even classify different insect species; they can then upload data to the cloud for in-depth research [90]. The same methods which are frequently employed for smart fertilisation, such as vehicle precision spray and automatic VRT chemigation, can also be applied for the treatment of diseases and other pesticide applications [91]. The IoT-based pest control technology offers various benefits, including the ability to lower total costs while also assisting in the climate's restoration.

4.5 *Yield Supervision, Crop Forecasting, and Harvest*

The technique known as "yield supervising" is used to examine several factors related to agricultural production, such as moisture content, grain mass flow, and harvested grain quantity. When it comes to precision farming, yield monitoring is regarded as being significant, not just during harvest, but also before that because yield quality monitoring is important. Crop forecasting is the technique of foreseeing the yield and production prior to the harvest. The farmer can plan and make decisions based on this prediction for the near future. For this monitoring, which spans many stages of development fruit characteristics such as colour, size, etc., are used. Predicting the ideal harvesting window not only maximises the crop quality and output, but also gives managers the chance to modify their management approach. Different mobile applications are developed for displaying harvest data that will be uploaded to the web platform of the manufacturer.

In this context, many researches have been carried out over the years, with sensors and IoT-based technologies assisting in the improvement of traditional agricultural processes to increase the crop output without or with minimal impact on its originality. To address the aforementioned problems, new advanced settings, which are more tightly controlled, are anticipated. With the help of IoT, the advance level of agricultural aspects like vertical farming, greenhouse farming, phenotyping, etc., are used to build the super smart agricultural system.

5 Cloud-Fog-Edge-Dew Computing in Agriculture

Along with IoT, cloud computing is utilised in the agricultural industry. Cloud storage is required because IoT devices produce more data than the classic database paradigm can store. Nowadays, along with cloud computing for better service provisioning fog, edge, and dew computing also become popular.

5.1 Cloud Computing

Cloud computing combines significant IT resources using the Internet on the backend and makes them accessible to the user community through clearly defined interfaces. Processors, storage, networks, specialised hardware resources, and other services are examples of cloud resources. These resources are made available to users in the cloud on a pay-per-use basis as needed. Today, cloud computing has applications in practically every field, including science, engineering, business, and social sectors.

The IoT and cloud-based big data analytics play a significant role in the feasibility study of smart agriculture. Smart farming maximises resource efficiency while minimising environmental effects. At the moment, sensors can provide incredibly precise measures of crop status. Actuators can control agricultural activities involving livestock, crops, greenhouses, irrigation, soil, and weather based on those values. This can lead to advancements in harvest forecasting, weather prediction, increase in productivity, water conservation, real-time data collecting, and production, as well as decreased operating costs, precise farm and field evaluation, equipment monitoring, and remote monitoring [92].

The area of agriculture is being benefited significantly from cloud computing over the past few decades. The major characteristics of cloud computing in agriculture include data collection and remote storage, low-cost access to information and communication technology (ICT) resources, online access to agricultural experts, automation of land records, and weather forecasting. Similar issues apply to the usage of cloud computing in the agriculture sector, including constant and fast network access, security, and privacy.

There are some major applications of cloud computing in the agriculture field:

- **Enormous amounts of information**: Individual farmers can easily store and access information from the cloud, including information about crops, weather, markets, farmers' experiences with agricultural procedures, and information about pesticides.
- **Inexpensive use of IT resources**: It is the more affordable and dependable way to get access to resources. The use of cloud computing allows farmers to access resources and services as needed.
- **Simple answer to farming problems**: Cloud computing has made it possible for farmers to easily solve issues, which may arise at many phases of their agricultural activities, from tilling to marketing and selling of their crops.
- **Tools for gathering data**: There are numerous efficient and trustworthy data collection methods available today. These methods are simple to combine with cloud computing applications.
- Applications for sensors include monitoring soil and water quality, forestry, and forecasting changing environmental conditions. The sensors may detect water in the soil, or they can measure humidity, liquid pH, or pressure.
- **Predicting the weather**: Farmers can choose their crops based on weather forecasting for a specified period of time. When the climate is unsuitable for a seasonal crop, a farmer can choose a different seasonal crop.

5.2 Fog Computing

Fog computing is a growing computing technique to enhance and support cloud computing. Fog computing platforms include a number of features to deliver services to users more quickly and improve the QoS. As a result, it is becoming a crucial strategy for IoT-based applications focusing on users and entails real-time operations [93, 94].

High processing power and extensive data storage are features of cloud and fog computing. The core concept of fog computing is a geographically dispersed architecture that connects several smart devices in an IoT environment similar to the cloud, but it is situated close to the end users. Hence, it may support latency-sensitive applications and services [3, 92]. The use of intermediate devices, between the end node and cloud, serves as fog devices. Precision agriculture, which is a developing field, is applied to the suggested fog computing approach. Additionally, using this framework, they were able to replicate and demonstrate how the two-tier fog computing technique may considerably minimise the quantity of data that is sent to the cloud [95].

Its purpose is to facilitate communication, computation, and storage between the user and the cloud. Instead of using cloud servers, computations, and data processing can be partially done by the intermediate nodes, which will lighten the overhead of the cloud servers. The services and applications of fog computing are dispersed, which has several benefits including speed, fault tolerance, and security. Fog is excellent for real-time processing since it is close to the edge, enabling real-time interactions between IoT devices [92, 95]. The development of creative farming methods is gradually increasing crop yield, increasing profitability, and lowering irrigation waste.

5.3 Edge Computing

Edge computing and related paradigms are at an early stage of developmental. Despite its advantages (cost savings, efficiency, scalability, and reliability), cloud computing faces significant difficulties when dealing with large amount of data. The challenges of cloud computing, such as low latency, real-time analytics, high network bandwidth, data management, energy usage, security, load balancing, and privacy, can be resolved using edge computing [96]. Edge computing is suggested as a way to enhance speed and solve cloud-related issues by enabling local data processing and storage at endpoints. Hence, edge computing is a computational approach that locates computing power and storage closer to the end user, at the edge of the network. It offers intelligent services with cloud computing [97–99]. Edge computing offers various advantages such as:

- Data streams from various sources are processed by the nodes before being sent to the cloud to conserve bandwidth and storage space and to filter out noise.

- Processing data close to its source of origin results in proximity and reduced latency.
- Decentralised processing and storage allow greater scalability.
- Each node of the network has privacy.

Edge computing is a strong component for enabling global smart environments in agriculture [100]. Edge-enabled services can be categorised as node-centric services and cloud-centric services. The node-centric services operate independently of the cloud. The cloud-centric services rely on at least one cloud service to function. The majority of the systems analysed in the preceding section might be categorised as cloud-centric. Reduced latency, better bandwidth usage, and work offloading are the three key advantages of edge computing [101–105], which were used to varied degrees. Local edge devices were employed for initial data processing, and core cloud services were used for second-level offloading, storage, and alert generating.

Rural broadband connection is still a major issue everywhere in the world. The consequences of intermittent Internet access can be partially mitigated by edge computing, but comprehensive analytics-driven service provision needs access to the cloud.

A third model based on delay tolerance should be taken into consideration for the agricultural domain, according to the intractability of this problem.

Delay-tolerant edge services may be used as a design template for services, which need a connection to the cloud, but only in circumstances where sporadic network access is the norm. The idea of delay/disruption tolerance is well-established in the networking field, where it is regarded as a workable solution to problems like excessive latency and uncertain network availability.

However, from the standpoint of service, delay tolerance must be taken into account from the very beginning of the service's design. When accustomed to immediate Internet access, services that do not respond in near real-time may initially appear out of the ordinary. However, many agricultural services can wait; the IoT's requirement for constant Internet access is frequently superfluous because many environmental parameters change rather slowly. Other wireless technologies may be used to cover the farm and link to the edge node there in situations where an urgent alert needs to be brought to the farmer's notice. In this scenario, the service's underlying logic must reside on the node. Global food security, the present state of smart agriculture, and the prevalent problem of internet connectivity are considered in [106]. In [106], the use of the edge computing model in agriculture has been illustrated.

The use of cloud, fog, and edge computing in the existing literature on agriculture are summarised in Table 1.

Table 1 Existing literature on the use of cloud, fog, and edge computing in agriculture

References	Service	Proposed Work
Gurado et al. [11]	Fog-based IoT	Fog-based IoT framework to produce a newly growing field of precision agriculture
Lavanya et al. [85]	IoT	An IoT-based system by developing Nitrogen-Phosphorus-Potassium (NPK) sensor along with Light Dependent Resistor (LDR) and Light Emittimg Diodes (LED)
Venkatesan et al. [89]	IoT	A framework for autonomous pesticide distribution that utilises an android-based application with a built-in mobile module
Kakamoukas et al. [107]	Cloud-edge	A climate-smart design to encourage and promote integrated agriculture systems
Almalki et al. [108]	Cloud-edge	Environmentally conscious IoT-based agricultural monitoring systems
Chew et al. [109]	Cloud-edge	IoT-based remote monitoring system for organic Black Soldier Files (BSF) farming
Li et al. [110]	Edge	Edge computing-based data collection approach for critical events in smart farming
Roopaei et al. [111]	Cloud-fog-edge	Cloud of Things (CoT)-based system that explores the main requirements, technical challenges, and legal concerns to manage data connected to water sources
Puri et al. [112]	AI and IoT	A solution based on AI and IoT to examine growth and health of plant (Ref. missing)

5.4 Dew Computing

The "micro-service(s)" concept is anticipated to be used by dew computing in a very diverse, extremely vertical, and highly distributed hierarchy. Data dispersal into low-end devices like smartphones, tablets, laptops, e-book readers, etc. becomes viable in this new area of centralised-virtualisation-free computing paradigm. As a result, it opens up a brand-new possibility for data accessibility without constant Internet access. As a result, dew computing encompasses all current network technologies as well as a wide range of essential traits like independent, mobile data aggregation, cooperative applications, and hybrid network behaviour [113]. Dew computing has the following advantages:

- It is lightweight.
- It maintains the information that is used frequently in its databases.
- The cloud server allows the regeneration of this data.
- All tasks, such as DBMS control, identity mapping, raw data synchronisation, rule-based data collecting, dew script analysis, etc., will fall under its purview. It will eventually become apparent to the user as a Personal Information Centre.
- The user will always have access to its services, whether or not they have an Internet connection.

Independence and collaboration are two pillars of dew computing architecture [114]. Independence refers to a computer's ability to function without cloud services or a network connection. In other words, it indicates that this application is not a cloud service or an entirely online application. Collaboration requires that, while in use, the dew computing automatically communicates data with cloud services. Synchronisation, correlation, or other forms of interoperation are examples of this collaboration [17]. In order to continue storing data when a connection is sporadic, dew computing seeks to replicate data close to sensors or users. In case of connection failure, it enables access to a local copy of the data. It enhances false tolerance and reliability [115]. The use of dew computing in agriculture will be demonstrated in Sect. 6.

6 Dew Computing in Agriculture

In the dew computing-based model, the sensor data is accumulated at the dew layer. When internet connectivity is available the data is transmitted to the cloud through the edge/fog nodes. Figure 1 presents the dew computing-based architecture for smart agriculture. The proposed paradigm contains:

- Sensors
- Microcontrollers
- Edge/fog nodes
- Cloud servers.

The four-layer architecture is described as follows:

Layer 1: Layer 1 contains the sensor nodes. A sensor node (S) is mathematically defined as

$$S = \langle \mathrm{ID}_s, O_s \rangle$$

where ID_s denotes the ID of the sensor node, and O_s denotes the type of the object.

The sensors are used to collect the object data such as soil temperature, soil moisture, nitrogen, phosphorous, potassium level, pH level, environmental temperature, humidity, etc. The sensors are connected to the microcontroller present at layer 2. Relay nodes can be used in layer 1 if required. The sensors at layer 1 transmit the collected data to the microcontroller at layer 2.

Layer 2: Layer 2 contains microcontrollers. A microcontroller (M) is mathematically defined as

$$M = \langle \mathrm{ID}_m, C_m \rangle$$

where ID_m denotes the ID of the microcontroller, and C_m denotes the configuration of the microcontroller.

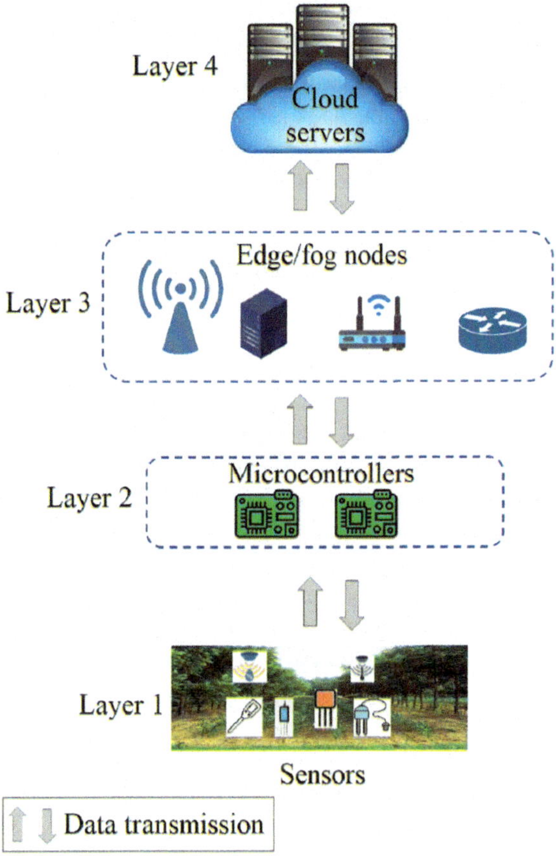

Fig. 1 Dew computing-based architecture for smart agriculture

The microcontroller receives the data from the sensor and stores the data. The microcontroller can pre-process the data. The microcontroller here serves the purpose of a dew layer. If Internet connectivity is not available, it holds and pre-processes the data. The microcontroller sends the data to the edge/fog node at layer 3 when Internet connectivity is available.

Layer 3: Layer 3 contains the edge/fog nodes. An edge/fog node (N) is mathematically defined as

$$N = \langle \mathrm{ID}_n, C_n \rangle$$

where ID_n denotes the ID of the edge/fog node, and C_n denotes the configuration of the edge/fog node.

The edge/fog node performs further processing on the data and forwards it to the cloud at layer 4.

Layer 4: Layer 4 contains the cloud servers. A cloud computing instance (I) is mathematically defined as

$$I = \langle \mathrm{ID}_i, C_i \rangle$$

where ID_i denotes the ID of the cloud component and P_i denotes the set containing the processing unit IDs.

The cloud servers after receiving the data store it if required for further analysis.

The total latency of the dew computing-based paradigm is computed as the sum of the data collection latency (L_1), data processing latency (L_2), and data transmission latency (L_3), given as

$$L = L_1 + L_2 + L_3 \tag{2}$$

Using MATLAB 2021, we have simulated the dew computing-based, edge computing-based, and only cloud computing-based architecture. Figure 2 presents the latency using the proposed dew computing-based paradigm, existing edge-based paradigm, and cloud-only paradigm. We observe that the use of dew computing reduces the latency by approximately 40 and 55% than the edge-based and cloud-only paradigms respectively.

As the dew layer holds the collected data and pre-processes the data before forwarding it to the next layer, the data transmission latency is reduced. Consequently, the overheads on the next layer as well as on the cloud are reduced. The use of dew computing moreover provides the advantage of accessing the data even when not connected to the Internet.

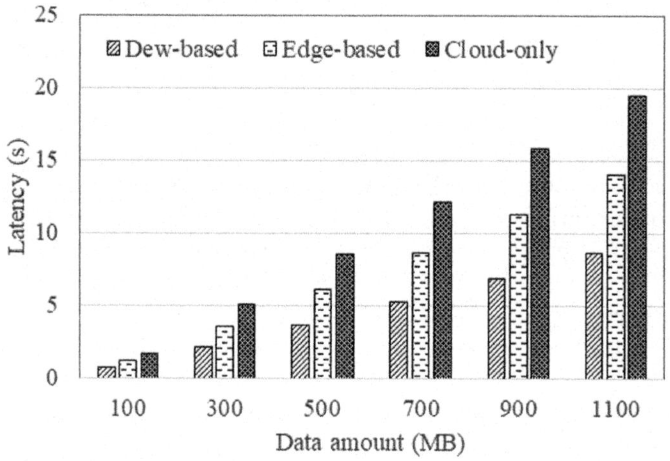

Fig. 2 Comparison of latency between dew computing-based, edge computing-based, and cloud-only paradigms

7 Conclusions

The use of the Internet of Things in smart agriculture has gained significant research interest in the last few years. The sensors are used to collect soil-related data, crop-related data, and the analysis of the collected data takes place inside the cloud. Based on the data analysis, the farmers can be guided for better crop production. However, the use of cloud-only paradigm has several shortcomings such as increase in latency, network bandwidth, network connectivity, huge overheads on the cloud, etc. To deal with these problems, fog and edge computing have come. The use of intermediate devices in data processing and brining the resources to the network edge solves various issues related to latency, bandwidth, etc. Nevertheless, the edge and fog computing scenarios require good Internet connectivity. But in the remote location, the Internet connectivity may not be always good. To resolve this issue, dew computing has come. The use of dew computing provides the facility of accessing the data even when Internet connectivity is unavailable. When the network connectivity resumes, the data transmission to the upper layer takes place, and finally the data is sent to the cloud if further analysis is required. In this chapter, we have discussed the architecture of dew computing-based smart agriculture, along with a brief discussion on the use of cloud, fog, and edge computing in smart agriculture. Various machine learning algorithms used for data analysis are also discussed. The simulation results present that the use of dew computing reduces the latency by approximately 40% and 55% than the edge-based and cloud-only paradigms, respectively. There are varied future research avenues. With the advancement of deep learning-based models, an efficient quality monitoring framework of agricultural products, greenhouse automation, and crop management system can be developed. Leveraging the dew-edge-cloud paradigm, we can develop a latency-aware smart farm management system to enhance the overall quality of life.

References

1. Elijah, O., Rahman, T.A., Orikumhi, I., Leow, C.Y., Hindia, M.N.: An overview of internet of things (IoT) and data analytics in agriculture: benefits and challenges. IEEE Internet Things J. **5**(5), 3758–3773 (2018)
2. Marjumin, N.H., Sidek, S., Hassan, M.A., Rajikon, M., Kamalrudin, M.: The challenges and contribution of internet of things (Iot) for smart living. Int. J. Recent Technol. Eng. **8**, 162–166 (2019)
3. Atzori, L., Iera, A., Morabito, G.: The internet of things: a survey. Comput. Netw. **54**(15), 2787–2805 (2010)
4. Fang, S., Da Xu, L., Zhu, Y., Ahati, J., Pei, H., Yan, J., Liu, Z.: An integrated system for regional environmental monitoring and management based on internet of things. IEEE Trans. Ind. Inf. **10**(2), 1596–1605 (2014)
5. Ray, P.P.: Internet of things for smart agriculture: technologies, practices and future direction. J. Ambient Intell. Smart Environ. **9**(4), 395–420 (2017)

6. Sadeeq, M.M., Abdulkareem, N.M., Zeebaree, S.R., Ahmed, D.M., Sami, A.S., Zebari, R.R.: IoT and Cloud computing issues, challenges and opportunities: a review. Qubahan Acad. J. **1**(2), 1–7 (2021)
7. Kaur, C.: The cloud computing and internet of things (IoT). Int. J. Sci. Res. Sci. Eng. Technol. **7**(1), 19–22 (2020)
8. Biswas, A.R., Giaffreda, R.: IoT and cloud convergence: opportunities and challenges. In: 2014 IEEE World Forum on Internet of Things (WF-IoT), pp. 375–376. IEEE (2014)
9. Goraya, M.S., Kaur, H.: Cloud computing in agriculture. HCTL Open Int. J. Technol. Innov. Res. (IJTIR) **16**, 2321–1814 (2015)
10. Deng, R., Lu, R., Lai, C., Luan, T.H., Liang, H.: Optimal workload allocation in fog-cloud computing toward balanced delay and power consumption. IEEE Internet Things J. **3**(6), 1171–1181 (2016)
11. Guardo, E., Di Stefano, A., La Corte, A., Sapienza, M., Scatà, M.: A fog computing-based iot framework for precision agriculture. J. Internet Technol. **19**(5), 1401–1411 (2018)
12. Yousefpour, A., Fung, C., Nguyen, T., Kadiyala, K., Jalali, F., Niakanlahiji, A., et al.: All one needs to know about fog computing and related edge computing paradigms. J. Syst. Archit. (2019)
13. Chen, J., Ran, X.: Deep learning with edge computing: a review. Proc. IEEE **107**(8), 1655–1674 (2019)
14. Sittón-Candanedo, I., Alonso, R.S., Corchado, J.M., Rodríguez-González, S., Casado-Vara, R.: A review of edge computing reference architectures and a new global edge proposal. Fut. Gener. Comput. Syst. **99**, 278–294 (2019)
15. Zhang, X., Cao, Z., Dong, W.: Overview of edge computing in the agricultural internet of things: key technologies, applications, challenges. IEEE Access **8**, 141748–141761 (2020)
16. Xu, L., Collier, R., O'Hare, G.M.: A survey of clustering techniques in WSNs and consideration of the challenges of applying such to 5G IoT scenarios. IEEE Internet Things J. **4**(5), 1229–1249 (2017)
17. Rindos, A., Wang, Y.: Dew computing: the complementary piece of cloud computing. In: 2016 IEEE International Conferences on Big Data and Cloud Computing (BDCloud), Social Computing and Networking (SocialCom), Sustainable Computing and Communications (SustainCom) (BDCloud-SocialCom-SustainCom), pp. 15–20. IEEE
18. Wang, Y.: Definition and categorization of dew computing. Open J. Cloud Comput. (OJCC) **3**(1), 1–7 (2016)
19. Van Klompenburg, T., Kassahun, A., Catal, C.: Crop yield prediction using machine learning: a systematic literature review. Comput. Electron. Agric. **177**, 105709 (2020)
20. Chlingaryan, A., Sukkarieh, S., Whelan, B.: Machine learning approaches for crop yield prediction and nitrogen status estimation in precision agriculture: a review. Comput. Electron. Agric. **151**, 61–69 (2018)
21. Xu, X., Gao, P., Zhu, X., Guo, W., Ding, J., Li, C., et al.: Design of an integrated climatic assessment indicator (ICAI) for wheat production: a case study in Jiangsu Province, China. Ecol. Indicat. **101**, 943–953 (2019)
22. Reddy, K.S.P., Roopa, Y.M., LN, K.R., Nandan, N.S.: IoT based smart agriculture using machine learning. In: 2020 Second International Conference on Inventive Research in Computing Applications (ICIRCA), pp. 130–134. IEEE (2020)
23. Jagtap, S.T., Phasinam, K., Kassanuk, T., Jha, S.S., Ghosh, T., Thakar, C.M.: Towards application of various machine learning techniques in agriculture. Mater. Today: Proc. **51**, 793–797 (2022)
24. Pearson, K.: LIII. On lines and planes of closest fit to systems of points in space. London Edinburgh Dublin Philos. Mag. J. Sci. **2**(11), 559–572
25. Fisher, R.A.: The use of multiple measurements in taxonomic problems. Ann. Eugen. **7**(2), 179–188 (1936)
26. Mishra, S., Mishra, D., Santra, G.H.: Applications of machine learning techniques in agricultural crop production: a review paper. Indian J. Sci. Technol. **9**(38), 1–14 (2016)

27. Uno, Y., Prasher, S.O., Lacroix, R., Goel, P.K., Karimi, Y., Viau, A., Patel, R.M.: Artificial neural networks to predict corn yield from compact airborne spectrographic imager data. Comput. Electron. Agric. **47**(2), 149–161 (2005)
28. Veenadhari, S., Mishra, B., Singh, C.D.: Soybean productivity modelling using decision tree algorithms. Int. J. Comput. Appl. **27**(7), 11–15 (2011)
29. Bhargavi, P., Jyothi, S.: Applying Naive Bayes data mining technique for classification of agricultural land soils. Int. J. Comput. Sci. Netw. Secur. **9**(8), 117–122 (2009)
30. Rainville, D., Durand, A., Fortin, F.A., Tanguy, K., Maldague, X., Panneton, B., Simard, M.J.: Bayesian classification and unsupervised learning for isolating weeds in row crops. Pattern Anal. Appl. **17**(2), 401–414 (2014)
31. Sharma, B., Yadav, J.K.P.S., Yadav, S.: Predict crop production in India using machine learning technique: a survey. In: 2020 8th International Conference on Reliability, Infocom Technologies and Optimization (Trends and Future Directions) (ICRITO), pp. 993–997. IEEE (2020)
32. Khoshnevisan, B., Rafiee, S., Omid, M., Mousazadeh, H., Rajaeifar, M.A.: Application of artificial neural networks for prediction of output energy and GHG emissions in potato production in Iran. Agric. Syst. **123**, 120–127 (2014)
33. Babu, M.P.: A web based tomato crop expert information system based on artificial intelligence and machine learning algorithms (2010)
34. Umair, S.M., Usman, R.: Automation of irrigation system using ANN based controller. Int. J. Electr. Comput. Sci. IJECS-IJENS **10**(02), 41–47 (2010)
35. Li, Y., Chao, X.: ANN-based continual classification in agriculture. Agriculture **10**(5), 178 (2020)
36. Dahikar, S.S., Rode, S.V.: Agricultural crop yield prediction using artificial neural network approach. Int. J. Innov. Res. Electr. Electron. Instrum. Control Eng. **2**(1), 683–686 (2014)
37. Kantanantha, N.: Crop decision planning under yield and price uncertainties. Georgia Institute of Technology (2007)
38. Suresh, K.K., Krishna Priya, S.R.: A study on pre-harvest forecast of sugarcane yield using climatic variables. Stat. Appl. **7&8**(1&2), 1–8 (New Series) (2009)
39. Horie, T., Yajima, M., Nakagawa, H.: Yield forecasting. Agric. Syst. **40**(1–3), 211–236 (1992)
40. Liakos, K.G., Busato, P., Moshou, D., Pearson, S., Bochtis, D.: Machine learning in agriculture: a review. Sensors **18**(8), 2674 (2018)
41. Narkhede, U.P., Adhiya, K.P.: Evaluation of modified K-means clustering algorithm in crop prediction. Int. J. Adv. Comput. Res. **4**(3), 799 (2014)
42. Mahesh, B.: Machine learning algorithms-a review. Int. J. Sci. Res. (IJSR) **9**, 381–386 (2020); Noble, W.S.: What is a support vector machine?. Nat. Biotechnol. **24**(12), 1565–1567 (2006)
43. Pradhan, A.: Support vector machine-a survey. Int. J. Emerg. Technol. Adv. Eng. **2**(8), 82–85 (2012)
44. Chandra, M.A., Bedi, S.S.: Survey on SVM and their application in image classification. Int. J. Inf. Technol. **13**(5), 1–11 (2021)
45. Mierswa, I.: Controlling overfitting with multi-objective support vector machines. In: Proceedings of the 9th Annual Conference on Genetic and Evolutionary Computation, pp. 1830–1837 (2007)
46. Mohamed, A.E.: Comparative study of four supervised machine learning techniques for classification. Int. J. Appl. **7**(2), 1–15 (2017)
47. Gupta, A., Katarya, R.: Social media based surveillance systems for healthcare using machine learning: a systematic review. J. Biomed. Inform. **108**, 103500 (2020)
48. Shakoor, M.T., Rahman, K., Rayta, S.N., Chakrabarty, A.: Agricultural production output prediction using supervised machine learning techniques. In: 2017 1st International Conference on Next Generation Computing Applications (NextComp), pp. 182–187. IEEE (2006)
49. Kataria, A., Singh, M.D.: A review of data classification using k-nearest neighbour algorithm. Int. J. Emerg. Technol. Adv. Eng. **3**(6), 354–360 (2013)
50. Karthikeya, H.K., Sudarshan, K., Shetty, D.S.: Prediction of agricultural crops using KNN algorithm. Int. J. Innov. Sci. Res. Technol **5**, 1422–1424 (2020)

51. Cutler, A., Cutler, D.R., Stevens, J.R.: Random forests. In: Ensemble Machine Learning, pp. 157–175. Springer, Boston, MA (2012)
52. Ali, J., Khan, R., Ahmad, N., Maqsood, I.: Random forests and decision trees. Int. J. Comput. Sci. Issues (IJCSI) **9**(5), 272 (2012)
53. Tan, K., Ma, W., Wu, F., Du, Q.: Random forest–based estimation of heavy metal concentration in agricultural soils with hyperspectral sensor data. Environ. Monit. Assess. **191**(7), 1–14 (2019)
54. Sharma, N., Juneja, A.: Combining of random forest estimates using LSboost for stock market index prediction. In: 2017 2nd International Conference for Convergence in Technology (I2CT), pp. 1199–1202. IEEE (2017)
55. Dargan, S., Kumar, M., Ayyagari, M.R., Kumar, G.: A survey of deep learning and its applications: a new paradigm to machine learning. Arch. Comput. Methods Eng. **27**, 1071–1092 (2020)
56. LeCun, Y., Bengio, Y., Hinton, G.: Deep learning. Nature **521**(7553), 436–444 (2015)
57. Bengio, Y., Lecun, Y., Hinton, G.: Deep learning for AI. Commun. ACM **64**(7), 58–65 (2021)
58. Sak, H., Senior, A.W., Beaufays, F.: Long short-term memory recurrent neural network architectures for large scale acoustic modeling. Interspeech (2014)
59. Zhu, N., Liu, X., Liu, Z., Hu, K., Wang, Y., Tan, J., et al.: Deep learning for smart agriculture: Concepts, tools, applications, and opportunities. Int. J. Agric. Biol. Eng. **11**(4), 32–44 (2018)
60. Gu, J., Wang, Z., Kuen, J., Ma, L., Shahroudy, A., Shuai, B., et al.: Recent advances in convolutional neural networks. Pattern Recogn. **77**, 354–377 (2018)
61. Kamilaris, A., Prenafeta-Boldú, F.X.: A review of the use of convolutional neural networks in agriculture. J. Agric. Sci. **156**(3), 312–322 (2018)
62. Schmidhuber, J.: Deep learning in neural networks: an overview. Neural Netw. **61**, 85–117 (2015)
63. Oquab, M., Bottou, L., Laptev, I., Sivic, J.: Learning and transferring mid-level image representations using convolutional neural networks. In: Proceedings of the IEEE Conference on Computer Vision and Pattern Recognition, pp. 1717–1724 (2014)
64. Long, J., Shelhamer, E., Darrell, T.: Fully convolutional networks for semantic segmentation. In: Proceedings of the IEEE Conference on Computer Vision and Pattern Recognition, pp. 3431–3440 (2015)
65. Chen, Y., Lin, Z., Zhao, X., Wang, G., Gu, Y.: Deep learning-based classification of hyperspectral data. IEEE J. Sel. Top. Appl. Earth Observ. Remote Sens. **7**(6), 2094–2107 (2014)
66. Grinblat, G.L., Uzal, L.C., Larese, M.G., Granitto, P.M.: Deep learning for plant identification using vein morphological patterns. Comput. Electron. Agric. **127**, 418–424 (2016)
67. Lee, S.H., Chan, C.S., Wilkin, P., Remagnino, P.: Deep-plant: plant identification with convolutional neural networks. In 2015 IEEE International Conference on Image Processing (ICIP), pp. 452–456. IEEE (2015)
68. Luus, F.P., Salmon, B.P., Bergh, F.V., Maharaj, B.T.: Multiview deep learning for land-use classification. IEEE Geosci. Remote Sens. Lett. **12**, 2448–2452 (2015)
69. Kuwata, K., Shibasaki, R.: Estimating crop yields with deep learning and remotely sensed data. In: 2015 IEEE International Geoscience and Remote Sensing Symposium (IGARSS), pp. 858–861. IEEE (2015)
70. Xinshao, W., Cheng, C.: Weed seeds classification based on PCANet deep learning baseline. In: 2015 Asia-Pacific Signal and Information Processing Association Annual Summit and Conference (APSIPA), pp. 408–415. IEEE (2015)
71. Yalcin, H.: Plant phenology recognition using deep learning: deep-pheno. In: 2017 6th International Conference on Agro-Geoinformatics, pp. 1–5. IEEE (2017)
72. Ienco, D., Gaetano, R., Dupaquier, C., Maurel, P.: Land cover classification via multitemporal spatial data by deep recurrent neural networks. IEEE Geosci. Remote Sens. Lett. **14**(10), 1685–1689 (2017)
73. Rußwurm, M., Körner, M.: Multi-temporal land cover classification with long short-term memory neural networks. Int. Arch. Photogramm. Remote Sens. Spatial Inf. Sci. **42** (2017)

74. Abbasi, A.Z., Islam, N., Shaikh, Z.A.: A review of wireless sensors and networks' applications in agriculture. Comput. Stand. Interfaces **36**(2), 263–270 (2014)
75. TongKe, F.: Smart agriculture based on cloud computing and IOT. J. Converg. Inf. Technol. **8**(2), 210–216 (2013)
76. Ngu, A.H., Gutierrez, M., Metsis, V., Nepal, S., Sheng, Q.Z.: IoT middleware: a survey on issues and enabling technologies. IEEE Internet Things J. **4**(1), 1–20 (2016)
77. Raza, U., Kulkarni, P., Sooriyabandara, M.: Low power wide area networks: an overview. IEEE Commun. Surv. Tutor. **19**(2), 855–873 (2017)
78. Vågen, T.G., Winowiecki, L.A., Tondoh, J.E., Desta, L.T., Gumbricht, T.: Mapping of soil properties and land degradation risk in Africa using MODIS reflectance. Geoderma **263**, 216–225 (2016)
79. Santhi, P.V., Kapileswar, N., Chenchela, V.K., Prasad, C.V.S.: Sensor and vision based autonomous AGRIBOT for sowing seeds. In: 2017 International Conference on Energy, Communication, Data Analytics and Soft Computing (ICECDS), pp. 242–245. IEEE (2017)
80. Williams, M.: What Percent of Earth is water. Universe Today 2016 (2014)
81. Water Facts_Worldwide Water Supply.: https://www.usbr.gov/mp/arwec/water-facts-ww-water-sup.html. Accessed 15 Apr. 2019
82. *Water for Sustainable Food and Agriculture by FAO.*: https://www.fao.org/3/a-i7959e.pdf. Accessed 15 Apr. 2019
83. Hassan, Q.F.: (Ed.). (2018). *Internet of things A to Z: technologies and applications.* John Wiley & Sons.
84. LaRue, J., Fredrick, C.: Decision process for the application of variable rate irrigation. Am. Soc. Agric. Biol. Eng. (Dallas, TX, USA, Tech. Rep.) (2012)
85. Lavanya, G., Rani, C., GaneshKumar, P.: An automated low cost IoT based fertilizer intimation system for smart agriculture. Sustain. Comput.: Inform. Syst. **28**, 100300 (2020)
86. Shi, J., Yuan, X., Cai, Y., Wang, G.: GPS real-time precise point positioning for aerial triangulation. GPS Solut. **21**(2), 405–414 (2017)
87. Colaço, A.F., Molin, J.P.: Variable rate fertilization in citrus: a long term study. Precision Agric. **18**(2), 169–191 (2017)
88. Khan, N., Medlock, G., Graves, S., Anwar, S.: GPS guided autonomous navigation of a small agricultural robot with automated fertilizing system (No. 2018-01-0031). SAE Technical Paper (2018)
89. Venkatesan, R., Kathrine, G.J.W., Ramalakshmi, K.: Internet of things based pest management using natural pesticides for small scale organic gardens. J. Comput. Theor. Nanosci. **15**(9–10), 2742–2747 (2018)
90. Suma, V.: Internet-of-things (IoT) based smart agriculture in India-an overview. J. ISMAC **3**(01), 1–15 (2021)
91. Oberti, R., Marchi, M., Tirelli, P., Calcante, A., Iriti, M., Tona, E., et al.: Selective spraying of grapevines for disease control using a modular agricultural robot. Biosyst. Eng. **146**, 203–215 (2016)
92. Kalyani, Y., Collier, R.: A systematic survey on the role of cloud, fog, and edge computing combination in smart agriculture. Sensors **21**(17), 5922 (2021)
93. Abbas, N., Zhang, Y., Taherkordi, A., Skeie, T.: Mobile edge computing: a survey. IEEE Internet Things J. **5**(1), 450–465 (2018). https://doi.org/10.1109/JIOT.2017.2750180
94. Shi, W., Cao, J., Zhang, Q., Li, Y., Xu, L.: Edge computing: vision and challenges. IEEE Internet Things J. **3**(5), 637–646 (2016)
95. Sarhan, A.: Fog computing as solution for IoT-based agricultural applications. In: Smart Agricultural Services Using Deep Learning, Big Data, and IoT, pp. 46–68. IGI Global (2021)
96. Mukherjee, A., De, D., Ghosh, S.K., Buyya, R.: Introduction to mobile edge computing. In: Mobile Edge Computing, pp. 3–19. Springer, Cham (2021)
97. Zhang, J., Chen, B., Zhao, Y., Cheng, X., Hu, F.: Data security and privacy-preserving in edge computing paradigm: survey and open issues. IEEE Access **6**, 18209–18237 (2018)
98. Satyanarayanan, M.: The emergence of edge computing. Computer **50**(1), 30–39 (2017)

99. Zamora-Izquierdo, M.A., Santa, J., Martínez, J.A., Martínez, V., Skarmeta, A.F.: Smart farming IoT platform based on edge and cloud computing. Biosyst. Eng. **177**, 4–17 (2019)
100. Sengupta, A., Gill, S.S., Das, A., De, D.: Mobile edge computing based internet of agricultural things: a systematic review and future directions. Mob. Edge Comput. 415–441 (2021)
101. Ghosh, S., Mukherjee, A., Ghosh, S.K., Buyya, R.: Mobi-iost: mobility-aware cloud-fog-edge-IoT collaborative framework for time-critical applications. IEEE Trans. Netw. Sci. Eng. **7**(4), 2271–2285 (2019)
102. Mukherjee, A., Ghosh, S., De, D., Ghosh, S.K.:. MCG: mobility-aware computation offloading in edge using weighted majority game. IEEE Trans. Netw. Sci. Eng. (2022)
103. Ghosh, S., Mukherjee, A.: STROVE: spatial data infrastructure enabled cloud–fog–edge computing framework for combating COVID-19 pandemic. Innov. Syst. Softw. Eng. 1–17 (2022)
104. Das, J., Ghosh, S., Mukherjee, A., Ghosh, S.K., Buyya, R.: Rescue: enabling green healthcare services using integrated IoT-edge-fog-cloud computing environments. Softw.: Pract. Exp. (2022)
105. Ghosh, S., Ghosh, S.K.: Mobility driven cloud-fog-edge framework for location-aware services: a comprehensive review. Mob. Edge Comput. 229–249 (2021)
106. O'Grady, M.J., Langton, D., O'Hare, G.M.P.: Edge computing: a tractable model for smart agriculture? Artif. Intell. Agric. **3**, 42–51 (2019)
107. Kakamoukas, G., Sarigiannidis, P., Maropoulos, A., Lagkas, T., Zaralis, K., Karaiskou, C.: Towards climate smart farming—A reference architecture for integrated farming systems. In: Telecom, vol. 2, no. 1, pp. 52–74. MDPI
108. Almalki, F.A., Soufiene, B.O., Alsamhi, S.H., Sakli, H.: A low-cost platform for environmental smart farming monitoring system based on IoT and UAVs. Sustainability **13**(11), 5908 (2021)
109. Chew, K.T., Jo, R.S., Lu, M., Raman, V., Then, P.H.H.: Organic black soldier flies (BSF) farming in rural area using Libelium Waspmote smart agriculture and internet-of-things technologies. In: 2021 IEEE 11th IEEE Symposium on Computer Applications & Industrial Electronics (ISCAIE), pp. 228–232. IEEE (2021)
110. Li, X., Ma, Z., Zheng, J., Liu, Y., Zhu, L., Zhou, N.: An effective edge-assisted data collection approach for critical events in the SDWSN-based agricultural internet of things. Electronics **9**(6), 907 (2020)
111. Roopaei, M., Rad, P., Choo, K.-K.R.: Cloud of things in smart agriculture: intelligent irrigation monitoring by thermal imaging. IEEE Cloud Comput. **4**(1), 10–15 (2017). https://doi.org/10.1109/MCC.2017.5
112. Puri, B.: IoT and AI-based Plant Monitoring System. International J. Mach. Learn. Netw. Collab. Eng. **4**(3), 135–142 (2021)
113. Ray, P.P.: An introduction to dew computing: definition, concept and implications. IEEE Access **6**, 723–737 (2017)
114. Roy, S., Sarkar, D., De, D.: DewMusic: crowdsourcing-based internet of music things in dew computing paradigm. J. Ambient. Intell. Humaniz. Comput. **12**(2), 2103–2119 (2021)
115. Skala, K., Davidovic, D., Afgan, E., Sovic, I., Sojat, Z.: Scalable distributed computing hierarchy: cloud, fog and dew computing. Open J. Cloud Comput. (OJCC) **2**(1), 16–24 (2015)

Dew Computing Enabled Consumer Electronics for Sustainable Internet of Agricultural Things

Satabdwi Sarkar, Anirbit Sengupta, Abhijit Das, Debashis De, and Nurzaman Ahmed

1 Introduction

Food and fiber, two of humanity's most basic needs, are provided by agriculture, which is a significant source of income for many nations. Due to technical advancements like the green movement, agriculture has seen a fundamental transformation in recent decades. The management of livestock, commodities, and groundwater level are only a few of the subjects that are the focus of agricultural research. The Green Revolution, is also known as the third invention of agriculture, involved the improvement of varieties of crop, irrigation, synthetic fertilizers, and pesticides, which has increased crop food production and food stability between the 1960s and 1980s, especially in developing nations. Consequently, even though the global population has doubled and food production has increased threefold since the 1960s, agricultural productivity has already increased by 25% to meet the demand. Consumer electronics, often known as home electronics, are gadgets designed for regular use,

S. Sarkar (✉)
Computer Science & Engineering, RCC Institute of Information Technology, Kolkata, India
e-mail: satabdwi.rcciit@gmail.com

A. Sengupta
Electronics and Communication Engineering, Dr. Sudhir Chandra Sur Institute of Technology and Sports Complex, Kolkata, India

A. Das
Information Technology, RCC Institute of Information Technology, Kolkata, India

D. De
Department of Computer Science and Engineering, Maulana Abul Kalam Azad University of Technology, Kolkata, West Bengal, India
e-mail: debashis.de@makautwb.ac.in

N. Ahmed
Donald Danforth Plant Science Center, St. Louis, MO, USA
e-mail: nurzaman.ahmed@dartmouth.edu

D. De and S. Roy (eds.), *Dew Computing*, Internet of Things,
https://doi.org/10.1007/978-981-99-4590-0_15

usually in private homes. Devices for communication, entertainment, and amusement are all included in the consumer electronics category. Today's cutting-edge consumer gadgets are made with semiconductor solutions, including tablets, mobile phones, wearables, and wireless charging platforms. In 1999, the Auto-ID center at MIT and its pertinent market research articles helped popularize the idea of IoT. IoT is an amalgamation of many devices that use the embedded technology it contains to communicate, detect, and react to internal and external conditions. Multiple devices are combined into the Internet of Things (IoT), which uses embedded technology to provide communication, sensing, and interaction with internal and external states [1].

The effects of IoT on precision agriculture are so severe that they have the potential to alter the national economy radically. This is what took place with the Netherlands, a tiny, heavily populated lowland nation that currently ranks second among exporters of vegetables worldwide. The novel approach to agricultural landscapes and the widespread application of IoT in agricultural processes are the keys. In the Netherlands, glass covers the most agricultural area, and self-contained greenhouses cultivate vegetables. To maintain a sustainable climate in the greenhouses, the Dutch use linked technology and sensors to track plant health, CO_2 levels, LED light, humidity, and data analytics. As a result, they produce ten times the typical production from an open field. Besides being incredibly efficient, growing food in modern greenhouses is also environmentally friendly because it utilizes much less water and chemicals.

The industry's leading products for electrostatic discharge (ESD) circuit protection, wireless charging, and capacitive touch controller integrated circuits (ICs) are those with high-performance silicon and a small footprint. Dew computing is a developing field of study with lots of application potential. We observed a different concept of dew computing in this chapter. The revised meaning is: In the context of cloud computing, "dew computing" refers to an on-premises computer software-hardware architecture paradigm in which the on-premises computer provides functions both independently of cloud services and in collaboration with them.

A model called "dew computing" was developed from the original idea of cloud computing. Cloud computing has also given rise to various models, such as fog computing, edge computing, dew computing, and others. Advocates contend that cutting-edge methods like dew computing can provide users with a better experience. Here, agriculture and technology are combined in this effort. There are two forms of agriculture: organic and inorganic. "Organic agriculture is a comprehensive approach to production management that fosters and improves the health of the Agroecosystem, including biodiversity, biological cycles, and soil biological activity." Farmers drive organic farming. Some farmers have established alternative production methods to increase their family's health, farm economies, and independence because they feel that conventional agriculture is not sustainable. Many developing nations use organic agriculture as a growing method.

The concept "Internet of Things" has gained popularity over the past five years. An IoT network is a collection of Internet-connected machines, cars, and appliances that can gather and share data without the need for human contact. IoT devices collect data and transfer it to a central data server, where it is processed, compiled, distilled, and used to facilitate a variety of operations. The advantages of IoT are enjoyed by the business community, the government, organizations, and the average consumer.

Mobile phones, computers, coffee makers, refrigerators (mine automatically orders replacement water filters), Apple Watches, Google Home, Fitbits, and others are examples of specific IoT-enabled gadgets. Any gadget with sensors and an Internet connection can be connected to the IoT.

IoT is used to automate operations in various applications, including smart grids, smart cities, and innovative industries. For example, as the global population increases too quickly, new IoT techniques are being employed in agriculture to produce food quickly, in large quantities, and with high quality while using less water, fertilizer, and space.

IoT is helping agriculturalists and technologists identify solutions to farmers' concerns, such as cost management, water shortages, and productivity problems.

1.1 WSN

Before the IoT era, wireless sensor networks (WSN) were used to create an intelligent environment and monitor and regulate several environmental elements. They can be utilized in agricultural wireless sensor networks, particularly in greenhouses. We can utilize WSN in various IoT applications thanks to its multiple capabilities, including integrating sensors and actuators, low-power consumption, digital transmission, security, and scalability. Operations include irrigation, farm monitoring, fertilizer use management, soil monitoring, intruder detection, water quality monitoring, etc.

The IoT and Big Data are topics that are frequently brought up together since they both produce the same kinds of enormous amounts of data. A detailed examination of IoT in agriculture is required to understand the state of the research. A fair lot of work has been done in this area. This study addresses many difficulties and trends in IoT smart farming to transform agriculture technology through IoT innovation [1].

Specific sensors assist farmers in understanding how the sun's effect and light adjustment affect plant growth. Additionally, there are weather stations that, when used in conjunction with other sensors, software, and tools, aid in monitoring the various aspects of the conditions in farming fields. They support farmers, growers, and landowners in their endeavors.

The wireless transmission provides the data that sensors have collected to the web server. Wireless communication between the field and the web server is accomplished using several sensor modules and transceivers. WSN modernizing agriculture may be achieved by fusing established practices with cutting-edge innovations like Internet of things and wireless sensor networks. Through a wireless protocol, the Wireless Sensor Network gathers data from various sensors and transmits it to the central server. The impact of numerous additional factors on productivity is significant. Insect and pest attacks, which can be prevented by applying the appropriate insecticides and pesticides, as well as attacks by wild animals and birds as the crop matures, are among the factors. Because of erratic monsoon rainfalls, water scarcity, and inefficient water utilization, crop output is dropping [2].

1.2 Sensors

The architecture of IoT devices includes sensors. Sensors are used to detect objects, machinery, and other things. A gadget that responds to a particular measurement by producing usable output. The sensor collects physical data and transforms it into a signal that can be used to analyze (e.g., electrically, mechanically, optically) the properties of any object or substance to identify the presence of a specific physical quantity. The sensor's output is a signal that's been transformed into something humans can understand, such as changes in characteristics, resistance, capacitance, and impedance.

In agriculture, sensors are used for everything from automated irrigation to weather monitoring. By utilizing inexpensive sensors, designers can build a hardware environment prototype for the data collecting and mining process. The impact of each sensor's data for agricultural monitoring varies; however, by providing reliable observations even with fewer sensors, a system's cost can be reduced (Fig. 1).

2 Cache Memory

It is a computer component which is chip-based that makes retrieving data from the computer's memory more efficiently. The computer's CPU can quickly retrieve data from it, serving as temporary storage. The computer's primary memory source, which is often some DRAM, is less accessible to the processor than this temporary

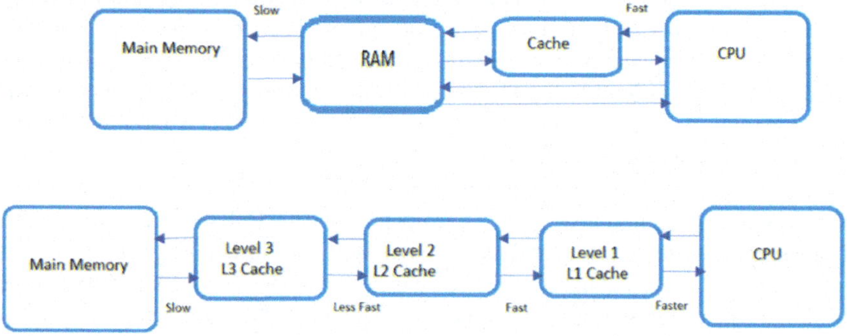

Fig. 1 Block diagram showing cache memory

storage space, known as a cache. As because it is frequently built directly into the central processing unit (CPU) chip or mounted on a different chip with a separate bus link from the CPU, cache memory is also referred to as CPU (central processing unit) memory [1]. Therefore, it can increase efficiency because it's physically close and is more accessible to the processor. Figure 2 shows the circuit diagram of Raspberry Pi4, and Fig. 3 shows the position of cache memory in computer architecture.

Fig. 2 Circuit diagram of raspberry Pi4

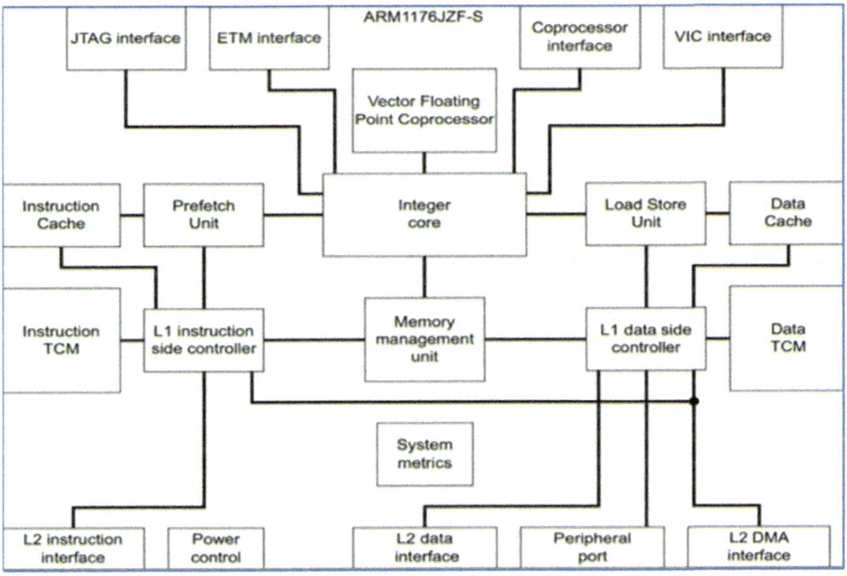

Fig. 3 Block diagram of raspberry Pi4

3 Flash Memory

One example of nonvolatile memory is flash memory, commonly referred to as flash storage, which erases blocks or data in units called and rewrites data at the byte level. In consumer electronics, business software, and commercial applications, flash memory is frequently utilized for data storage and transfer. Whether a flash-equipped device is powered on or off, flash memory keeps data for a significant time.

We imagine single-level storage systems that can theoretically host and map many IoT devices on a single, shared address space, allowing user programs to access those devices without being aware of the precise storage kinds or physical device locations. NAND flash memory is appropriate for IoT devices, which are frequently compact and mobile.

Figure 4 shows the proposed system of storage allowing for load/store access to the flash memory with the aid of a storage memory management unit (SMMU), making the flash memory appear as huge main memory, with the prototype board corresponding to the dotted rectangle part.

Figure 5a shows the overall structural design from a hardware perspective. Figure 5b describes the schematic diagram of an Unmanned Air Vehicle (UAV)/ Drone, and Fig. 5c illustrates the operational block diagram of a UAV.

The key problem is to disguise flash memory's restrictions, such as a finite number of write cycles. This is similar to memory-mapped IO and virtual memory, but our goal is to use flash memory as the main memory [3]. When a write-back is required, the SMMU compresses the data instead of writing the cached DRAM data back to the flash memory.

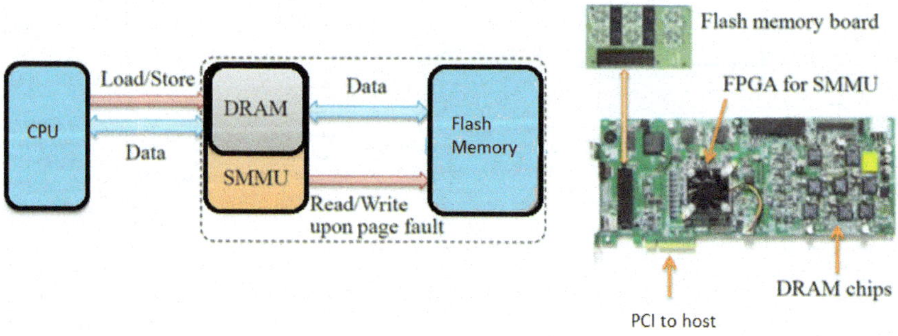

Fig. 4 Block diagram and circuit diagram flash memory

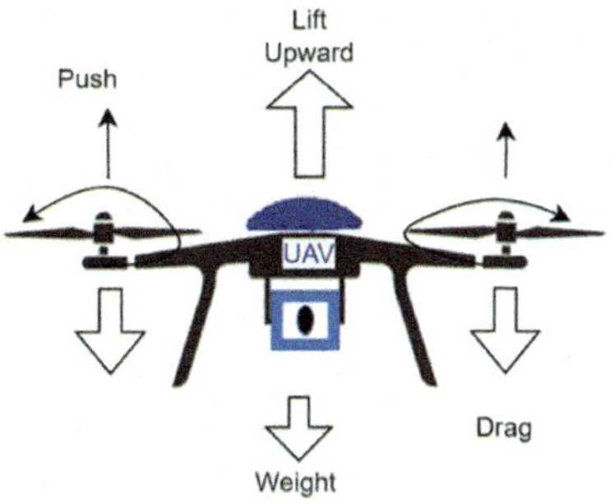

(A) Drone System Overall hardware structural Design

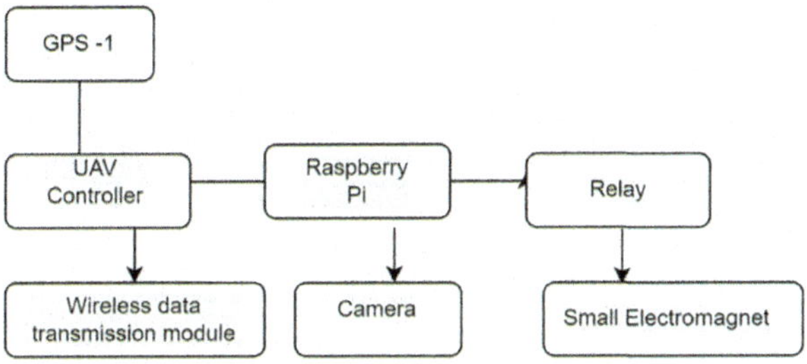

(B) Schematic diagram of Unmaned Air Vehicle

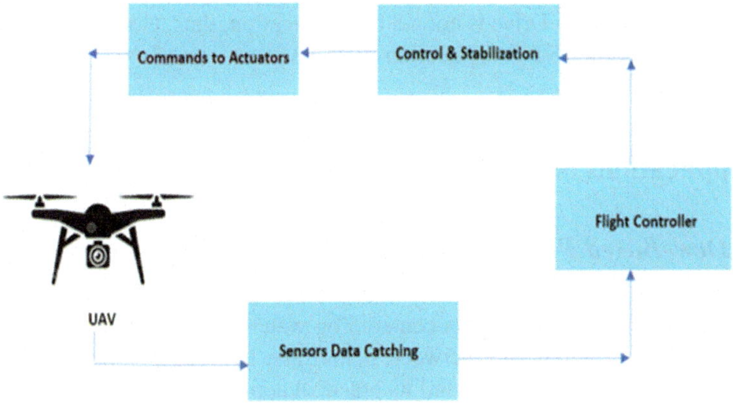

(C) Operational Block diagram of UAV

Fig. 5 **a** Unmanned aerial vehicle (UAV)/drone system overall hardware structure diagram; **b** schematic diagram of UAV. **c** Operational block diagram of UAV

4 Dew Computing

According to Wang [4], the evolution of dew computing is based on two keywords: (i) collaboration and (ii) independence. Another concept given by Skala et al. [5] defined that dew computing actually keeps the virtual infrastructure far away from the central system.

A model called Dew Computing was developed from the original idea of cloud computing. Fog computing, edge computing, and other models are examples of new ones that have come out of cloud computing. The word "dew" describes a phenomenon that occurs in nature: the dew is on the ground, fog is closer to the earth, and clouds are far above it. Like fog computing is next to the user, dew computing is on the user's end. Cloud computing is a faraway service. Dew computing can be thought of as physical or ground.

An on-premises computer's storage is copied in a cloud service under the dew computing category known as "storage in Dew," and the storage of that computer is automatically synced with its cloud counterpart. Dropbox is a typical example of a current STiD service [6]. The independent aspect of Dropbox is first satisfied by the fact that users always have access to its files and folders. Additionally, these files and folders are automatically synchronized with cloud services to support the collaboration capability. There are a lot more apps that are comparable. We also wish to review Google Drive [7, 8] quickly. A file satisfies the collaboration criterion but not the independence feature because it cannot be viewed if the Google Drive service is unavailable. Likewise, the file satisfies the cooperation criterion but not the independence feature because it can only be opened if the Google Drive service is available [2]. Google Drive is not a STiD application, thus. Google Drive Offline, on the other hand, is a STiD application [9].

5 Applications of Dew Computing

5.1 Dew-Based Precision Farming

This farming management idea is centered on watching, quantifying, and reacting to crop variability within and between fields. The late 1980s saw the creation of the first theoretical works on PA and its practical applications. Precision agriculture research aims to define a decision support system (DSS) over whole farm management to improve input returns while safeguarding resources. Figure 6 shows dew-based Precision farming, where the farmer gets the information in their mobile phone through satellite communication.

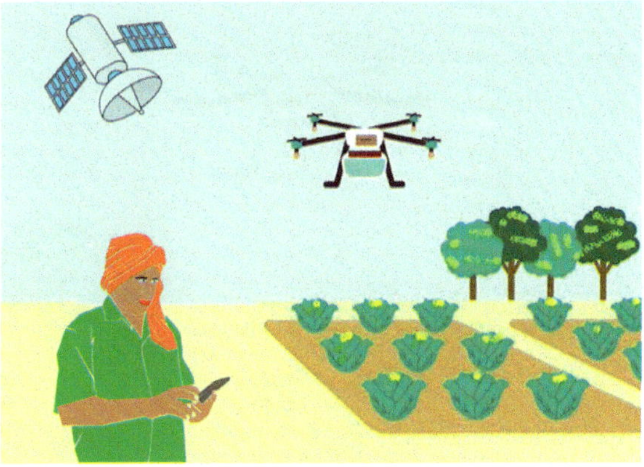

Fig. 6 Dew-based precision farming

5.2 *Dew Powered Agricultural Drones*

IoT enables monitoring and managing microclimate conditions for boosting output. For example, IoT-enabled devices can sense soil moisture and nutrients in contrast to weather information better to manage smart irrigation and fertilizer systems for outdoor planting. For instance, this prevents resource waste if sprinkler systems only release water when necessary.

Unmanned aerial vehicles (UAVs) or drones are currently employed in precision agriculture to monitor harvests or to identify diseases, parasite infections, or information on plant development. They may also measure soil moisture and release fertilizer precisely and accurately inside the ground. UAVs are frequently used for crop insurance surveys, dusting, and surveillance. UAVs are commonly utilized for monitoring, dusting, and crop insurance surveys. A UAV mounted with a high-resolution camera is frequently used to survey a target area. In this, Fig. 7 represents dew-based agricultural drones, by which the farmer can easily spread pesticides and fertilizers. Figure 8 shows how cache-enabled agriculture can monitor and control temperature, irrigation systems, and weather via a wireless sensor network.

5.3 *Smart Dew Greenhouse Management*

Smart dew greenhouse management: Increasing the yields of fruits, vegetables, and other agricultural products is a goal of greenhouse farming. Greenhouses can regulate environmental conditions through personal intervention or a relative control mechanism. An IoT-powered smart greenhouse can intelligently monitor and regulate the climate without the need for human involvement. Manual intervention, however, results in lost production, wasted energy, and labor expenditures. This renders the idea of greenhouses as a whole useless. Consequently, intelligent greenhouses are a superior option. In this, Fig. 9 shows how the Internet of Things can be used to build

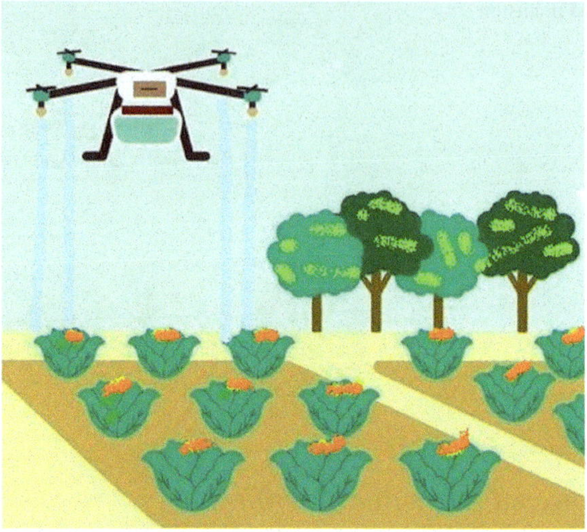

Fig. 7 Dew-based agricultural drones

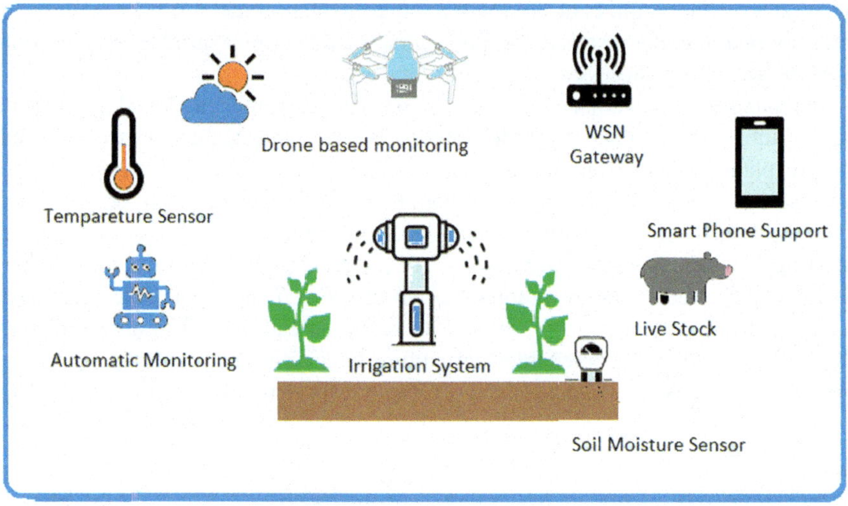

Fig. 8 Cache enabled IoT for agriculture

a smart greenhouse. Without needing personal assistance, the farmer can control, monitor and manage the climate, light, security system, and crop harvesting process through these intelligent greenhouses.

In a smart greenhouse, several sensors are utilized to evaluate environmental conditions and determine whether they are suitable for plants. Remote access is created by using IoT to connect the system to a cloud. As a result, continuous manual

Fig. 9 Dew-based smart greenhouse management

monitoring is no longer necessary. Instead, the cloud server implements a control action inside the greenhouse, which also manages data processing. The greenhouse's IoT sensors offer critical data on temperature, humidity, and more [5]. Through a Wi-Fi signal, these sensors operate everything from lighting and window opening to temperature control and cooling.

5.4 *Livestock Management Using Dew Computing*

In this chapter, we have discussed that the Internet of Thing's involvement in smart farming is possible through Livestock management and data processing using dew computing.

A system designed to monitor the crop field using sensors is what we mean when we talk about IoT-based smart farming. The irrigation system is automated while these sensors monitor every factor necessary for crop production, including soil moisture, humidity, light, temperature, etc.

Farmers may check the state of their fields from anywhere through this system. Wireless IoT applications can be used by farmers in the area of livestock management also. Here in this Fig. 10, we can see that the farmer can gather information about their cattle movements, comfort, and health through a wireless sensor network. This knowledge lowers labor expenses and aids in the prevention of sickness. Individually examining livestock for indications of illness or damage is a traditional approach to livestock monitoring. This approach is both expensive and incredibly unreliable. Comparing IoT-based farming to conventional farming, it is far too efficient.

Solutions for managing livestock that is IoT-enabled remove uncertainty regarding herd health. Battery-powered sensors worn on an animal's collar or tag track the

Fig. 10 Dew-based livestock management

animal's location, temperature, blood pressure, and heart rate and wirelessly transmit the information to farming equipment in almost real time.

IoT-based smart farming emphasizes organic farming, family farming (complicated or confined spaces, particular animals and/or cultures, maintenance of specific or high-quality varieties, etc.), and farming that is very transparent in addition to modernizing traditional farming practices. The advantages of IoT-based smart farming also extend to environmental considerations. For example, it can help farmers use water efficiently and get the most out of their treatments and inputs.

Key advantages include:

- Real-time monitoring of livestock vitality and health enables farmers to give the animals prompt medical attention and stop the spread of illness.
- To avoid theft and to spot grazing trends, track grazing animals.
- Compile and examine historical data to find patterns in the health of cattle or to monitor the spread of disease.
- Keep an eye on an animal's preparation for mating or giving birth to avoid losing any new-born calves and to improve breeding techniques.

5.5 *Dew-Enabled Plant Healthcare Monitoring*

Most crucially, these gadgets allow hospitals to keep track of their patient's health in the intensive care unit and occasionally at home. All day long, it can keep an eye on the patients. Hence shortening hospital stays and still giving current real-time information that might save lives. Smart beds reduce the wait time for available space in hospitals by informing the person of availability. The addition of IoT sensors can make the difference between life and death by reducing malfunctions and boosting reliability. Rajalakshmi et al. [10] described monitoring the crop field using soil moisture, temperature, humidity, light sensors, and an automated irrigation system. To maintain the server database, the data from sensors is encoded in JSON format and wirelessly transmitted to the web server. When an agricultural field's temperature and moisture content is at risk, the irrigation system will automatically turn on. Farmers can check their fields' status from anywhere by regularly using the notifications provided to their mobile devices [11–13].

5.6 Dew Controlled Forest-Fire Monitoring

Real-time environmental analysis of factors like weather and harmful gases could generate pertinent environmental data that could aid in averting or identifying a disaster. IoT (Internet of Things) devices and sensors now enable it to monitor various environmental factors, including pressure, temperature, humidity, and the concentrations of polluting gases like carbon monoxide and carbon dioxide. Radical changes and combinations of these factors may be signs of unfavorable weather conditions that could result in a natural disaster like a forest fire.

Castro et al. [14] have emphasized the significance of the "rule of 30" [1], which is based on three essential variables that are connected to the identification of forest fires: temperatures above 30 °C, humidity levels below 30%, and wind speeds above 30 km/h in the exact location. Applying this rule as a guideline for preventing forest fires can assist in determining areas with a high risk of fires, allowing the deployment of protective measures and contingency planning that may be advantageous for environmental preservation.

A deep learning-based strategy termed LBFFPS has been proposed in [15]. It gives the suggested technique the capacity to keep track of the forest area using an intelligent surveillance system called SFMK. A controller with IoT capabilities, several smart sensors, and a camera portion make up the SFMK surveillance kit. The suggested approach can quickly detect the forest fire based on these parameters and report it to the appropriate authority. In addition, the sensors connected to the SMOKE can detect smoke and monitor temperature and humidity. To monitor the forest zone and automatically communicate any changes to the server entity, a high-definition 1020-megapixel camera is installed in the SFMK [16].

5.7 Dew Server-Based Monitoring of Climate Condition

Crop output is significantly influenced by the climate. Varied crops require different climate conditions to grow, and any lack of understanding about climate seriously impairs crop output in terms of quantity and quality. In the following Fig. 11, we can see that the farmer can now access real-time weather information thanks to IoT technology.

Farmers can select a crop that will thrive in a specific climate using the environmental data collected by the sensors installed in agricultural fields.

The entire Internet of Things (IoT) ecosystem is made up of sensors that identify current meteorological variables, including humidity, rainfall, and temperature, all of

Fig. 11 Dew-based climate monitoring condition

which are essential for crop development. These sensors can predict any significant climate change that can have an impact on productivity. The necessity for physical presence is reduced thanks to a warning delivered to the server about the difference in the weather. Higher yields are the result in the end [11].

5.8 Dew-Based Crop Protection Against Animal and Insect Intrusion

The production of crops is seriously threatened by animal encroachment, which also lowers farmer profits and jeopardizes food security.

Solutions based on the Internet of Things and machine learning techniques are available to solve this issue. The ESP8266 Wireless Fidelity module, Pi Camera, Buzzer, and LED are all connected to the Raspberry Pi, which controls the machine algorithm. Machine learning technologies like the Region-based Convolutional Neural Network and Single Shot Detection technology are crucial to identify objects in photos and categorizing the animals. The testing shows that the Single Shot Detection method works better than the Region-based Convolutional Neural Network algorithm. Finally, the programs that integrate with the Twilio API decimate the information to the farmers so they can act swiftly in their farm fields [39, 12, 17]. In the above diagram, Fig. 12 shows how the farmer can easily detect and monitor the crop condition and predict the attack of insects or animal intrusion with the help of technology and WSN.

Fig. 12 Dew-enabled crop protection against animal intrusion

5.9 Dew-Facilitated Home and Office Management with Indoor and Rooftop Garden

Rooftop gardens have emerged as a new trend in metropolitan areas due to the lack of green space.

Small-scale home greenhouse management systems with IoT capabilities track and automate operations. Smart gardening programs track soil and ambient conditions, automate irrigation, and check on the health of plants to increase productivity and reduce the amount of labor needed to maintain the greenhouse. These solutions are similar to those used in agriculture and smart farming. We may turn our roof into a tranquil outdoor area where we can eat organic, fresh food by following simple guidelines. So, here are a few layout suggestions for a rooftop garden. A rooftop garden, also known as a roof garden, is a planted green area on the top of a house or structure. Terraced gardens are situated in metropolitan areas, where they cultivate fruits and vegetables and provide a play area or peaceful green space. The plans incorporate asphalt, decking, seating, flowerbeds or raised planters, cultivated gardens, or urban farms. Buildings throughout the world's major cities have a significant number of green roofs. Both extensive and intense vegetation kinds are possible for green roofs. Intensive vegetation on a rooftop consists of large shrubs and tiny trees, whereas extensive vegetation refers to herbs, small plants, and shrubs. The following Fig. 13a shows the prototype model of Rooftop Gardening with smart irrigation, and Fig. 13b indicates the technic of Smart Indoor Gardening. Figure 14 describes the detailed Tree Health monitoring system and Green House management system process.

The advantages of rooftop or indoor gardening are significant. Some of them can be listed as:

- These autonomously control the temperature and humidity of our indoor gardening environment to produce and maintain the best possible growing conditions for your plants.
- Compile and examine previous information on the greenhouse environment for improved testing, management, and optimization of ideal growing conditions.

Fig. 13 Dew-enabled smart home farming: **a** Rooftop garden with smart irrigation, **b** smart indoor garden

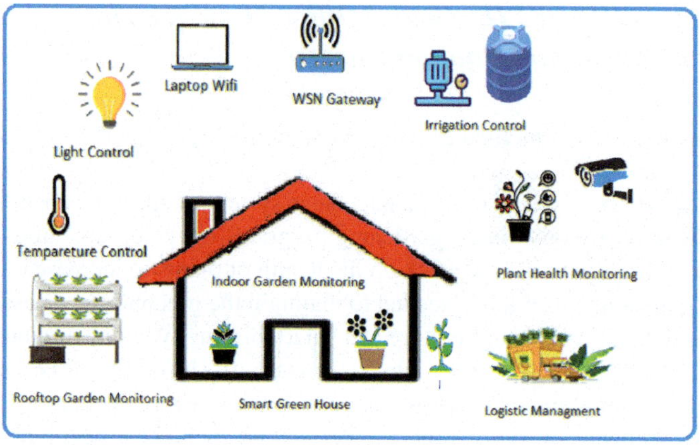

Fig. 14 Tree health monitoring and green house management

- Automate irrigation in greenhouses to prevent plant over-or under-watering, decrease water waste, and reduce the need for labor-intensive manual labor.

5.10 Dew-Based Disaster Management to Protect Crop

We have seen many natural calamities, such as salinity, river erosion, cyclones, droughts, and floods. By causing harm to the crops, cattle, fisheries, and Agro-forestry, these devastating events significantly impede the nation's agricultural production systems, social development, and economic growth. Because of the limitations of climate change on coastal agriculture, CSA is crucial for managing disaster risk reduction in that area. These natural catastrophes can affect anybody, anywhere. Still, they wreak the most havoc on the most defenseless, impoverished communities—often those belonging to underrepresented groups, those residing in remote places, or those residing in fragile ecosystems.

5.11 Dew-Enabled Logistic for Food Supply Chain Management

We can manage your worldwide assets quickly and cut costs with Smarter Technologies' innovative logistics management systems. With real-time data on all critical assets and personnel, your business can reduce costs and increase productivity.

- Real-time asset tracking
- Affordable technology
- Robust security.

5.12 Dew-Based Autonomous Driving in the Agricultural Sector

Global food chains are being impacted by shortages in agriculture. This efficiency is considerably increased by using smart vehicles, which also automate conventional farming. As autonomous farming promises to produce more crops with less work and less damage to the environment, inventions like agricultural drones, self-driving tractors, and seed-planting robots could be crucial to the future of food supplies. Figure 15 shows the process of autonomous driving in agriculture. These electronic devices have sensors, computer vision, extremely accurate (less than one inch) GPS, and machine learning capabilities to enable self-driving and precision farming. These machines occasionally still need an operator, but owing to technology, that operator doesn't necessarily need to be a highly skilled ag truck driver.

- Autonomous tractors could help solve farming labor shortages.
- Drones and seed-planting robots are among other innovations.

Fig. 15 Dew-enabled auto-vehicle in agriculture

- Autonomous farming equipment.

5.13 Dew-Based Agricultural Waste Management

Dew aids in our daily efforts to develop a trustworthy, sustainable water source for everyone. The industry may boost profitability while complying with regulations and conserving the environment by employing advanced management and wastewater resource-conversion technologies.

The mining and ore sectors benefit from technological advancements in dew management that enable them to control water levels without releasing pollutants produced during the process into the environment. Dew effectively produces fresh water from the resulting impure wastewater while preserving the environment [16, 18].

6 Benefits of IoT in Agriculture

IoT enables smart farming to improve the sector. Landowners and producers may make better overall decisions thanks to the collection and monitoring of pertinent environmental, meteorological, and chemical land use data and other sophisticated IoT solutions. This gives us the flexibility to deal with any situation and respond quickly. In addition to increasing agricultural output and saving crops and labor, it warns us of environmental concerns and helps us take appropriate action.

Figure 16 gives an overall view of the benefits of IoT in the Agriculture field. IoT-based remote sensing uses sensors installed along farms, like weather stations, to collect data that is then sent to analytical tools for study. Farmers can use analytical dashboards to monitor their crops and act on the information they gain. The saving actions are: (a) using caution when allocating resources, (b) enabling land users to take prompt action, (c) enhancing food security, agricultural product health, production, volume, and quality, (d) improving any administrative procedures, (e) it has enormous potential to alleviate harmful climate impacts, (f) crop Evaluation.

Aggregation of Data: For centralized data gathering in sizable farming operations, 5G technology holds considerable promise. A substantial corporate farm might gather the data from micro-monitored crop management devices via a private 5G network. These systems contain soil moisture sensor density, which may be hundreds of times higher than what is supported by current technologies. With this network, a more effective real-time monitoring system that includes triggers for slowing irrigation and other crop support systems may be possible. Here, Fig. 17 describes the overall Aggregation of IoT-based agriculture benefits.

Statistical Analysis: It is seen that large industrial farms can better integrate predictive analytics as a result of 5G technology's data aggregation capabilities.

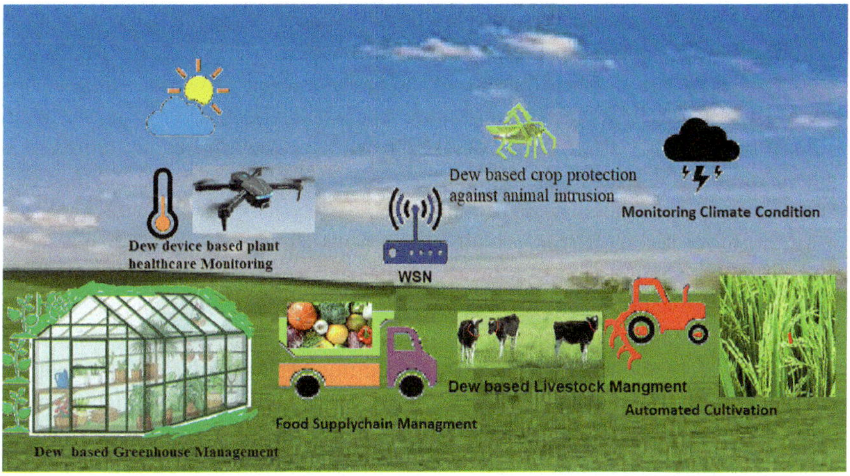

Fig. 16 IoT-based agricultural scenario

Fig. 17 Aggregation of agriculture benefits

Analytics software considers current and historical data on conditions (such as soil moisture and pesticide use).

Operation of Drones: We have observed that farmers increasingly use drones to check their crops. Drones offer more precise data about crop damage and other factors while being less expensive than driving tractors through fields. In addition, drones will be able to gather better-quality video data and transmit it more quickly because of 5G's enormous bandwidth. Real-time reporting and the development of AI drone technology will be made possible by this high-speed data transfer capacity [16].

Animal Tracking and Monitoring: Animal monitoring sensors will likely continue to be connected through Wi-Fi, Bluetooth, or LTE LPWAN until REL 17 boosts the viability of 5G low-power and denser sensor networks. One example is significant, centrally located farms where it is possible to install 5G infrastructure over a restricted region (such as a chicken farm) and track individual animals. To track an animal's whereabouts and health, ag-tech developers have developed herd management sensors, such as smart collars and ear tags.

Autonomous Agricultural Vehicles: Farm tools will benefit from advancements in autonomous vehicle technology in other industries. The spacing between seed rows and the pressure applied to them when they are planted in the ground are just two examples of the numerous little agricultural task specifics that operators of tractors with onboard computers can now handle. In addition, driverless farm equipment will advance to give farmers more flexibility and efficiency while reducing labor costs.

IoT sensors can also benefit trucks used to deliver crops. These sensors can monitor the cargo's temperature and notify if it gets too warm or cold (i.e., cold chain). Asset trackers and other small mobile sensors will probably continue to use high-latency technologies like LPWAN. However, autonomous vehicles with more powerful onboard computers will be able to send and receive more significant, ultralow-latency data streams thanks to 5G.

Weather Stations: The weather has complete control over farming operations. Large percentages of a farmer's crop may be lost to avoidable diseases and damage. By giving farmers information about the state of their fields, linked farm monitoring stations can help alleviate this problem.

IoT Based Decomposition Management Process for Organic Materials: Here in this chapter, the authors have discussed a fertilizer kit for computing manure creation technique and inevitably supplies that natural manure to the field as per the need of the crops. Intelligent irrigation system uses components like LM35 MQ6, different motors for installation and is also set up to deliver the exact quantity of water and nutrition at the correct time. The complete manure formation and distribution is automated, reducing the need for human intervention. In the paper, B. P. Chaudhury et al. discussed the IoT tools used in industrial applications but also found their occurrence in monitoring of climatic health. To be precise, it can measure the volume of carbon content in the air. The researchers have designed a cost-effective IoT-based wireless monitoring system that will help to measure air quality. In this device, they have put three sensors to measure carbon dioxide concentrations, temperature, relative humidity, and dust level [16, 19]. Here Fig. 18 describes how the worker can manage the dew-based air pollution and waste management process.

7 Existing Dew-Enabled IoT Devices

The dew server uses device control to keep track of the IoT device's status and manage its operation. This module's fundamental capabilities include the following:

Fig. 18 Dew-based air pollution and waste management

- determine sensor status.
- alter a particular variable.
- cause particular actions.
- activate a function.

To provide relevant information and to enable the dew server to respond to the received status, the IoT device broadcasts its status to a nearby dew server. Due to the dew server's control over the IoT sensor, arranging a particular aspect to start a more precise detection and quantification of the required signal is possible. When an IoT device is reset, the dew server may, for instance, begin a new measuring of the IoT sensor. In the case of an IoT actuator, it may start a specific action that regulates the actuator's performance.

Additional pipelined stream processing is required for the streaming data sent by the IoT sensor. Data collection, storage, and transfer to higher level servers are the three crucial duties the dew server does with the incoming streaming data. The dew server's primary function is data collection. IoT sensors typically broadcast data without verifying whether it has been gathered. Higher sampling rates may also cause data to be permanently lost because IoT sensors cannot store them. The dew server realizes the data storage function of the IoT sensor, especially when it is battery operated and is a light mobile device, even without wires which can provide a more secure connection, such as the wearable.

The dew server's primary function is data collection. IoT sensors typically send data without verifying whether it has been gathered. Additionally, more excellent sampling rates may result in data loss for all time because IoT sensors cannot store the information. In reality, the dew server implements the data storing function of the IoT sensor, particularly in the scenario where it is battery operated and is a light mobile device, without any connections that can provide a more secure connection, as are the wearable sensors. Since the dew server buffers the received data stream on a low-power radio link and produces a signal with more extensive power, it also serves as a digital repeater in some ways.

Data processing is complementary to data collection and storage. Depending on the size of the dew server, data processing may include at least the following:

- pre-processing.
- analyzing.
- classification (decision-making).
- visualization.

Data preparation typically uses digital filters or comparable techniques to prepare data for further processing because the incoming data stream may be noisy. Each received sample must undergo data processing to estimate the perceived signal and draw additional inferences accurately.

On most devices, one may monitor and start specific operations for the IoT device via the dew server's user interface. At a minimum, this human–computer interface does the following tasks:

- activation instructions (via menu options and control buttons).
- establishing different criteria.
- real-time visualization.
- browsing via stored info.

Users can manage the dew server even though it may execute particular algorithms and operate independently.

Generally, dew-enabled devices are cache memory-based devices. Such as in different agriculture devices based on Raspberry Pi have a cache memory of 16 KB, whereas Arduino Nano R3 has 512 bytes of Flash, and some devices have an in-direct cache and flash memory. Here in the following table, we can observe different types of sensors, their uses in various related fields of agriculture and the involvement of cache and flash memory. Table 1 shows the list of devices capable of implementing Dew functionalities. It compares the type of processor and size of cache and flash memory. Identifying services and installing them in a device as per the capabilities of process and memory is a challenge.

8 Challenges and Future Research Directions

Invent of IoT: The basic idea of a network of smart devices was explored as early as 1982, when a modified Coca-Cola vending machine at Carnegie Mellon University became the first ARPANET-connected device, reporting its inventory and whether freshly loaded drinks were cold or not. The modern idea of the IoT was created by academic conferences like UbiComp and PerCom, as well as Mark Weiser's 1991 paper on ubiquitous computing, "The Computer of the twenty-first Century [13, 25, 33]."

The term "Internet of Things" was first used in 1999 by MIT's Executive Director of Auto-ID Labs, Kevin Ashton. Although he was the first to define the IoT in a demonstration for Procter & Gamble, the definition has changed over time.

Table 1 Comparison among existing dew-enabled devices

IoT application domain	IoT enable devices	Sensors	Microcontroller or microprocessor	Cache memory	Flash	Comments
A. Agriculture devices	A.1 Smart irrigation [3, 10, 20, 21]	PIR Sensor *Sensor S5851A*	Raspberry PI	16 KB cache memory	No	
	A.2 Smart Greenhouse using sensors [22, 23]	CO2 level				
		Temperature *DS18B20* PIR, URD Repeller LM2596 DC			EPROM	
		Moisture level				
	A.3 Temperature humidity [24]	DHT11, TMP007 LDR, HDC1010				[24]
	Camera		CC3200		256 KB 256 KB RAM	
	Soil health monitoring [21]	Soil temperature				
		Air humidity				
		Soil moisture				
		Leaf moisture				
		Precipitation, barometric pressure				
B. Healthcare devices	B.1 Plant diseases monitoring [13, 25]					

(continued)

Table 1 (continued)

IoT application domain	IoT enable devices	Sensors	Microcontroller or microprocessor	Cache memory	Flash	Comments
	B.1 Blood pressure [21, 26]	BP sensor	Intel Quark SoC X1000		8 MB	
	B.2 Heart rate [4]	URD Sensor HC-SR04				
	B.3 Temperature [21, 26]	LM35 Body temperature sensor				
C. Smart home farming-organic farming	C.1 Pollution monitoring and reporting [27, 28]		Arduino NRF24L01 HSC-04		32 KB SRAM	
	C.2 Smart lighting on streets [13]	Noise/sound, vibrations, light intensity	Raspberry Pi	16 KB cache memory	No	
	C.3 Leaf disease detection system [23, 19]	Soil moisture sensor, ultra-sonic senor	Raspberry Pi	16 KB cache memory	No	
	C.4 Crop health monitoring [27–30]	Soil moisture	Arduino Nano R3		32 KB SRAM	
		Soil temperature DS 18D20,				
		Soil quality				
		Soil pH				
		Turbidity sensor				
		2 Axis Joy stick				

(continued)

Table 1 (continued)

IoT application domain	IoT enable devices	Sensors	Microcontroller or microprocessor	Cache memory	Flash	Comments
	C.5 Greenhouse smart security and monitoring devices for agriculture [12, 31, 32]	Raspberry PI and GPIO	Raspberry PI		32 KB SRAM	
		Flex or bend sensor	MSP 430 F5529		128 KB 8 KB RAM	
		PIR sensor	Raspberry PI		32 KB SRAM (Manual Cache)	
		Vibration sensor				
		Temperature sensor LM35				
D. Industrial manufacturing	Monitoring of supply chain and inventory management	Motion detection sensor				
	Energy saving sensors	Outdoor air quality sensor				
	Process monitoring	Indoor air quality sensor				
	Logistic management					
		Water quality sensor				
		Water level sensor				

(continued)

Table 1 (continued)

IoT application domain	IoT enable devices	Sensors	Microcontroller or microprocessor	Cache memory	Flash	Comments
E. Home/building automation	Home/building *automation* [31] Industrial control monitoring	Texas *instruments*' CC2530			32/64 KB	
	Micro electromechanical system (MEMS)	Sensor GY-83				
F. Crop protection against animal intrusion	Crop protection against animal intrusion [17]	ESP 8266	Raspberry Pi	16 KB cache memory	No	

Invent of Dew: The decade phases of the ICT development and its connections to the distributed computing hierarchy and g-phenomena are described, including cloud computing (cc), fog computing (fc), and dew computing (dc). Cellular-connected drones and robots equipped with cameras that can quickly scan large areas of land to monitor crop fields and livestock carefully are other examples of smart technology altering farming and could benefit from 5G. Farmers can use the time saved to do less time-consuming duties while also creating new opportunities for future, more robust crops by automating the mass collection of comprehensive information that was previously done by hand. The farmers make decisions by developing models and projections. Analytics will improve, optimizing farm production, as 5G delivers denser real-time data.

9 Conclusions

IoT-enabled gadgets and sensors are being developed to improve agriculture-related management and benefit humanity. This article has discussed devices based on Raspberry Pi, Arduino, and several flash and cache memory types. Several difficulties were encountered when developing this solution, including using data transfer protocols, contact with hardware, incorporating sensors, and converting recorded data into relevant information for user presentation. In addition, the synchronization of all system data across various platforms, such as new alerts and measurements, the integration of multiple technologies, mobile devices, Web services, and IoT devices, and further considerations will also be illustrated.

References

1. Farooq, M.S., Riaz, S., Abid, A., Abid, K., Naeem, M.A.: A survey on the role of IoT in agriculture for the implementation of smart farming. IEEE Access **7**, 156237–156271 (2019). https://doi.org/10.1109/ACCESS.2019.2949703
2. Prathibha, S.R., Hongal, A., Jyothi, M.P.: IoT based monitoring system in smart agriculture. In: 2017 International Conference on Recent Advances in Electronics and Communication Technology (ICRAECT), pp. 81–84. IEEE (2017)
3. Thorpe, J., van Oorschot, P.C.: Graphical dictionaries and the memorable space of graphical passwords. In: USENIX Security 2004, San Diego (2004); Williams, J.: Narrow-band analyzer. Doctoral dissertation, Ph.D. dissertation, Dept. of Electrical Eng., Harvard Univ., Cambridge, Mass., (Thesis or dissertation)) (1993)
4. Wang, Y.: Definition and categorization of dew computing. Open J. Cloud Comput. (OJCC) **3**(1), 1–7 (2016)
5. Skala, K., et al.: Scalable distributed computing hierarchy: cloud, fog and dew computing. Open J. Cloud Comput. (OJCC) **2**(1), 16–24 (2015)
6. Google Inc., Google Drive.: https://drive.google.com. Accessed 22 May 2016
7. Drago, I., et al.: Inside dropbox: understanding personal cloud storage services. In: Proceedings of the 2012 Internet Measurement Conference (2012)

8. Google Inc., Google Drive Offline.: https://support.google.com/drive/answer/2375012?co=GENIE.Platform%3DDesktop&hl=en. Accessed 22 May 2016
9. Balakrishna, K., et al.: Application of IOT and machine learning in crop protection against animal intrusion. Glob. Trans. Proc. **2**(2), 169–174 (2021)
10. Rajalakshmi, P., Mahalakshmi, S.D.: IOT based crop-field monitoring and irrigation automation. In: 2016 10th International Conference on Intelligent Systems and Control (ISCO). IEEE (2016)
11. Tripathy, P.K., et al.: MyGreen: an IoT-enabled smart greenhouse for sustainable agriculture. IEEE Consum. Electron. Mag. **10**(4), 57–62 (2021)
12. Baranwal, T., Pateriya, P.K.: Development of IoT based smart security and monitoring devices for agriculture. In: 2016 6th International Conference-Cloud System and Big Data Engineering (Confluence). IEEE (2016)
13. Kassim, M.R.M.: Iot applications in smart agriculture: issues and challenges. In: 2020 IEEE Conference on Open Systems (ICOS). IEEE, pp. 19–24 (2020)
14. Toledo-Castro, J., Santos-González, I., Hernández-Goya, C., Caballero-Gil, P.: Management of forest fires using IoT devices. Tenerife, Spain, UBICOMM (2017)
15. Rela, M., Rao, S.N., Patil, R.R.: Performance analysis of liver tumor classification using machine learning algorithms. Int. J. Adv. Technol. Eng. Explor. **9**(86), 143 (2022)
16. Sangeetha, K., Shamini, S.S., Devi, R.L.: IoT based controlled decomposition process for organic materials. Elsevier J. (2021)
17. Ananthi, J., et al.: Forest fire prediction using IoT and deep learning. Int. J. Adv. Technol. Eng. Explor. **9**(87), 246 (2022)
18. Saha, H.N., et al.: Waste management using internet of things (IoT). In: 2017 8th Annual Industrial Automation and Electromechanical Engineering Conference (IEMECON). IEEE (2017)
19. Lu, Y., et al.: Multimodal plant healthcare flexible sensor system. ACS Nano **14**(9), 10966–10975 (2020)
20. Martinez, J.M.P., et al.:Integrating data warehouses with web data: a survey. IEEE Trans. Knowl. Data Eng. **20**(7), 940–955 (2008). https://doi.org/10.1109/TKDE.2007.190746
21. Ma, S., et al.: Development of noncontact body temperature monitoring and prediction system for livestock cattle. IEEE Sens. J. **21**(7), 9367–9376 (2021)
22. Nicole, R.: The last word on decision theory. J. Comput. Vis. (2012)
23. Gurram, V.S.L., Koneti, P.K., Maddali, A., Dr. Omkumar, S.: Agricultural robot with leaf disease detection using raspberry PI", Int. J. Res. Publ. Rev. **3**(3), 1274–1277 (2022)
24. Chen, W.K.: Linear Networks and Systems. Wadsworth, Belmont, CA (1993); Duncombe, J.U.: Infrared navigation—Part I: an assessment of feasibility. IEEE Trans. Electron. Dev. **11**(1959), 34–39
25. Neupane, K., Baysal-Gurel, F.: Automatic identification and monitoring of plant diseases using unmanned aerial vehicles: a review. Remote Sens. **13**(19), 3841 (2021)
26. Lee, G., et al.: Wireless IC tag based monitoring system for individual pigs in pig farm. In: 2019 IEEE 1st Global Conference on Life Sciences and Technologies (LifeTech). IEEE (2019)
27. Sengupta, A., et al.: GrowFruit: an IoT-based radial growth rate monitoring device for fruit. IEEE Consum. Electron. Mag. **11**(3), 38–43 (2021)
28. Sengupta, A., et al.: AgriStick: an IoT-enabled agricultural appliance to measure growth of jackfruit using 2-axis joystick. IEEE Instrum. Meas. Mag. **25**(3), 58–62 (2022)
29. Bodake, K., et al.: Soil based fertilizer recommendation system using internet of things. MVP J. Eng. Sci. **1**(1), 13–19 (2018)
30. Sengupta, A., et al.: FarmFox: a quad-sensor-based IoT box for precision agriculture. IEEE Consum. Electron. Mag. **10**(4), 63–68 (2021)
31. Fei, X., Xiao-Long, W., Yong, X.: Development of energy saving and rapid temperature control technology for intelligent greenhouses. IEEE Access **9**(2021), 29677–29685 (2021)

32. Corkery, G., et al.: Monitoring environmental parameters in poultry production facilities. In: Computer Aided Process Engineering-CAPE Forum 2013. Institute for Process and Particle Engineering, Graz University of Technology, Austria (2013)
33. Gondchawar, N., Kawitkar, R.S.: IoT based smart agriculture. Int. J. Adv. Res. Comput. Commun. Eng. **5**(6), 838–842 (2016)

The manufacturer's authorised representative in the EU is Springer Nature Customer Service Centre GmbH, Europaplatz 3, 69115 Heidelberg, Germany. If you have any concerns regarding our products, please contact ProductSafety@springernature.com

Printed and bound by CPI Group (UK) Ltd, Croydon, CR0 4YY
01/10/2026
02225868-0009